WEED SCIENCE: PRINCIPLES AND PRACTICES

Weed Science:

PRINCIPLES AND PRACTICES
SECOND EDITION

GLENN C. KLINGMAN
Instructor, University of Nebraska
Professor, North Carolina State University
Lilly Research Laboratory, Greenfield, Indiana
Retired

FLOYD M. ASHTON
Professor of Botany
University of California at Davis
Davis, California

With the editorial assistance of

LYMAN J. NOORDHOFF
Communications Specialist, Extension Service
United States Department of Agriculture
Washington, D. C.
Retired

175 YEARS OF PUBLISHING
1807 1982

A WILEY-INTERSCIENCE PUBLICATION

JOHN WILEY & SONS,
New York • Chichester • Brisbane • Toronto • Singapore

***Library of Congress Cataloging in Publication Data*:**

Klingman, Glenn C.
Weed science.

"A Wiley-Interscience publication."
Includes index.
1. Weed control. 2. Herbicides. I. Ashton, Floyd M. II. Noordhoff, Lyman Judson, 1917– III. Title.

SB611.K55 1982 632'.58 82-2750
ISBN 0-471-08487-5 AACR2

Printed in the United States of America

10 9 8 7 6 5 4 3 2 1

Preface

Weed control is an expensive, but necessary, part of crop production—directly increasing the price of food. Weeds may poison or seriously slow weight gains of livestock. They cause human allergies such as hay fever and poison ivy. They infest home lawns and gardens. Weeds create problems in recreation areas such as golf courses, parks, and fishing and boating areas. They are troublesome along highways, railroads, in industrial areas, and in irrigation and drainage systems. Tillage to control weeds may seriously increase soil erosion. Human labor used to control weeds is not available for more-productive uses. *Weeds affect everyone.*

The science of weed control has advanced more during the last 35 years (since 1947) than in the previous 100 centuries. In this book old and reliable methods of weed control are integrated with the very newest chemical techniques. Together these practices result in more-effective and less-expensive weed-control programs. Together they help reduce the cost of food production.

Progressive farmers have rapidly accepted new herbicides. 2,4-D was first used on a large scale in 1947. Ten years later, 92% of the farmers (in progressive farming areas) depended regularly upon herbicides.

This textbook is designed principally for college classroom instruction in the principles and practices of weed science. The first third of the book deals with basic principles and methods of weed control, the second third with specific chemistry of herbicides, and the last third with weed-control practices in specific crops, pastures and range, brush and undesirable trees, aquatic areas, total vegetation control, lawns, turf, and ornamentals. Also, it deals with application techniques, control of spray drift, volatility, and similar basic principles. The broad coverage will be helpful to research scientists, extension specialists, county agents, vocational agriculture teachers, herbicide-development representatives, and advanced farmers.

It is impossible to acknowledge adequately the vast number of workers and the vast number of papers published in the field. This book brings together the modern philosophy and techniques of weed control. It is not intended as a complete review of the literature.

Trade names have been used in some cases along with chemical names and abbreviations. Trade names and common chemical names are cross indexed in the appendix. Trade names have been given as a convenience to the reader, *not* as an endorsement of any one product.

This book is intended for worldwide use. It cannot delve into the wide range of cultural practices and conditions encountered in different countries. Therefore, nothing in this book is to be construed as recommending or authorizing the use of any particular weed-control practice or chemical in any given area.

For each chemical, see the manufacturer's label for method and time of application, rates to be used, weeds controlled, and special precautions. *Label recommendations must be followed*–regardless of statements in this book. Use of any product name does not constitute recommendation of that product. Neither does omission constitute not recommending the product.

For our children, our grandchildren, and to other children of the world we commend the technology of *Weed Science* as it contributes to the production of a safe, healthful, and abundant food supply with a safe environment.

We are grateful to our wives Loree Klingman, Phyllis Ashton, and Ruth Noordhoff for their patience and assistance during preparation of this manuscript.

GLENN C. KLINGMAN
FLOYD M. ASHTON
LYMAN J. NOORDHOFF

March 1982

Contents

WEED SCIENCE: PRINCIPLES AND PRACTICES

1 Introduction

Weed control is as old as the growth of food crops. Slowly and with growing momentum, man has learned to use tools, horsepower, tractor power, and chemicals in the fight to control weeds.

During thousands of years man has achieved amazing advances. First he substituted a sharp stick for his fingers. Centuries later he discovered the metal hoe. Then man greatly reduced his labor by harnessing a horse or ox to drag the hoe or plow. In 1731 a major advance was proposed—planting crops *in rows* to permit "horse-hoeing." Jethro Tull explained his idea in *Horse-Hoeing Husbandry*. He was among the first to use the word *weed* with its present spelling and meaning. And less than 200 years later tractors started to replace horses.

All these methods battled weeds with brute force. But with the development of herbicides, chemical energy is replacing mechanical energy for weed control (see Figure 1-1).

Weed control by biological means (mainly insects and plant diseases) has succeeded somewhat; considerable research is under way in this area.

Improved agricultural technology produces more food as shown in Figure 1–2. Weed control is an important part of these gains.

Before 10,000 BC, weeds were removed from crops by hand. One person could hardly feed himself, and starvation was common. During centuries to follow, man slowly improved the uses of hand tools.

By 1000 BC, man used an animal to drag the hoe (as a crude plow), thus reducing human labor, mostly in seedbed preparation. Still, one person could produce only enough food for two people, and starvation continued to be widespread.

After 1731, when growing crops in *rows* with horse-hoeing was introduced, each person could then provide food for four persons.

In 1920 tractors were beginning to be widely used. This new-found power enabled each farmer to produce enough food for eight people.

About 1947 herbicide usage began spreading as a common practice. At that time one farmer could feed 16 people.

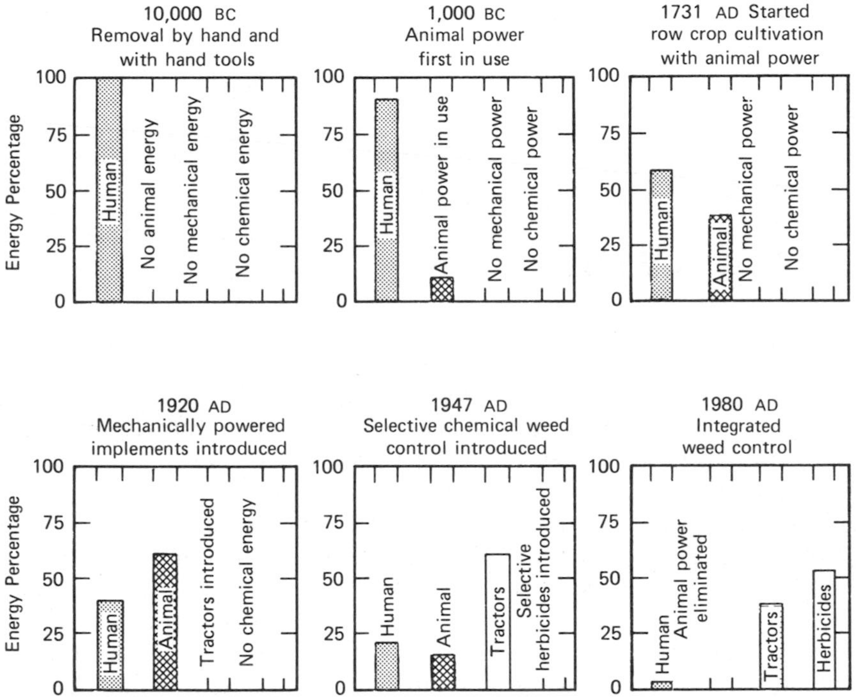

Figure 1-1. Energy sources providing weed control at different times. Data shown for 1920, 1947, and 1980 are for the United States.

During the 1950s, 1960s, and 1970s, many new herbicides were developed, along with other improved agricultural technology. Thus in 1980 one farmer could feed 38 persons. This means that fewer than three persons out of 100 are now needed directly to produce food. The other 97 can improve our standard of living by working as doctors, dentists, researchers, teachers, and skilled craftsmen to build better homes, schools, hospitals, refrigerators, automobiles, roads, agricultural equipment, agricultural chemicals—the list continues on and on.

With this technology the United States can produce far more food than we need for ourselves. In 1981 U.S. agriculture produced $43.8 billion from sale of agricultural products to other nations. Then U.S. citizens used this sizable income to buy imports such as oil, autos, TV sets, radios, and coffee. The U.S. food surplus also is a major source in helping to feed the hungry elsewhere in the world.

Several methods can be used to measure the effectiveness of agriculture technology in terms of cost and health effects. For example, in 1980 the U.S. population spent an average of 16% of total disposable income for food. This included production, shipment, processing, distribution, and sale costs through the grocery store. In comparison, people in many developing nations of the world with low agricultural

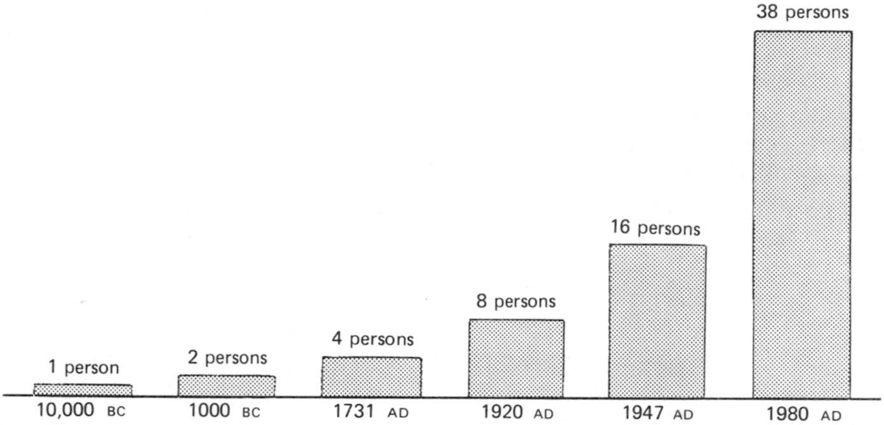

Figure 1-2. Crop energy output per farmer or the number of people fed by one farmer. Data for 1920, 1947, and 1980 are for the United States.

technology had to spend more than 50% of their disposable income for food.

Nutrition affects life expectancy in important ways. In 1920 in the United States the average life expectancy at birth was 54.1 years; by 1980 it had risen to 74.4 years. Also during this same time, in our nation the average height of the 20-year-old male increased by 2 in. An increase of 20.3 years in life expectancy during only 60 years is a tremendous testimonial to diet or nutrition and to medical science—during a period of the highest herbicide usage in history.

Weed scientists have been called "agricultural revolutionaries." Millions of hoe-hands have been replaced by modern weed-control techniques. There have been serious discussions on whether hoe-hands have a continuing right to work as hoe-hands. At what wages? Can we justify unemployment, even though it may be temporary? How could education help make the shift easier? How could government help? Has the change led to a higher standard of living for laborers? For farmers? For consumers? How has the change affected food supply? What percentage of income in the United States is used to pay for food? How important is it to find more productive use of labor than hoeing weeds? The authors propose these and similar questions for further thought and discussion. Answers to these questions will be decisive in determining "policy" in "developing" parts of the world.

DEFINITIONS OF A WEED, PESTICIDE, AND HERBICIDE

A weed is a *plant growing where it is not desired,* or a *plant out of place.* Therefore, rye in a wheat field is a weed; so is corn in a peanut field.

Weeds encompass all types of undesirable plants—trees, broadleaf plants, grasses, sedges, aquatic plants, and parasitic flowering plants (dodder, mistletoe, witchweed).

Weeds are also classified as *pests.* Other important pests are insects, plant diseases, nematodes, and rodents. A chemical used to control a pest is called a *pesticide.* A chemical used specifically for *weed control* is known as a *herbicide.*

TYPES OF WEED COSTS

Weeds affect almost everyone. Specific damages include the following:

Lower crop and animal yields.
Less-efficient use of land.
Higher costs to control insects and plant diseases.
Poorer-quality products.
More water-management problems.
Lower human efficiency.

Lower Plant and Animal Yields

Weed control is an expensive but necessary part of agricultural production, directly affecting the price of food. However, without modern weed science technology, food would be less abundant and cost much more. Weeds reduce yields of field crops, vegetables, fruits, pastures, meat, milk, and wool. Weeds reduce crop yields not only by releasing chemicals into the soil (*allelopathic effects,* as shown in Figures 1-3 and 3-2), but also by competition with the crop for soil water, soil nutrients, carbon dioxide, and light.

Livestock yields are often reduced through less pasture forage, or by poisonous or toxic plants that cause slower animal growth or death (see Figure 1-4).

Less-Efficient Use of Land

Weeds cause the following problems:

Costs are increased through cultivation, hoeing, mowing, and spraying.
Harvesting costs—both hand and machine—may be increased.
Root damage may result from cultivation.
Crop choice may be limited. Some crops don't compete effectively against weeds.

Figure 1-3. Weeds often reduce the yield of crops through allelopathic effects. *Left*: Alfalfa growing in weed-free soil. *Center*: Alfalfa watered with a filtrate drained from ground quackgrass rhizomes. *Right*: Alfalfa grown in soil containing ground quackgrass rhizomes (T. Kommendahl, University of Minnesota.)

Reforestation costs are increased and tree growth is slowed.

Costs are increased on rights-of-way for railroads, highways, and power and telephone transmission lines.

Land values may be reduced.

Higher Costs To Control Insects and Plant Diseases

Weeds harbor insect and disease organisms that attack crop plants. For example, the carrot weevil and carrot rust-fly may be harbored by the wild carrot, only later to attack the cultivated carrot. Aphids and cabbage root maggots live in mustard, and later attack cabbage, cauliflower, radish, and turnips. Onion thrips live in ragweed and mustards and may later prey on the onion crop. The disease of curly top on sugar beets is carried by insect vectors that live on weeds in wastelands. Many insects overwinter in weedy fields and field borders.

Disease organisms such as black stem rust may use the European barberry, quackgrass, or wild oat as hosts prior to attacking wheat, oats, or

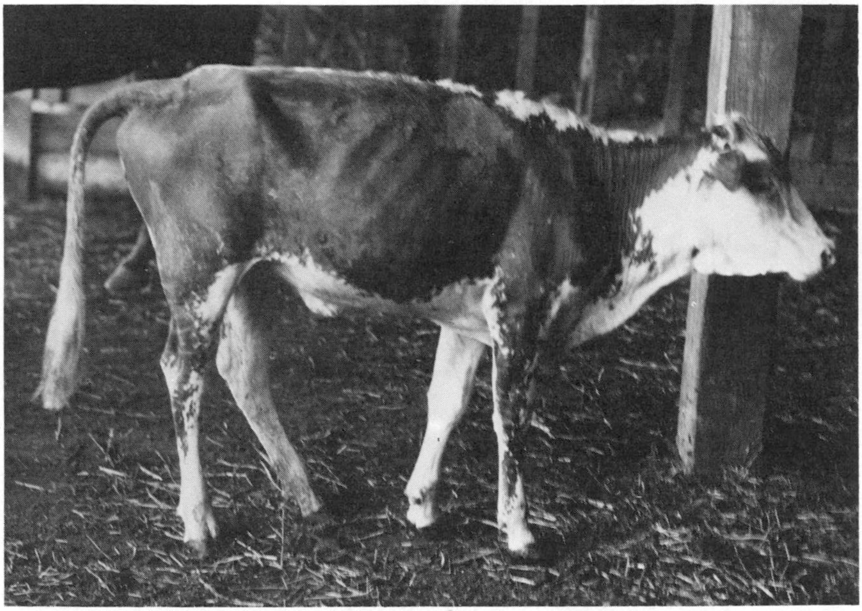

Figure 1-4. Weeds may reduce livestock yields. This animal was fed prostrate spurge. The abnormal stance and digestive disturbances became more pronounced as the spurge feeding continued (O. E. Sperry, Texas A&M University.)

barley. Some virus diseases are propagated on members of the weedy nightshades. For example, the virus causing "leaf roll" of potatoes lives on black nightshade. It is thought that aphids carry the virus to potatoes. A three-way harboring and transmission of a mycoplasma disease from weeds to citrus has been discovered in California. Leaf hoppers transmitted the disease organism, citrus stubborn disease (*Spiroplasma citri*), to and from diseased periwinkle and to and from London rocket (*Sisymbrium irio*). These weedy plants act as a source of the disease organism to infect citrus trees.

Poorer-Quality Products

All types of crop products may be reduced in quality. Weed seeds and onion bulblets in grain and seed, weedy trash in hay and cotton, spindly "leaf crops," and scrawny vegetables are a few examples.

Livestock products may be lower priced or unmarketable because of weeds; for example, onion, garlic, or bitterweed flavor in milk, and cocklebur in wool reduce the quality of the products. Poisonous plants may kill animals, slow down their rates of growth, or cause many kinds of abnormalities (see Figure 1-4).

More Water-Management Problems

Weeds are important in irrigation and drainage systems. They affect recreation and fishing on ocean beaches, lakes, and farm ponds (see Figure 1-5). They may give off undesirable flavors and odors in public water supplies, as well as affecting the shipping use of inland waterways and harbors.

Lower Human Efficiency

Weeds reduce human efficiency through allergies and poisoning. Hay fever, caused principally by pollen from weeds, alone accounts for tremendous losses in human efficiency every summer and fall. Poison ivy, poison oak, and poison sumac cause losses in terms of time and human suffering; children occasionally die from eating poisonous plants or fruits.

Weed control constitutes a large share of a farmer's work required to produce a crop. This effort directly affects the cost of crop production and thus the cost of food. In other words, it affects all of us, whether we farm or not.

Farming efficiency has permitted the release of agricultural workers for other means of livelihood. In 1850, in the United States about 65%

Figure 1-5. In St. Cloud, Florida, The State Mosquito Control Board sprayed ditches with diuron to eliminate weeds and improve drainage. The weed-control program reduced the expenses of mosquito control enough to permit a one-third savings in the total budget. *Left*: One year after hand cleaning. *Right*: One year after chemical treatment with diuron. (E. I. du Pont de Nemours and Company.)

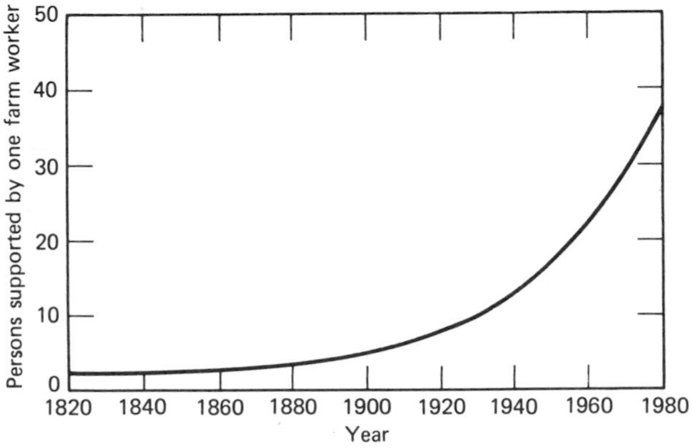

Figure 1-6. Each farmer in the United States produced enough agricultural products to provide for the number of nonfarmers shown in the graph. (Adapted from *USDA Agricultural Handbook* Nos. 423 and 574.)

of the work force was employed in agriculture. By 1870, this number had dropped to 52%, by 1890 to 42%, by 1910 to 31%, by 1930 to 11%, by 1970 to about 5%, and by 1980 to about 3%. This is possible only because the 97% are no longer needed to grow food. A nation's economy is directly related to the ratio of agricultural to industrial workers.

Farm output has risen steadily since 1910, the fastest rise occurring after 1940. A farm worker in 1980 produced, on an average, as much in 1 hr as he did in 3 hr in 1950, 5 hr in 1940, and 7 hr in 1910. In 1980, one agricultural worker produced enough agricultural products for 38 persons (see Figure 1-6). It has been said that one worker in a chemical factory producing 2,4-D is equivalent to the work efficiency of 100 hoe-hands in a cornfield.

The total number of hours worked in agriculture has continued to decline, despite increased production (see Figure 1-7). This increase in production efficiency is a result of modern agricultural technology involving research as well as educational programs. Weed control is an important part of this technology.

Weeds have been a plague to man since he gave up a hunter's life. Traveling in developing nations, one may feel that half the world's population work in the fields, stooped, moving slowly, and silently weeding. These people are a part of the great mass of humanity that spends a lifetime simply weeding. Many young people doing such work in Africa, Asia, and Latin America can never attend school; women do not have time to prepare nutritious meals or otherwise care for their families. Modern weed-control methods integrated into the economies and cultures of developing nations provide relief from this arduous chore and give nations the opportunity to improve their standards of living through more-productive work.

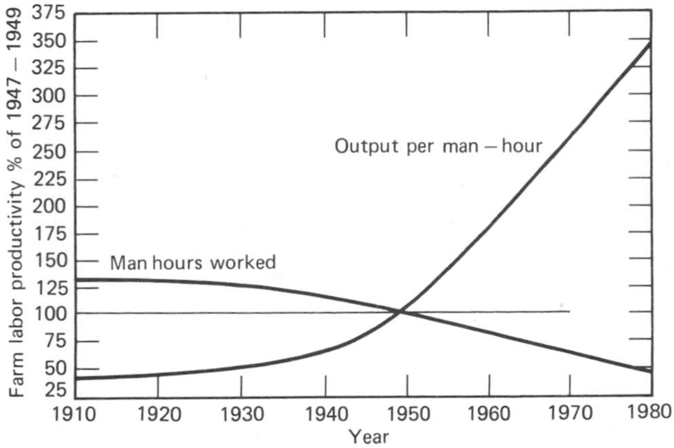

Figure 1-7 Farmers in the United States are working fewer hours and producing more products per hour of work than ever before. (Adapted from *USDA Agricultural Handbook* Nos. 423 and 574.)

FARM LOSSES FROM WEEDS

Farm losses from weeds are much higher than generally recognized. They represent a major cost item in food production. Because weeds are so common and so widespread, people do not fully appreciate their significance in terms of losses and control costs. Weeds are common on all 485 million acres of U.S. cropland and almost one billion acres of range and pasture. The whole system of weed control is so intermixed with so-called standard practices that separate accounting is difficult. For example, the original reason for planting crops in rows has long since been forgotten by the average person. Just how much of plowing and crop cultivation is for weed control, and weed control alone?

In 1971, the total cost to agriculture as a result of pests (Table 1-1) was slightly over $12 billion/year. Of this amount, the annual cost re-

Table 1-1. Annual costs[1] of Plant Pests of Crops (Adapted from USDA *Pesticide Review*)

Pest	Losses (X 1000)	Control (X 1000)	Total (X 1000)	Total (%)
Diseases	$3,152,815	$ 115,000	$ 3,267,815	27.1
Insects	2,965,344	425,000[1]	3,390,344	28.1
Nematodes	372,335	16,000	388,335	3.2
Weeds	2,459,630	2,551,050	5,010,680	41.6
Total	8,950,124	3,107,050	12,057,174	100.0

[1] Includes control costs for man, animals, and households as well as crops.

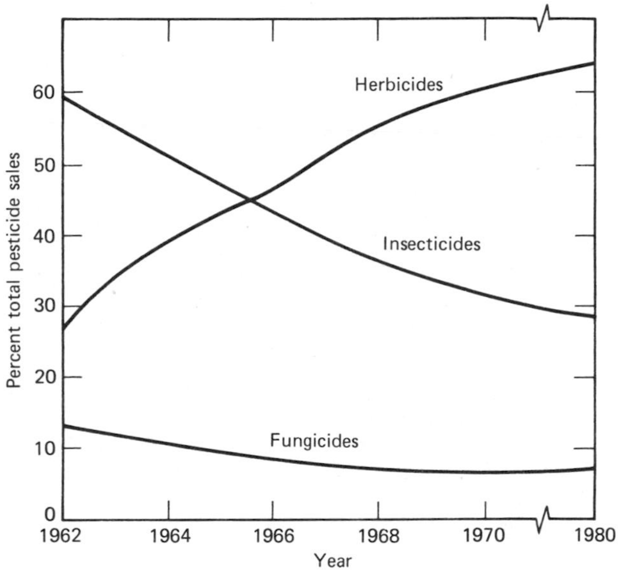

Figure 1-8. Value of pesticide sales. (Adapted from *USDA Pesticide Review*, 1971, and *USDA Agric. Econ. Rpt.* No. 461., 1980.)

sulting from plant diseases was about 27%, insects 28%, nematodes 3%, and weeds 42%. It is believed that these percentages are still essentially the same.

Since the early 1960s there has been a major shift in relative sales of various types of pesticides (see Figure 1-8). Since about 1968 the value of herbicide sales has exceeded the combined value of insecticide and fungicide sales.

The weed-control losses noted above are agricultural. If we add the losses to suburban lawns and gardens, public parks and other recreation areas, rights-of-way of highways, railroads, telephone and electric power transmission lines, industrial plants, lakes, ponds, ditches, tidewater areas, and public health, the figures become almost unbelievable.

SOCIOLOGICAL BENEFITS OF WEED-CONTROL CHEMICALS

These sociological benefits grow out of the use of agricultural chemicals:

1. The physical burden of farm work is lessened because the drudgery of hand labor is reduced.

2. Fewer people are required to produce food. Sons and daughters of farmers are now surgeons, teachers, researchers, legislators, manufacturers, writers, or labor-union leaders.

3. The percentage of money spent on food is reduced, leaving a larger share of income for other uses.
4. Farm children are freed and can attend schools.
5. Land use is improved.
6. There are health benefits.

HISTORY OF WEED CONTROL

For centuries man fought weeds with his hands, sharp sticks, and hoes (see Figure 1-1). Only recently has man used animal-powered cultivators and mechanical power. Sea salt was probably the first chemical used to kill plant life. About 1900, purified chemicals were used for selective weed control. Wide-scale use of selective herbicides began in 1947, after the discovery of 2,4-D [(2,4-dichlorophenoxy)acetic acid].

The science of weed control has advanced more since 1947 than in the previous million years. Major advances are still ahead. Most of us will see the day when a specific chemical will be used for a specific crop. The chemical or chemical combinations will inhibit all growth except the desired crop.

Here are the major figures and events that have laid the foundations for our present weed-control programs.

Julius Sachs, a German botanist, conducted many experiments (1859–1887) to learn the factors involved in the rooting and flowering of plants. He proposed that minute amounts of root-forming and flower-forming substances were moving either upward or downward in the plant. He wrote that "chemical messengers" were related to the flowering behaviors of begonias and squash. This is one of the earliest articles on translocation of growth-regulating substances.

Charles Darwin, an English biologist better known for his theory of evolution, wrote the book *Power of Movement in Plants,* published in 1900. In it he reported plant movement in response to light. Darwin germinated oats and *Phalaris* in darkness and then exposed the coleoptiles to light coming from one direction. The coleoptiles bent toward the light. By covering with tinfoil or by painting different areas of the coleoptile with india ink, he determined that the *coleoptile tip produced the growth-regulating substance.* This substance moved downward, causing curvature below. He found that the stem tips of Cruciferae (mustards) and Chenopodiaceae (goosefoot) also produced growth-regulating substances. Darwin wrote: "In several respects, light seems to act on plants in nearly the same manner as it does on animals by means of the nervous system . . . the effect . . . is transmitted from one part to another."

1908: Bolley (United States) reported successful weed control in wheat using common table salt, iron sulfate, copper sulfate, and sodium arsenite. In his words: "When the farming public has accepted this method (selective weed control) of attacking weeds . . . the gain to the country at large will be much larger in monetary consideration than that which has been afforded by any other single piece of investigation applied to field work in agriculture."

1941: R. Pokorny (United States) reported chemical synthesis of 2,4-D.

1942: P. W. Zimmerman and A. E. Hitchcock (United States) first reported 2,4-D to be a growth substance.

1944: P. C. Marth and J. W. Mitchell (United States) established the selectivity of 2,4-D. Dandelion, plantain, and other broadleaved weeds were removed from a bluegrass lawn. C. L. Hamner and H. B. Tukey (United States) successfully used 2,4-D in field weed control.

1945: W. G. Templeman (England) established the preemergence principle of soil treatment for selective weed control.

1951: *Weeds,* the journal of the Association of Regional Weed Control Conferences, was published.

1956: Weed Science Society of America was organized and assumed publication of *Weeds.* The publication was renamed *Weed Science* in 1968.

1961: *Weed Research,* official journal of the European Weed Research Council, was published.

1968: *Weed Control, Principles of Plant and Animal Pest Control,* Vol. 2, Publication 1597, was published by the prestigious National Academy of Sciences—thus confirming the importance of weed science as a discipline.

1970: The International Conference of Weed Control was organized by the Food and Agricultural Organization (FAO) of the United Nations. This meeting showed the international concern for solutions to the world's weed problem.

1976: International Weed Science Society formed.

PREVENTION, CONTROL, AND ERADICATION

Prevention

Prevention means stopping a given species from contaminating an area. Prevention is often the most practical means of controlling weeds. This is best accomplished by (1) making sure that new weed seeds are not carried onto the farm in contaminated crop seeds, feed, or on machinery; (2) preventing weeds on the farm from going to seed; and (3) preventing the spread of perennial weeds that reproduce vegetatively.

Control

Control is the process of limiting weed infestations. In crops the weeds are limited so that there is a minimum of weed competition. The amount of control is usually balanced between the costs involved and the amount of possible injury to the crop. Control is the usual aim when dealing with annual weeds competing with farm crops.

Eradication

Eradication is complete elimination of all live plants, plant parts, and seeds from an area. Two main problems are involved: (1) eliminating the living plants and (2) exterminating seeds in the soil. Because seeds may remain dormant in the soil for many years, it is usually easier to eliminate living plants than seeds. For eradication, both must be exterminated.

Usually serious perennial weeds are first located in small areas on the farm. Effective soil sterilants applied at this time may prevent them from spreading.

Most biological control programs using a highly specific insect or disease organism as the control agent can be expected to result in control, but not eradication. Dormant seeds in the soil may germinate long after the control agent has disappeared.

WEEDS CLASSIFIED

The plant's life span, season of growth, and its methods of reproduction largely determine the methods needed for control or eradication. In temperate climates there are three principal groups: annuals, biennials, and perennials.

Annuals

Annual plants complete their life cycle in less than one year. Normally they are considered easy to control. This is true for any one crop of weeds. However, because of an abundance of dormant seed and fast growth, annuals are very persistent, and they actually cost more to control than perennial weeds. Most common field weeds are annuals. There are two types—summer annuals and winter annuals.

Summer Annuals

Summer annuals germinate in the spring, make most of their growth during the summer, and the plants mature and die in the fall. The seeds lie dormant in the soil until the next spring. Summer annuals include such weeds as cockleburs, morning glories, pigweeds, common lambsquarters, common ragweed, crabgrasses, foxtails, and goosegrass. These weeds are most troublesome in summer crops like corn, sorghum, soybeans, cotton, peanuts, tobacco, and many vegetable crops.

Winter Annuals

Winter annuals germinate in the fall and winter and usually mature seed in the spring or early summer before the plants die. The seeds often lie dormant in the soil during the summer months. In this group, high soil temperatures (125°F or above) have a tendency to cause seed dormancy—inhibit seed germination.

This group includes such weeds as chickweed, downy brome, hairy chess, cheat, shepherdspurse, field pennycress, corn cockle, cornflower, and henbit. These are most troublesome in winter-growing crops such as winter wheat, winter oats, or winter barley.

Biennials

A biennial plant lives for more than one year but not over two years. Only a few troublesome weeds fall in this group. Wild carrot, bull thistle, common mullein, and common burdock are examples.

There is confusion between the biennials and winter annual group, because the winter annual group normally lives during two calendar years and during two seasons.

Perennials

Perennials live for more than two years and may live almost indefinitely. Most reproduce by seed and many are able to spread vegetatively. They are classified according to their method of reproduction as *simple* and *creeping.*

Simple Perennials

Simple perennials spread by seed. They have no natural means of spreading vegetatively. However, if injured or cut, the cut pieces may produce new plants. For example, a dandelion or dock root cut in half longitudinally may produce two plants. The roots are usually fleshy and may grow very large. Examples include common dandelion, dock, buckhorn plantain, broadleaf plantain, and pokeweed.

Creeping Perennials

Creeping perennials reproduce by creeping roots, creeping aboveground stems (stolons), or creeping belowground stems (rhizomes). In addition, they may reproduce by seed. Examples include red sorrel, perennial sow thistle, field bindweed, wild strawberry, mouseear chickweed, ground ivy, bermudagrass, johnsongrass, quackgrass, and Canada thistle.

Some weeds maintain themselves and propagate by means of tubers, which are modified rhizomes adapted for food storage. Nutsedge (nutgrass) and Jerusalem artichoke are examples.

Once a field is infested, creeping perennials are probably the most difficult group to control. Cultivators and plows often drag pieces about the field. Herbicides applied and mixed into the soil may reduce the chances of establishment of such plant pieces. Continuous and repeated cultivations, repeated mowing for one or two years, or persistent herbicides are often necessary for control. Cultivation in combination with herbicides is proving effective on some weeds. An eradication program requires the killing of seedlings as well as dormant seeds in the soil.

SUGGESTED ADDITIONAL READING

Weed Science published by Weed Science Society of America, 309 West Clark St., Champaign, IL 61820. See index at the end of each volume.

Weeds Today. Same availability as above.

Herbicide Handbook of the WSSA. Same availability as above.

Agrichemical Age. A commercial publication available from 83 Stevenson St., San Francisco, CA 94105.

Weed Research. Official journal of the European Weed Research Society. Available from EWRS, P.O. Box 14, Wageningen, The Netherlands.

The Economics of Agriculture Pest Control, an Annotated Bibliography, 1960–80, USDA, Econ. and Stat. Ser., Bib. and Lit. of Agric. No. 14, 1981.

2 Methods of Weed Control

Efficient weed control is basic to efficient and profitable agriculture. Weeds injure crops very slowly and in subtle ways. "Good" farmers commonly lose 10–25% of crop yields because of weeds. Farmers using poorer weed control easily lose 25–50% of their crops. Both groups of farmers would be shocked if they were to lose the same amount of money from the sudden death of livestock, from a hailstorm, or from a sudden insect attack. They'd call these losses catastrophic. Yet losses due to weeds are seldom viewed as catastrophic. In fact they're often excused as due to bad weather, poor soil, poor seed, poor season, and so forth.

CROP YIELD LOSSES DUE TO WEEDS

These four typical examples emphasize the actual cost of weeds.

Soybeans yields dropped 20, 39, 63, and 87% when infested by 1620, 3240, 6480, and 12,960 cocklebur plants per acre, as reported by Waldrep and McLaughlin in North Carolina.

In Mississippi, cotton fields kept weed-free averaged 2050 lbs of cotton per acre for three years. Robinson, working in Mississippi, stated that when weeds remained in the cotton row but were removed from the middles, average yields dropped 758 lb/acre. Where weeds were not controlled, the cotton crop failed completely.

For cotton grown in Alabama, according to Crawley and Buchanan, yields were cut up to 88% by one morningglory plant per 1½ ft of row.

In Brazil, William and Warren reported vegetable yield losses due to purple nutsedge. Irrigation was provided so that water was not limiting. Crop yield losses from nutsedge were as follows: garlic, 89%; okra, 62%; carrots, up to 50%; and transplant tomatoes, 53%.

Weeds cause similar losses in other U.S. crops such as corn, wheat, oats, barley, rice, fruits, and vegetables, and with pasture or range areas.

Most crops can be managed to effectively control annual weeds, and to a lesser extent perennial weeds. This management program usually calls for integration of several weed-control methods that provide maximum crop yields of high quality.

EARLY-SEASON WEED CONTROL

Effective early-season weed control is usually more important than late-season weed control. Buchanan and Burns, working in Alabama, reported that cotton must be weed-free for the first eight weeks to avoid yield loss. They also found that addition of nitrogen fertilizer did not basically alter the crop–weed relationship. In Oklahoma, Hill and Santelmann found that peanuts need to be weed-free for the first six weeks to avoid lower yields. Working in Iowa, Staniforth studied the effect of soil nutrients on corn and weeds. He reported that even with an abundance of nitrogen, competition between crop and weeds did not cease.

Working in Mississippi, Barrentine found that cocklebur permitted to grow in soybeans for the first 4, 6, 8, 10, and 16 weeks slashed soybean seed yields by 10, 36, 40, 60, and 80%. Friesen, working with tomatoes in Canada, reported that for the largest yields the crop had to be kept weed-free for 36 days after transplanting. If weeds were allowed to grow to the 48th day after transplanting, and then removed, the yield could be cut by 50%. If 5% of the weeds escaped removal 36 days after transplanting, the tomato yield was reduced by 60%.

In summary, if weeds are controlled until the crop *forms a canopy* overtop the weeds, the crop has a considerable competitive advantage. Annual weeds *germinating after the complete crop canopy has developed* seldom cause serious harm. Most crops develop a canopy in four to eight weeks after crop emergence. Weed-control programs should aim to provide nearly 100% weed control until the crop canopy develops. Weeds that escape before the canopy forms may seriously reduce crop yields.

As for crop yields, usually it makes little or no difference whether weeds are removed by hand, hoe, tractor, cultivator, biological control, selective herbicide, or a combination of these.

The following crop-management practices favor early canopy development:

1. Select a cultivar (variety) with rapid seed germination.
2. Select a cultivar that quickly forms a canopy over weeds.
3. Use a maximum number of crop plants per acre without exceeding the number of crop plants that reduce yields.
4. Use narrow rows for row crops, Need for cultivation may require row widths wide enough to permit cultivation.

5. Delay planting spring crops until soil is warm enough for rapid germination of crop seed. This delay may also give time to destroy one or two flushes of weed growth before planting the crop.

6. Use agronomic practices favorable to rapid growth of the crop, such as proper soil fertility, good seedbed preparation, proper depth of planting, adequate water, and tillage as needed to control weeds.

7. Use a selective herbicide or combination of herbicides that will control weeds with minimum or no crop injury. If the herbicide or herbicide combination is sufficiently selective, the first five items mentioned above will be less important.

METHODS OF WEED CONTROL

Often the best weed control with lowest cost comes from a program that combines two or three methods into an integrated, season-long use for both mechanical (tillage) and chemical methods.

Six methods of controlling weeds are the following:

1. Mechanical.
2. Crop competition.
3. Crop rotation.
4. Biological.
5. Fire.
6. Chemical.

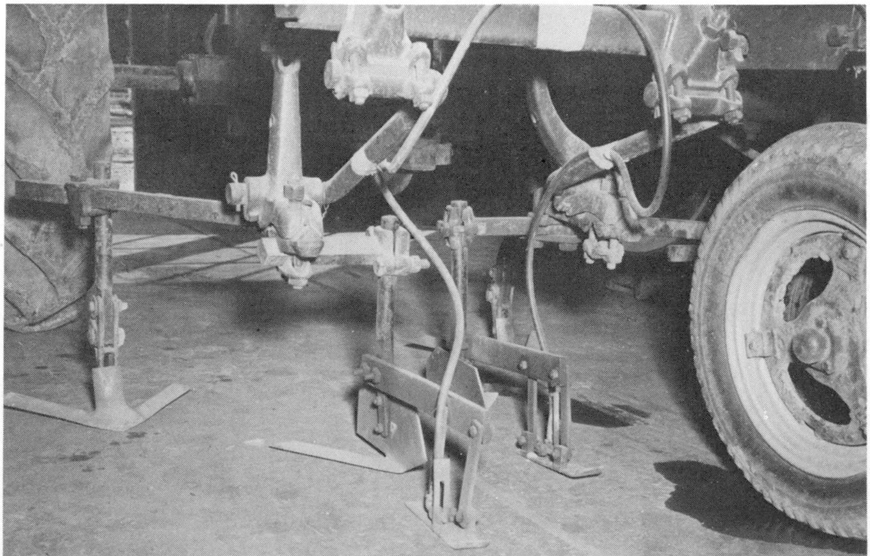

Figure 2-1. Shielded sweeps kills weeds between the rows, and a parallel-action spray applicator applies the herbicide to the weeds in the row. (O. B. Wooten, Delta Branch Experiment Station, Stoneville, MS.)

Mechanical

Mechanical control involves two well-known weed-control techniques–tillage and mowing.

Tillage

One type of tillage is *burial.* This is effective on most small annual weeds. If all growing points are buried, most annual weeds are killed. Burial is only partly effective on weeds with underground stems and roots that are capable of sprouting; examples are field bindweed, Canada thistle, quackgrass, bermudagrass, johnsongrass, and nutsedge. For control, such perennials must be repeatedly cut off or buried until the underground parts are killed by carbohydrate starvation (see Figure 2-3).

The second method of tillage is the *disturbed rooting system.* Shallow cultivation equipment such as sweeps, knives, harrows, finger weeders, and rotary hoes is used. The objective is to loosen or cut the root system so the plant dies from desiccation (drying out) before it can reestablish its roots. Small weeds are easily controlled by tillage, and it is most effective in hot, dry weather with dry soils. In moist soils, or if it rains soon after tillage, the roots may quickly reestablish themselves. In effect the weed is transplanted with little or no injury. Some herbicides, such as the dinitroanilines, reduce the ability of the weed's root system to reestablish itself, greatly increasing the effectiveness of cultivation.

Most serious perennial weeds are easily destroyed by tillage as seedlings, but are difficult to kill after they develop rhizomes, stolons, tubers, or reproductive roots. Combination of tillage with certain chemicals often increases the degree of control.

One of the most important reasons for growing crops in rows, instead of broadcast, is to permit tillage or cultivation. Only one benefit from cultivation stands out as distinct and clear. This is the benefit of weed control.

Other benefits for cultivation are occasionally claimed, such as increased soil aeration, the breaking of surface crusts, and increased rainfall penetration. In some cases these may be enhanced, in other cases hindered.

Cultivation, especially late deep cultivation, may injure crop roots. It may also cause rapid moisture loss and rapid drying of the soil. Cultivation is nearly always associated with a decreasing organic-matter content in the soil.

Experiments with corn have shown that if weeds are controlled without cultivating, at least on certain soils, the yields will equal those of plots cultivated as often as four times during the season. In New York,

Sturtevant and Lewis compared different cultivation treatments for corn as early as 1886. Here are their results:

Treatment	Yield per Acre (bushels)
Free weed growth	18.1
Weed removal, no cultivation	70.5
Ordinary cultivation plus hoeing	56.8

The New York report states, "Strangely enough, we have, during the existence of this station, not been able to obtain decisive evidence in favor of cultivation." Many later studies have provided data to support this statement.

Kiesselbach et al. stated that weed control is the main consideration in corn cultivation. In their experiments in Nebraska, late cultivation tended to reduce the yield.

In these studies investigators had to disturb the soil to some extent by scraping, hoeing, or hand-pulling the weeds. These actions may produce the effect of shallow cultivation.

Today herbicides make it possible to grow weed-free corn without scraping and hoeing. Research work in North Carolina was conducted by the senior author on three distinct soil types. They were Tidewater soil high in organic matter, which remains moderately loose and friable when dry; a Coastal Plain sandy loam, which remains loose and friable even though dry; and a Piedmont clay loam, which becomes hard and cracked when dry.

There were three treatments on each soil type:

1. No cultivation and sprays—gave weed-free corn.
2. One cultivation and sprays to control weeds.
3. Three cultivations.

These tests showed there is no benefit from cultivation, beyond weed control, on the Coastal Plain sandy loam soil. Yields of corn were the same with or without cultivation, provided the weeds were controlled (see Figure 2-2).

Results were similar on the Tidewater soil, with one related finding: There appeared to be a trend toward favoring one early, shallow cultivation, especially in a dry year.

The Piedmont clay loam soil showed a small advantage, above and beyond weed control, from one early cultivation in a dry season. But in seasons with favorable moisture, these differences disappeared. In

Figure 2-2. Corn averaged 131 bushels/acre for four years without cultivation. The experiment was performed on North Carolina sandy loam soil, and weeds were controlled by the use of herbicides. (North Carolina State University.)

all cases, one cultivation when the corn was 12–15 in. tall was as effective in favoring high corn yields as three cultivations.

To generalize, corn grown on porous, well-aerated soils show no benefit from soil tillage beyond weed control. If the weeds are controlled without corn injury, yields on noncultivated soil will equal those of cultivated corn. Late cultivation, which injures the corn roots, is likely to reduce corn yields.

Increased corn growth and corn yields may be expected from one early cultivation on soils that become hard and compacted when dry—in a dry year. There is no advantage, other than weed control, in giving three cultivations compared to one.

Mowing

Mowing is effective on tall-growing plants, but not on short-growing plants. Tall annual weeds are mowed primarily to reduce competition with crop plants and to prevent seed production.

Repeated mowing not only prevents seed production of perennial weeds, but also may starve the underground parts. To control tall perennial weeds, repeated and frequent cutting may be required for one to three years. At no time must the plant be permitted to replenish its underground stored food supply. The best time to start mowing or cultivation is usually when the underground root reserves are at a low ebb.

For many species this is between full leaf development and the time when flowers appear during late spring (see Figure 2-3).

Stem tips produce growth-inhibiting substances. These substances hinder growth of buds on the lower stem and on belowground stems and roots. Thus the stem apex (tip) produces substances that tend to inhibit the growth of lateral buds. This is known as *apical dominance.*

Following mowing or cultivation, apical dominance is no longer present. The dormant buds may start to grow, with more new stems developing than were originally present. This may appear to thicken the stand; however, it is desirable—if you mow this new growth repeatedly. The new stems grow at the expense of the belowground stored food, and repeated cutting hastens food depletion and death of the plant.

Annual weeds are often mowed to prevent seed production. Mow when the first flowers appear because some weed seeds will germinate even though you cut the plant soon after pollination. (See Chapter 3.)

Some annual weeds sprout new stems below the cut. This growth can often be controlled by cutting rather high at the first mowing and enough lower with the second mowing to cut off the sprouted stems. By the second mowing the stem is often hard and woody and cannot put out new sprouts below the cut. This procedure is effective on bitter sneezeweed, horseweed, and many others.

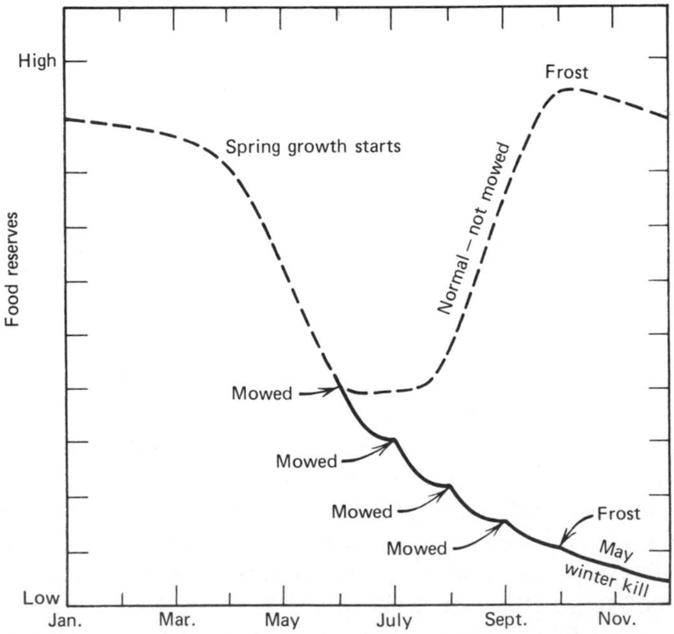

Figure 2-3. Food reserves of a perennial unmowed plant compared with reserves of a repeatedly mowed plant.

Weeds that form rosettes, mats, or a close-growing sod cannot be controlled by mowing. In fact, mowing tends to favor such weeds because it removes competition. Also, weeds that produce seed heads close to the ground are not controlled by mowing. That is why mowing does not kill established plants such as common dandelion, buckhorn plantain, dock, bermudagrass, crabgrass, goosegrass, and the foxtails.

Crop Competition

Crop competition is one of the cheapest and most useful methods available to farmers. It means using the best crop production methods—methods so favorable to the crop that weeds are crowded out. The formation of a crop canopy over weeds, discussed earlier, gives the crop a strong competitive advantage over weeds. Actually, competition makes full use of one of the oldest laws of nature—*survival of the fittest.*

Carried one step further a certain "balance of nature" is implied—a sort of "utopia" of peaceful harmony. Of course, this is not the case, because every living organism in nature carries on the most ruthless kind of competition. "Nature's balance" changes day by day. It is truly a survival of the fittest.

Man is the most disturbing influence of all when his wishes conflict with his environment. He uses his intelligence to bend most factors of the environment to his own advantage. Otherwise we could not feed the 4 billion people on this earth.

Herbicides are a relatively new tool in man's fight to control plant competition in his environment. If used properly, chemicals can do the job better and more efficiently than other methods—and safely too.

Weeds are naturally strong competitors. If not, they would fail nature's test of *survival of the fittest.* Those weeds that can best compete always tend to dominate. Domination may involve quick germination and very rapid early growth. For example, quick-growing annual weeds tend to dominate in desert areas having a moist spring but a dry summer and fall.

Weeds compete with crop plants for *light, soil moisture, soil nutrients,* and *carbon dioxide.* One mustard plant requires twice as much nitrogen and phosphorus, four times as much potassium, and four times as much water as a well-developed oat plant. The average common ragweed requires three times as much water as a corn plant.

Weeds, as a group, have much the same requirements for growth as crop plants. For every pound of weed growth, the soil produces about 1 lb less of crops. The competition for nutrients can be easily seen from chemical analysis of weeds growing in corn and competing with corn.

Table 2-1 shows that the weeds had an average of approximately twice as much nitrogen, 1.6 times as much phosphorus, 3.5 times as much potassium, 7.6 times as much calcium, and 3.3 times as much magnesium as corn.

Early weed competition usually reduces crop yields far more than late-season weed growth. Late weed growth may not seriously reduce yields, but it makes harvesting difficult, reduces crop quality, reinfests the land with seeds, and harbors insects and diseases during the fall and winter.

In planning a control program, it is important to know the weed's life cycle. Possibly the cycle can be interrupted to gain easy but very effec-

Table 2-1. Chemical Composition of Corn and Corn Weeds in September (*Agronomy Journal,* 1953. 45, 213)

		Mean Percentage Composition, Air-Dry Basis				
Plant	Growth Stage	Nitrogen	Phosphorus	Potassium	Calcium	Magnesium
Corn	Total plant, early milk stage	1.20	0.21	1.19	0.18	0.15
Pigweed	Green plants, 25–50% seeds ripe	2.61	0.40	3.86	1.63	0.44
Lambsquarters	Green plants, 25–50% seeds ripe	2.59	0.37	4.34	1.46	0.54
Smartweed	Green plants, 25–50% seeds ripe	1.81	0.31	2.77	0.88	0.56
Purslane	Lush plants, seeds partly ripe	2.39	0.30	7.31	1.51	0.64
Galinsoga	Lush plants, seeds partly ripe	2.70	0.34	4.81	2.41	0.50
Ragweed	Green plants, seeds partly ripe	2.43	0.32	3.06	1.38	0.29
Crabgrass	After bloom, seeds not ripe	2.00	0.36	3.48	0.27	0.54
Average for Weeds		2.36	0.34	4.23	1.36	0.50

tive control. In crop production this may be a well-timed chemical spray or a shift in planting date; thus the crop gets the upper hand or competitive advantage.

Smothering with plastic, tar paper, old linoleum, straw, sawdust, or any other similar material is largely a matter of competition for light. Most weed seedlings cannot penetrate these thick coverings and are killed because of lack of light.

Crop Rotations

Certain weeds are more common in some crops than in others. For example, pigweed, lambsquarters, common ragweed, and crabgrass are often found in corn and in other summer-cultivated crops. Wild mustard, wild oat, wild garlic, cornflower, thistles, and corn cockle are a few that occur frequently in the grain crops. Dock, bitter sneezeweed, ironweed, thistles, mullein, and weedy bromegrasses are often found in pastures.

Rotation of crops is an efficient way to reduce weed growth. A good rotation for weed control usually includes strong competitive crops grown in each part of the rotation involving both (1) summer row crops and (2) winter or early spring grain crops, drilled or broadcast.

Crop rotation with herbicide rotation is effective for the control of most weeds. For example, herbicides used in corn are usually different from those used in soybeans or cotton. A rotation of these crops with herbicide rotation usually provides efficient and effective weed control.

Biological Weed Control

Biological control is defined as control of the population of an organism through its natural enemies. For weed control, a natural enemy of the weed is introduced. This enemy must be otherwise harmless. It must not attack the crop. Natural enemies of weeds are primarily (1) insects, (2) plant diseases, and (3) certain other plants with allelopathic effects. Other biological controls include weed-eating fish and selective grazing by livestock (sheep and goats).

In general, biological control works best on large areas, like rangeland or waterways, infested by a mono-weed species (one weed). The method has special utility when one aggressive weed devastates thousands of square miles. It has been successful when the weed had been introduced, likely as weed seed, and in the move had been freed of its natural enemies. The predator or disease organism is then brought from areas of the world where the weed is native and is being controlled by its natural enemies.

Insects for Biological Control

To provide effective biological weed control, an insect must

1. Severely stunt or kill the weed.
2. Injure only the intended species–and no other.
3. Be mobile enough to locate its intended weedy host.
4. Reproduce faster than the intended weedy host.
5. Be adapted to an area or region similar to that of the weedy host.

The outstanding example of biological weed control is the prickly pear or cactus (*Opuntia* spp.) in Australia. The prickly pear was originally planted for ornamental purposes, but it spread rapidly from 1839 to 1925, covering 60 million acres and threatening much of the cultivated land. In seeking control measures research scientists found insects that attacked the prickly pear and *no other plants.* A moth borer (*Cactoblastis cactorum*) from Argentina was the most effective. It tunneled through the cladodes, underground bulbs, and roots, and became even more effective after several bacterial root organisms were accidentally introduced into the wounds caused by the insect.

By 1931 the moth borer had multiplied to such numbers that nearly all the prickly pears were destroyed by midseason. With little or no food supply, many of the moth borers starved. Then followed several years when prickly pear increased, with a later increase in the moth borer. Several "waves" of each type of growth occurred before an equilibrium or "balance of nature" was reached. For this reason, biological methods *control* rather than *eradicate.* This is especially true of weed species reproducing from seeds that may lie dormant in the soil for years.

There are other cases of effective biological control. In the western United States, St. Johnswort (*Hypericum perforatum*), a poisonous range weed, is being controlled by leaf-eating beetles (*Chrysolina* spp.) (see Figure 2-4). In Hawaii, pamakani (*Eupatorium adenophorum*), a range weed, is being controlled by a stem gallfly (*Procecidochares utilis*), and in Fiji, curse (*Clidemia hirta*), another range weed, is being controlled by a shoot-feeding thrip (*Liothrips urichi*).

Other more recent biological weed-control research has indicated control of musk thistle by a weevil (*Rhinobyllus conicus*), purple nutsedge by a moth (*Bactra verutana*), alligatorweed by three different insects, and puncturevine by seed weevils (*Microlarinus lareynii*).

Plant Diseases or Plant Pathogens

Plant pathogens attacking weeds also have shown their usefulness in weed control. Usually a disease organism is chosen that naturally attacks the weed but is harmless to desirable plants. These "warriors"

Figure 2-4. St. Johnswort or Klamath weed control by *Chrysolina quadrigemina* at Blocksburg, California. *Top*: Photograph taken in 1946. The foreground shows weeds in heavy flower, whereas the rest of the field has just been killed by beetles. *Middle*: Portion of the same location in 1949 when heavy cover of grass had developed. *Bottom*: Photograph taken in 1966 showing the degree of control that has persisted since 1949. Such results were reported throughout the state. (C. B. Huffaker, University of California, Berkeley.)

include certain species of bacteria, fungi, viruses, protozoa, and nematodes. The organism must be artifically propagated and at the appropriate time applied as a spray. Depending upon the organism and environmental conditions, one to six sprays may be required.

Several disease organisms hold promise of controlling weeds. Northern jointvetch is a serious leguminous weed of rice. Spray application of an anthraconose fungus (*Colletotrichum gloeosporiodes*) controls jointvetch without injuring rice. Another pathogenic fungus gives promise of controlling winged waterprimrose in rice. With cotton a pathogenic fungus (*Alternaria macrospora*) looks hopeful for control of spurred anoda without crop injury.

The disease organism must develop fast enough to quickly reduce the weed's competitive effect against the crop. Also, the disease organism must be suitable for large-scale production, packaging, storage, distribution, and application.

Allelopathic Effect

Some plants stunt or inhibit the growth of other plants in ways other than direct competition for light, nutrients, CO_2, and water. For many years this effect was confused with competitive effects. Now it is known that *some plants have a harmful effect on other plants through the release of a chemical compound(s) into the environment.* This reaction is called the *allelopathic effect.* It is likely caused by a biochemical substance, possibly similar to a growth regulator, that stunts, inhibits, or kills plants. The toxic substance may be released by the living plant, most likely through the roots, or it may be released from nonliving plant residues. Allelopathy is believed to be one way that weeds affect crop growth, and vice versa. The allelopathic effect of quackgrass on alfalfa is shown in Figure 1-3 and on wheat seedlings in Figure 3-2.

Stachon and Zimdahl, working in Colorado, ground or macerated foliage and roots of Canada thistle. Both foliage and roots, when added to soil, reduced later growth of pigweed and foxtail. Addition of nutrients did not mask the effect. Extracts from quackgrass in petri dishes gave similar reduced growth of the two weeds.

De Frank and Putnam, working in Michigan, confirmed that there is an allelopathic activity in both living plants and plant residues. Residues of sorghum and sudangrass as a soil mulch reduced a crabgrass population by at least 98% and common purslane by 50%.

Drost and Doll, working with yellow nutsedge in Wisconsin, reported that extracts both from the weed and from its residues have an allelopathic effect on corn and soybean. Soybean is the most sensitive.

Lockerman and Putnam, working in Michigan, reported differences in allelopathic effects among cucumber cultivars (varieties). Some cucumber cultivars grown in the field with proso millet suppressed the number of millet plants by 84% and reduced the fresh weight of the millet by 58%. Also, growth of pigweed and barnyardgrass was suppressed. Other cucumber cultivars showed no allelopathic reduction.

Fay and Duke, working in New York, reported that different oat cultivars possessed different levels of allelopathic activity. The ability of wild oat to stunt other plants growing nearby has been recognized for many years.

Research on allelopathic effects is limited. There is evidence of allelopathy in sorghum, rye, wheat, barley, oats, tobacco, and cucumber.

Many weeds have shown allelopathy. Later research surely will add to this list. At this time allelopathy is either strongly suspect or it has been proven in quackgrass, bermudagrass, johnsongrass, wild oat, Canada thistle, field bindweed, leafy spurge, purple nutsedge, yellow nutsedge, velvetleaf, pigweed, and sunflower.

Residues of the above crops and weeds usually show an allelopathic effect. Toai and Linscott, working in New York, determined the length of time that quackgrass residues remained toxic to alfalfa. The quackgrass was killed and left in or on the soil. With soil temperatures of 77°F, alfalfa could be seeded without injury after 21 days; but at 68°F and wet soil conditions, a delay of 42 days was required to avoid allelopathic effects.

Aquatic Weeds

Several fish species have been tried for aquatic weed control. One of the more interesting and more effective is the white amur, or grass carp (*Ctenoparyngodon idella*) (see Figure 2-5). It is native to the northern rivers of China and Siberia. Although native to cold water, it also thrives in warmer water.

It has been called the "hoe-hand with fins." It grows rapidly, at times up to 100 lb. It eats aquatic plants almost exclusively. Game fish apparently coexist very well in the same water. It has been used on a large scale in the state of Arkansas. It can also be a valuable source of protein for people in developing countries.

In some areas people are concerned that the white amur will reproduce naturally. To overcome this objection, research has developed methods of producing and releasing fish only of one sex, thus preventing reproduction.

Control vs Eradication by Biological Methods

Biological control methods do not eradicate weeds that reproduce by dormant seed. Dormant seed will germinate and produce seed in later years (up to 100 years) in a place where the biological control agent is no longer present.

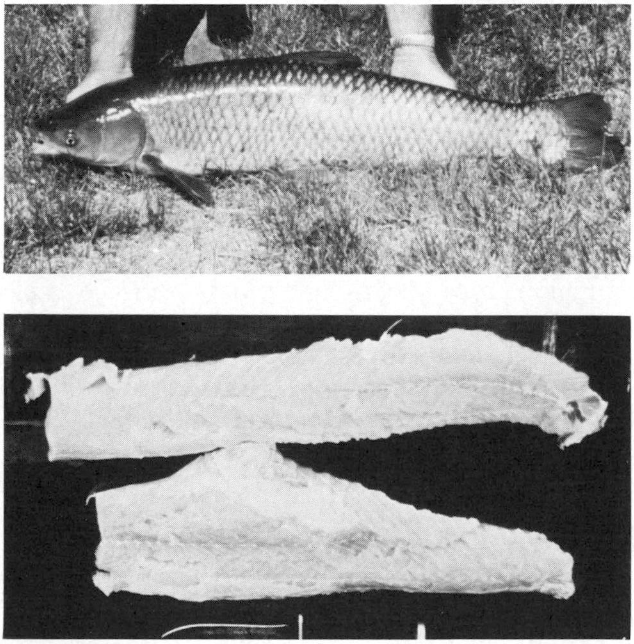

Figure 2-5. *Top*: The Chinese grass carp, or white amur, grows rapidly by grazing on water weeds. *Bottom*: A valuable by-product of this activity is a highly nutritious fish flesh. (Richard R. Yeo, USDA, SEA, ARS and LeRoy Holm, Madison, WI.)

Biological Weed Control in Low-Cost Control Areas

Biological weed control is well adapted to areas that will support only low-cost control methods. This includes extensive range areas, many forested areas, and many aquatic areas. Also, if a single problem weed causes trouble, there is a good chance of success with this method. If the one weed is highly objectionable—for example, poisonous to livestock—then biological control may represent a major accomplishment. But if many weeds are growing together, when one species is controlled, others (unchecked) will quickly replace the controlled weed.

Biological Weed Control in Cultivated Crops

In cultivated fields, a complex of 8 to 10 different weed species is usually present. Control of one weed may give little or no increase in crop yield. Disappearance of one weed may only let other weeds occupy that area. The major exception to this is the few perennial weeds that are extremely serious in cultivated fields. Thus efficient biological removal of such weeds as Canada thistle, quackgrass, bermudagrass, johnsongrass, purple nutsedge, yellow nutsedge, and field bindweed would be a major contribution to more-efficient crop production.

Fire

Fire can be used to remove undesirable plants from ditch banks, roadsides, and other waste areas; to remove undesirable underbrush and broadleaved species in conifer forests; and for annual weed control in some row crops. Burning must be repeated at frequent intervals if it is to control most perennial weeds. In alfalfa, weeds and some insects can be controlled by burning.

In waste areas, if the vegetation is green, a preliminary searing will usually dry the plants enough so that they will burn by their own heat 10–14 days later. Large trucks with maneuverable booms equipped with oil-burning nozzles are occasionally used for the preliminary sear.

When you want to favor pine trees over hardwoods, proper burning is usually successful in removing many of the hardwoods. Undesirable underbrush can also be removed by *controlled burning* at regular intervals. Burn often enough so that an excessively hot fire does not develop. Also, burn when there is a slight but steady breeze to sweep the fire along and when the debris close to the ground is slightly moist. *Controlled burning* reduces the waste and hazard of the destructive, uncontrolled forest fires (see Chapter 27).

Fire from special burners is effective in *flaming* or *sizz-weeding* cotton and corn to remove small annual weeds, but is much less effective on perennial weeds. The flame weeder must be used when the weeds are very small, and thus must be used at regular intervals. Proper adjustment and proper speed are important to avoid injuring the crop. The recent increase in the price of petroleum has reduced the economic attractiveness of this method.

Burning dried vegetation seldom kills the weed seeds; thus the practice has little value for this purpose.

Chemical

The use of chemicals for weed control has rapidly increased since 1947, when 2,4-D was first used on a broad scale as a herbicide (weed-killing chemical).

Time of Chemical Treatment

The time of chemical application may largely determine its usefulness in various crops. The time of application may be given with respect to the *crop*, or with respect to the *weed*.

Preplanting treatment is any treatment made before the crop is planted. For example, methyl bromide may be used in seedbeds to kill most weed seeds and vegetative reproductive parts before planting. It also controls many plant disease organisms, nematodes, and insect pests.

Preplant soil-incorporation applications are commonly used to place the herbicide in the soil area of the germinating weed seeds. This greatly increases the reliability of weed control regardless of additional rainfall, but the crop seed must have "true tolerance" to the herbicide. Several herbicides, such as EPTC and trifluralin, are soil-incorporated.

Preemergence treatment is any treatment made prior to emergence of a specified crop or weed. The treatment can be applied preemergence to both the crop and weeds, or to just the weeds. Therefore, a statement as to "preemergence to the crop," "preemergence to the weeds," or "preemergence to both crop and weeds" will clearly establish the timing of the treatment. The treatment is usually applied to the soil surface.

Postemergence treatment is any treatment made after emergence of a specific crop or weed. For example, 2,4-D gives effective postemergence control for most broadleaved weeds in corn, sorghum, small grains, and grass pastures.

The chemical may be applied postemergence to the crop, but preemergence to the weeds. For example, corn may be cultivated when it is 24–30 in. tall, leaving the field free of weeds. A herbicide sprayed on the soil surface between the rows at this time may inhibit weed-seed germination; thus the treatment is postemergence to the corn and preemergence to the weeds. Also, the chemical may be applied and incorporated into the soil between the crop rows.

Area of Application

Chemicals are applied *broadcast,* as a *band,* as a *directed spray,* and as *spot* treatment.

Broadcast treatment or blanket application is uniform application to an entire area.

Band treatment usually means treating a narrow strip—usually directly over or in the crop row. The space between the rows is not chemically treated, but is usually cultivated for weed control. Without the chemical, however, cultivation effectiveness may be reduced.

This method reduces the chemical cost, because the treated band is often one-third of the total area with comparable savings in chemical cost. In addition, when the chemical has a long period of residual soil toxicity (remains toxic in the soil for a long time), the smaller total quantity of the chemical reduces the residual danger to the succeeding crop. However, lack of chemical treatment between the rows may result in less-effective weed control from cultivation between the rows.

Figure 2-6. A directed spray. Nozzles direct spray across the row, killing small weeds. The cotton stem is tolerant of the aromatic oil, but the leaves are easily killed. (Delta Branch Experiment Station, Stoneville, MS.).

Directed sprays are applied to a particular part of the plant, usually to the lower part of the plant stem or trunk. Such sprays are usually directed at or just above the ground line (see Figure 2-6).

Dropped nozzles to spray between row crops give a directed spray treatment. The height of the nozzles is influenced largely by the size of the crop and also the size of the weeds to be controlled.

Trees are often basally treated by directing the spray to the base of the trunk. The chemical may kill by chemically girdling the tree and by preventing sprout growth (Figure 27-6).

Spot treatment is treatment of a restricted area, usually to control an infestation of a weed species requiring special treatment. Soil sterilant treatments are often used on small areas of serious perennial weeds to prevent their spread.

INTEGRATED PEST MANAGEMENT (IPM)

This term means the use of two or more control methods, with appropriate integration, to control a given pest. The objective is to maintain the pest population just below the level that causes economic loss (economic threshold) with minimal adverse environmental effects.

The term IPM was initially coined by entomologists who favored biological control of insects over the use of insecticides. However, now the term is used with the broader meaning described above and applies to weeds and diseases as well as to insects. It does not necessarily mean less use of pesticides, and in some cases increased use of pesticides may be required.

The term IPM has considerable appeal to the general public, even though they may not completely understand the concept. The current policy of the U.S. Department of Agriculture is "to develop, practice, and encourage the use of integrated pest management (IPM) methods."

The concept of IPM is not new to weed scientists. They have never depended only on chemical control methods but rather have combined (integrated) various complementary methods to provide seasonlong weed control. Efficient weed control is seldom accomplished by using any one control method alone. To attain the efficiency needed in the production of food crops, integrated management is needed that will provide efficient seasonlong weed control. Such control helps greatly to provide abundant and safe food without undue hazard to the environment. For additional information, see CAST Rpt. no. 93.

SAFETY OF HERBICIDES TO THE USER AND TO THE PUBLIC

Before herbicides are sold to the public, they are thoroughly tested to assure that they are effective as well as safe. The main agency responsible for these regulations in the United States is the Environmental Protection Agency (EPA). Local government agencies may also add their requirements, more restrictive than those of EPA, for unique local conditions.

EPA's prime concern with pesticide regulations is to see that the food supply is free of harmful pesticide residues and that the environment is not adversely affected. To achieve these goals, federal laws require manufacturers of pesticides to conduct exhaustive research on potential products before EPA will approve their use. These studies include:

1. Efficacy—that the product effectively does the job as claimed on the label. However, EPA no longer attaches emphasis to efficacy.
2. Toxicology—both acute and chronic.
3. Determination of herbicide residues, or absence of residues, in food and feed crops. If there are detectable residues, a tolerance level must be established.
4. Fate in the environment (residues in soil, runoff, groundwater, and wildlife).
5. Impact on the environment (induced changes in natural populations of animals, plants, and soil microorganisms).

The precise research requirements are available from EPA. These guidelines involve research in many phases of toxicology, biology, chemistry, and biochemical degradation of the compound, and the chemical's effect on air, water, soil microorganisms, wildlife, birds, fish, and other aquatic organisms. The cost of doing the research to develop a new herbicide in a new area of chemistry is between 20 and 50 million dollars and requires 6–10 years of research, indicating the thoroughness and complexity of the required research.

This research examines not only the applied compound, but also breakdown products. As a result, we can be assured that when applied according to label instructions, herbicides are safe to use.

Toxicity values, determined from experimental animals, are very helpful in assessing the toxic effect of a chemical on humans or other animals. Toxicity is the inherent capacity of a known amount of a substance to produce injury or death. Often, high-dosage levels are used to establish an *effect.*

A biological system's response to chemical exposure is dose-related. At a low dosage rate, the chemical may serve as a stimulant; at a moderate dosage rate, the chemical may have no effect; whereas at a high dosage rate, the chemical may act as a serious biological depressant. At high rates of application, the animal biological system may completely fail in its normal response. Thus high dosage levels may cause biological responses completely different from those caused by low levels of chemical exposure. There may be no biological relationship between dosage levels required to produce an *effect* and exposure levels experienced in normal usage.

"Safety in use" is a function of toxicity and exposure (exposure level and length of time of exposure). It is a total estimate of the hazards of use. Thus "safety in use" usually indicates safety or hazard better than toxicity data alone.

Risks from toxic compounds may be reduced by the following means:

Using a very dilute formulation (where directly ingested).

Using a formulation not readily absorbed through the skin or inhaled.

Using the compound only occasionally and where humans are not directly exposed.

Using the compound only by knowledgeable applicators who are properly equipped.

Terms Used in Expressing Toxicity

LD means *lethal dose* and LD_{50} means the dose that will kill 50% of a population of test animals; LC means *lethal concentration* and LC_{50} means the concentration that will kill 50% of the animals.

Acute oral refers to a *single dose* taken by mouth or ingested.

Acute dermal refers to a *single dose* applied directly to the skin (skin absorption).

Inhalation refers to exposure through breathing or inhaling.

Parenteral refers to introduction otherwise than by the intestine. More specifically, it usually refers to injection beneath the skin, into the veins, into the muscle, or as a dermal treatment.

Toxicity values are expressed as *acute oral* LD_{50} in terms of milligrams of the substance per kilogram (mg/kg) of body weight of the test animal; or *acute dermal* LD_{50} in terms of mg/kg; or inhalation data LC_{50} in milligrams of mist or dust per liter (mg/l) of air or parts per million (ppm) by volume of gas or vapor.

SUGGESTED ADDITIONAL READING

Weed Science. Journal of the Weed Science Society of America. Available in land grant university libraries or by subscription from WSSA, 309 West Clark St., Champaign, IL. 61820 (index at the end of each volume).

Weeds Today. Same availability as above.

Herbicide Handbook of the WSSA. Same availability as above.

Agrichemical Age. A commercial publication available from 83 Stevenson St., San Francisco, CA 94105.

Rice, E. L., 1974, *Allelopathy,* Academic, New York.

Zimdahl, R. L., 1980, *Weed-Crop Competition,* International Plant Protection Center, Oregon State University, Corvallis.

CAST (Council for Agric. Science and Technology) Report No. 93. March, 1982. 105 pages. Available from, CAST, 250 Memorial Union, Ames, Iowa. 50011.

3 Biology of Weeds and Weed Seeds

The biology of weeds is concerned with the establishment, growth, and reproduction of weeds as well as the influence of environment on these processes. Heredity and environment are the master factors governing life. Heredity determines what an organism becomes–that is, a dog or a corn plant–by controlling the life form, growth potential, method of reproduction, length of life, and so on. The environment largely determines the extent to which these life processes proceed.

Use of weed biology through environmental management makes it possible to shift plant ecology in a desired direction. In cultivated crop fields, the environment is made favorable to the crop plants and unfavorable to competing weeds. In pasture and range areas, desirable plants are favored by elimination of yield-reducing weeds including certain grasses, broadleafs and brush species. In forest areas, desirable trees are favored by elimination of competing annual weeds, brush, and undesirable trees. The use of herbicides is an important environmental management tool.

ECOLOGY

Ecology is the relationship of organisms to their environment. The ecology of weeds is primarily concerned with the effects of *climatic, physiographic,* and *biotic* factors.

Climatic

Climatic factors include the following:

Light (intensity, quality, length of day).
Temperature (extremes, average, frost-free period).
Water (amount, percolation, runoff, evaporation).

Wind (velocity, duration).
Atmosphere (CO_2, O_2, humidity, toxic substances).

Physiographic

Physiographic factors include the following:

Edaphic (soil factors including pH, fertility, texture, structure, organic matter content, CO_2, O_2, water drainage).
Topographic (altitude, slope, exposure to the sun).

Biotic

Biotic factors include the following:

Plants (competition, diseases, toxins including allelopathy, stimulants, parasitism, soil flora)
Animals (insects, grazing animals, soil fauna, man).

Many of the most common weeds have a broad tolerance to ecological conditions. For example, common weeds such as lambsquarters, chickweed, and shepherdspurse grow on almost all types of soils. Rarer species such as saltgrass, halogeton, and alkali heath are usually found only on alkali soils.

Similar environmental requirements of certain weeds and selected crop species produce some rather common crop–weed associations. Some examples are mustard in small grain, barnyardgrass in tomatoes, burning nettle in lettuce, chickweed in celery, and pigweeds in sugarbeets.

REPRODUCTION OF WEEDS

Weeds multiply and reproduce by both sexual and vegetative (asexual) means. Sexual reproduction requires pollination of a flower, which in turn produces seed. Vegetative reproduction is carried out by stems, roots, leaves, or modifications of these basic organs such as rhizomes (underground horizontal stem), stolons (aboveground horizontal stem), tubers, corms, bulbs, and bulblets. The roles these organs play in weed dissemination are discussed in the last section of this chapter. Seed dissemination is emphasized and discussed first because weeds spread most widely by that method.

SEED DISSEMINATION

Seeds in general have no method of movement; therefore, they must depend on other forces for dissemination. Regardless of this fact, they are excellent travelers. The spread of seeds, plus their ability to remain viable in the soil for many years (dormant), poses one of the most complex problems of weed control. This fact makes "eradication" nearly impossible for many seed-producing weeds.

Weed seeds are scattered by (1) crop seed, grain feed, hay, and straw; (2) wind; (3) water; (4) animals, including man; (5) machinery; and (6) weed screenings.

Crop Seed, Grain Feed, Hay, and Straw

Weeds are probably more widely spread through crop seeds, grain feed, hay, and straw than by other means.

Studies of wheat, oats, and barley seed sown by farmers in six North Central states reveal the seriousness of the problem. Researchers took seed directly from drill boxes and analyzed for weed seeds. About 8% of the samples contained primary noxious weeds, and about 45% contained secondary noxious weed seeds. About 80% of the seed had gone through a "recleaning" operation. Much of the recleaning was of limited benefit, because only part of the weed seeds was removed.

In the above study, *certified, registered,* and *foundation* seeds were free of primary noxious weeds and contained only a small percentage of the common weeds (Furrer, 1954).

Remember the difference between the serious, difficult-to-control weeds and those that are common on the farm and easily controlled. If the weed is already common on the farm, the soils are probably already infested; a few more seeds will make little difference. However, if a farm is free of a serious weed, one seed may be enough to start an infestation.

Farmers often feel that a low percentage of weed seeds on the seed label means that the few weeds present are of little importance. This may be a serious mistake (see Table 3-1). One dodder plant may easily spread to occupy one square rod during one season; thus only 0.001% dodder seed is enough to completely infest a legume crop the first year!

The prevalence of weed seed in legume seed is clearly shown by data collected by official state seed analysts. In 3643 samples, weed seeds averaged from 0.10 to 0.38% by weight. These figures indicate that certain weed species can completely infest a field the first year, despite an extremely low percentage of weed seeds present in the crop seed. These percentages are often below legal tolerances, which make it necessary to state their presence on the seed label.

Weeds are commonly spread through grain feed, hay, and straw. Where straw is used for mulching, it is important that the straw be free of viable weed seeds as well as grain seeds. The grain seed in the straw may prove to be a weed under such circumstances. Most of the grain seed

Table 3-1. Field Dodder[1] and its Rate of Planting in Contaminated Legume Seed Sown at the Rate of 20 lbs/acre

Dodder Seed by Weight (%)	No. of Dodder Seeds (per lb of legume seed)	No. of Dodder Seeds Sown (per acre)	(per square rod)
0.001	8	160	1
0.010	80	1,600	10
0.025	200	4,000	25
0.050	400	8,000	50
0.100	800	16,000	100
0.250	2,000	40,000	250

[1] There are 550,000 to 800,000 dodder seeds per pound.

will germinate and die if the straw is kept moist for 30 days with temperatures favorable to germination.

As shown later in this chapter, an appreciable number of weed seeds in grain and hay are viable after passing through the alimentary canal of the animal. If the manure is allowed to become "well rotted," the weed seeds will be killed.

Wind

Weed seeds have many special adaptations that help them spread. Some are equipped with parachutelike structures (pappus) or cottonlike coverings that make the seed float in the wind. Common dandelion, sowthistle, Canada thistle, wild lettuce, some asters, and milkweeds are examples (see Figure 3-1).

Water

Weed seed may move with surface water runoff, in natural streams and rivers, in irrigation and drainage canals, and in irrigation water from ponds.

Some seeds have special structures to help them float in water. For example, curly dock has small pontoon arrangements on the winged seed covering. Other seeds are carried in moving water or along the river bottom. Flooded areas from river overflows are nearly always heavily infested with weeds.

Irrigation water is a particularly important means of scattering seed. In 156 weed-seed catches in Colorado, in three irrigation ditches, 81 different weed species were found. In a 24-hr period, several million seeds passed in a 12-ft ditch.

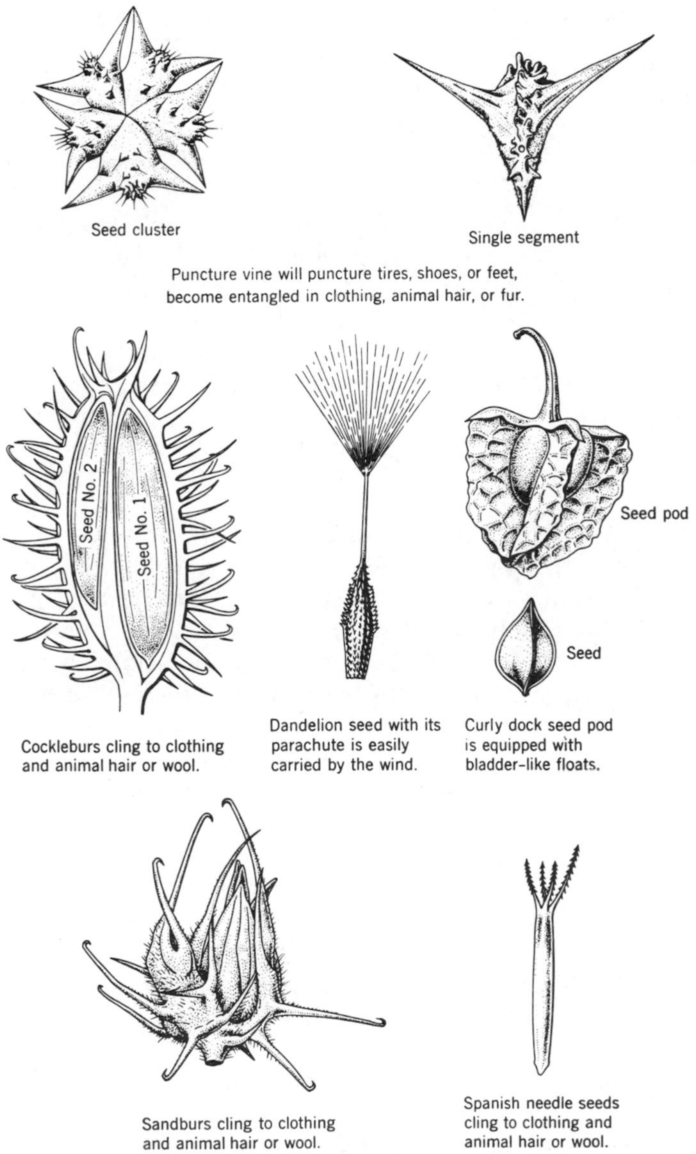

Seed cluster
Single segment
Puncture vine will puncture tires, shoes, or feet, become entangled in clothing, animal hair, or fur.
Seed No. 2
Seed No. 1
Cockleburs cling to clothing and animal hair or wool.
Dandelion seed with its parachute is easily carried by the wind.
Seed pod
Seed
Curly dock seed pod is equipped with bladder-like floats.
Sandburs cling to clothing and animal hair or wool.
Spanish needle seeds cling to clothing and animal hair or wool.

Figure 3-1. Examples of seeds that have special adaptations aiding in their spread.

Scientists have found great variation in the length of time that seeds remain viable in fresh water. Results in Table 3-2 are based on weed seed suspended in bags (luminite screen sewn with nylon thread) at 12- and 48-in. depths in a freshwater canal at Prosser, Washington. Water could circulate freely within the bags. The bags were removed periodically over a five-year period and germination counts were made.

Table 3-2. Germination of Weed Seed after Storage in Fresh Water

Field bindweed	After 54 months, 55% germinated
Canada thistle	After 36 months, about 50% germinated After 54 months, none germinated
Russian knapweed	After 30 months, 14% of seed still sound; none germinated after 5 years
Redroot pigweed	After 33 months, 9% still sprouted
Quackgrass	None sprouted after 27 months
Barnyardgrass	After 3 months less than 1% germinated; none germinated after 12 months
Halogeton	After 3 months less than 1% germinated; none germinated after 12 months
Hoary cress	After 2 months germination dropped to 5% or less, and dropped to zero after 19 months

There was little or no variation in germination between the 12- and 48-in. depths, but there was considerable difference in viability among species. Clearly, some seeds can be stored in fresh water for three to five years and still germinate. In some cases storage in water tended to break dormancy and increased the percent germination; this was especially true after two to four months in the water (Bruns and Rasmussen, 1958).

Every effort should be made to free irrigation channels and water storage facilities from weed-seed-producing plants that will be troublesome in irrigated fields.

Animals (Including Man)

Animals, including man, are responsible for scattering many seeds. They may carry the seed on their feet, clinging to their fur or clothes, or internally (ingested seed).

Many seeds have specially adapted barbs, hooks, spines, and twisted awns that cling to the fur or fleece of animals or to man's clothing. Sandburs, cockleburs, sticktights, and beggarticks are examples. Others may imbed themselves in the animal's mouth, causing sores. Examples are wild barley, downy brome, and various needlegrasses. Other seeds have cottonlike lint or similar structures that help the seed cling to fur or clothes. Annual bluegrass and bermudagrass seed will stick to the fur of rabbits and dogs. Mistletoe seeds are sticky and become attached to the feet of birds. Birds often carry away fleshy fruits containing seeds for food.

Many weed seeds pass through the digestive tracts of animals and remain viable. Often weed seedlings are found germinating in animal

Table 3-3. Percentage of Viable Seeds Passed by Animals Based Upon Total Number of Seeds Fed (Harmon and Keim, 1934)

	Percentage of Viable Seeds Passed by					
Kind of Seeds	Calves	Horses	Sheep	Hogs	Chickens	Average
Field bindweed	22.3	6.2	9.0	21.0	0.0	11.7
Sweetclover	13.7	14.9	5.4	16.1	0.0	10.0
Virginia pepperweed	5.4	19.8	8.4	3.1	0.0	7.3
Velvetleaf	11.3	4.6	5.7	10.3	1.2	6.6
Smooth dock	4.5	6.5	7.4	2.2	0.0	4.1
Pennsylvania smartweed	0.3	0.4	2.3	0.0	0.0	0.6
Average	9.6	8.7	6.4	8.8	0.2	6.7

droppings. Table 3-3 shows results of feeding weed seeds to different kinds of livestock and washing the seed free from the feces. The figures are given as a percentage of viable seeds recovered.

Machinery

Machinery can easily carry weed seeds, rhizomes, and stolons. Weed seeds are often spread by harvesting equipment, especially combines. Cultivation equipment, tractors, and tractor tires often carry dirt that may include weed seeds. Also, cultivation equipment may drag rhizomes and stolons, dropping them later to start new infestations.

Weed Screenings

Most weed seeds have a reasonably high feed value. Common ragweed seed, when chemically analyzed, had 20.0% crude protein, 15.7% crude fat, 18% nitrogen-free extract, and 4.83% ash. Because of their relative cheapness, weed-seed screenings are often included in livestock feeds (see Tables 2-1 and 28-1).

Over the years, considerable attention has been given to the problem of destroying the viability of weed seeds in screenings. Seeds are usually finely ground, or soaked in water and then cooked. Fine grinding with the hammer mill has reduced to a minimum the hazard of scattering live weed seeds.

Pelleting destroyed the seed's viability when high temperatures were used. Fish solubles with a high protein content were added to screenings after heating to about 200°F. The mixture was pressed into pellets.

Germination tests indicated that the seeds were no longer viable with that amount of heat.

NUMBER AND PERSISTENCY OF WEED SEEDS

Number in the Soil

Some species produce a remarkable number of living seeds per acre. Wild poppies seriously infested the Rothamsted (England) Experiment Station in 1930 with an estimated 113 million poppy seeds per acre.

In Minnesota, weed-seed counts at four different locations on 24 different plots showed from 98 to 3068 viable weed seeds per square foot of soil, 6 in. deep. Converted to an acre basis, this is between 4.3 million and 133 million seed per acre in the upper 6 in.

Soil samples were taken on 10 plantations in Louisiana heavily infested with johnsongrass. The average number of viable johnsongrass seeds per acre was 1,657,195. A sugar cane crop was grown for three years without permitting the addition of new johnsongrass seed; after three years the johnsongrass seed population in the upper 2.5 in. of soil dropped to 1.3% of the original number.

Number Produced by Plants

The persistence of annual and biennial weeds depends mainly on their ability to reinfest the soil. The first infestation of most perennial species depends on seed. Obviously, if we could control the production of seed, we could eventually eliminate many species.

One scientist, reporting on 245 species, found that the number of weed seeds produced by one plant ranges from 140 for leafy spurge to nearly a quarter of a million seeds for common mullein. Of these, 24 species have been selected and are listed in Table 3-4. Only plump, well-developed seeds were counted from well-developed plants growing with comparatively little competition. Those 24 species averaged 25,688 seeds per plant. A second report listed 263 species. One witchweed plant may produce as many as one-half million seeds.

Age of Seed and Viability

The length of time that a seed is capable of producing a seedling varies widely with different kinds of seeds and with different conditions. Certain seeds keep their viability for many years, whereas others die within a few weeks after maturing if they do not find a suitable environment for germination.

Table 3-4. Number of Seeds Produced per Plant, Number of Seeds per Pound, and Weight of 1000 Seeds (Stevens, 1932)

Common Name	Number of Seeds (per plant)	Number of Seeds (per lb[1])	Weight of 1000 Seeds g
Barnyardgrass	7,160[2,3]	324,286	1.40
Buckwheat, wild	11,900	64,857	7.0
Charlock	2,700	238,947	1.9
Dock, curly	29,500	324,286	1.4
Dodder, field	16,000[3]	585,806	0.77
Kochia	14,600	534,118	0.85
Lambsquarters	72,450	648,570	0.70
Medic, black	2,350	378,333	1.2
Mullein	223,200	5,044,444	0.09
Mustard, black	13,400[4]	267,059	1.7
Nutsedge, yellow	2,420[2]	2,389,474	0.19
Oat, wild	250[2]	25,913	17.52
Pigweed, redroot	117,400[2]	1,194,737	0.38
Plantain, broadleaf	36,150	2,270,000	0.20
Primrose, evening	118,500	1,375,757	0.33
Purslane	52,300	3,492,308	0.13
Ragweed, common	3,380[2]	114,937	3.95
Sandbur	1,110[2]	67,259	6.75
Shepherdspurse	38,500[2,3]	4,729,166	0.10
Smartweed, Pennsylvania	3,140	126,111	3.6
Spurge, leafy	140[4]	129,714	3.5
Stinkgrass	82,100[2,3]	6,053,333	0.07
Sunflower, common	7,200[2,3]	69,050	6.57
Thistle, Canada	680[2,3]	288,254	1.57

[1] Calculated from the weight of 1000 seeds.
[2] Many immature seeds also present.
[3] Many seeds shattered.
[4] Yield of one main stem.

A short-lived species, silver maple, has seeds with about 58% moisture when shed, and they will germinate immediately. However, when moisture content drops to 30–34%, they die. In nature this occurs within a few weeks.

Lotus (*Nelumbo nucifera*) seeds found in a lake bed in Manchuria were still viable after approximately 1000 years. The seeds were 1040 ± 210 years old, as measured by a radioactive carbon dating technique. The seeds had been covered in deep mud and very cold water.

To determine the longevity of seed, Duvel placed 107 species in porous clay pots and buried them 8, 22, and 42 in. deep in the soil. He removed samples at various intervals and left others undisturbed. His findings, reported by Toole and Brown, were as follows:

After 1 year, seed of 71 species germinated
After 6 years, seed of 68 species germinated
After 10 years, seed of 68 species germinated
After 20 years, seed of 57 species germinated
After 30 years, seed of 44 species germinated
After 38 years, seed of 36 species germinated

After 38 years:
- 91% of jimsonweed seed germinated
- 48% of mullein seed germinated
- 38% of velvetleaf seed germinated
- 17% of eveningprimrose seed germinated
- 7% of lambsquarters seed germinated
- 1% of green foxtail seed germinated
- 1% of curly dock seed germinated

The actual percentage of germination is not as important as the fact of survival. Programs to eradicate plants with long seed dormancy are doomed to failure until techniques are developed to break (100%) seed dormancy in soil.

The three depths did not greatly influence the data; however, seed at the 42-in. depth lived slightly longer. The 8-in. depth is far below the ideal depth for most seeds to germinate. If one sample had been buried 1 in. deep or less, greater differences caused by depth would have been likely.

In another test started in 1879, Beal mixed seeds of 20 different weed species with sand and buried the mixtures in uncorked bottles, with the opening tipped downward. After 20 years, 11 of the species were still alive. After 40 years, 9 species were still alive. These 9 were redroot pigweed, prostrate pigweed, common ragweed, black mustard, Virginia pepperweed, eveningprimrose, broadleaf plantain, purslane, and curly dock.

After 50 years, moth mullein germinated for the first time. Therefore, the seeds were still alive at 40 years but did not germinate because they were dormant. After 70 years, moth mullein had a germination rate of 72%, eveningprimrose 17%, and curly dock a few percent. After 80 years, moth mullein still had a germination rate of 70–80%, but with the other two species only a few seeds germinated. It appeared that these two species were nearing the end of their survival period.

Thus many weed seeds retain their viability for 40 years and longer when buried deep in the soil. Scientists believe that much of the longevity depends upon the seeds being buried deep, with a reduced oxygen supply available. If brought to the surface with other conditions favorable, many of the seeds will germinate. It is evident that repeated cultivation for several years without opportunity of reinfestation effectively reduces the weed-seed population in a soil.

Depending on climatic conditions, abandoned cultivated crop land will usually return to a forest or grassland vegetation. If cultivated after 30–100 years, the original field weeds immediately appear. The weed seeds have remained dormant while buried deep in the soil.

GERMINATION AND DORMANCY OF SEEDS

Germination includes several steps that result in the quiescent embryo's changing to a metabolically active embryo as it increases in size and emerges from the seed. It is associated with an uptake of water and oxygen, use of stored food, and, normally, release of carbon dioxide. For a seed to germinate, it must have an environment favorable for this process. This includes an adequate, but not excessive, supply of water, a suitable temperature and composition of gases (O_2/CO_2 ratio) in the atmosphere, and light for certain seeds. Specific requirements for seed germination differ for various species. Although these factors are optimal, a seed may not germinate because of some kind of dormancy.

Dormancy is a type of resting stage for the seed. Dormancy may determine the time of year when a seed germinates, or it may delay germination for years and thus guarantee the viability of the seed in later years. Five environmental factors affect seed dormancy: *temperature, moisture, oxygen, light, and the presence of inhibitors* including allelopathic effects. Other factors directly related to the seed and its dormancy include impermeable seed coats (to water, oxygen, or both), mechanically resistant seed coats, immature embryos, and after-ripening.

Temperature

The temperature that favors seed germination varies with each species. There is a *minimum* temperature below which germination will not occur, a *maximum* temperature above which germination will not occur, and an *optimum,* or ideal, temperature when seeds germinate quickest. Thus some seeds germinate only in rather cool soils, whereas others do so only in warm soils.

The temperature requirements for most crop seeds are well established, and farmers recognize these requirements and plant accordingly. Cotton, for example, requires relatively high temperatures for germination, whereas the small grains will germinate at relatively cool temperatures.

Russian pigweed seed has germinated in ice and on frozen soil. Wild oat may germinate at temperatures of 35°F. Comparatively low temperatures (between 40 and 60°F) are necessary before certain winter annuals will germinate. High temperatures may cause a secondary type of seed dormancy, especially with some winter annual weeds. Wormseed mustard was introduced to secondary dormancy by temperatures of 86°F, and many summer annuals require temperatures of 65–95°F to germinate. Alternating temperatures are often better than a constant temperature for seed germination.

When redroot pigweed seed (a summer annual) was placed in germinators at 68°F, some seed remained dormant for more than six years. The seeds could be induced to germinate at any time in three ways: (1) by raising the temperature to 95°F, rubbing with the hand, and replacing at 68°F; (2) by partial desiccation; or (3) by alternating the temperatures.

Temperature alone does not completely explain the periodicity of seed germination. Often the seeds have another form of dormancy that temporarily stops germination. This may be a survival mechanism to keep the plant from germinating immediately upon maturity in a season not suited to the plant.

Seeds may lie dormant for as little as several weeks or as much as several years. For example, cheat and hairy chess have a primary dormancy of four to five weeks after maturity. During this period they will germinate only if subjected to low temperatures (59°F or below). However, if the seeds are stored for four to five weeks, germination will then readily occur at temperatures of 68–77°F.

Moisture

Germination is normally a period of rapid expansion and high rates of metabolism or cell activity. Much of the expansion is simply an increase in water, expanding cell walls. If water content of the seed is reduced, the activity of enzymes–and consequently metabolism–slows down. The amount of moisture contained in seeds may determine their respiratory rate. During germination the seed respires at a very rapid rate. Many seeds cannot maintain this high rate of respiration until they reach a moisture content of 14% or more. Thus in dry soils the seed remains dormant.

Dry seeds can tolerate severe conditions; some have been kept in boiling water for short intervals without injury and others in liquid air (-310°F). When moist enough for germination, the same seeds may be killed by cold temperatures of 30°F or warm temperatures of 105°F.

Oxygen

In addition to the right temperature and sufficient moisture, germination depends on oxygen. Aerobic respiration requires more free oxygen than anaerobic respiration; thus some seeds start germination under anerobic conditions, then shift to aerobic respiration when the seed coat ruptures.

The percentage of oxygen found in the soil varies widely, depending on soil porosity, depth in the soil, and organisms in the soil that use oxygen and release CO_2 (microorganisms, roots, etc.). The percentage of oxygen in the soil is usually inversely proportional to the percentage of CO_2. In swampy rice land there may be less than 1% oxygen in the soil atmosphere; in freshly green-manured land, 6–8%; and where corn is growing rapidly, 8–9%, compared with about 21% in a normal atmosphere. The percentage of carbon dioxide may range from 5 to 15% under such conditions, compared with 0.03% for normal air.

Different species vary considerably in the amount of oxygen needed for seeds to germinate. Wheat seed germinated well when the replenished oxygen supply of the soil was 3.0 $mg/m^2/hr$ or more. It failed to germinate when the rate was below 1.5 mg. Rice seed germinated at 0.5 mg. Broadleaf cattail and some other acquatic plants germinate better at low oxygen concentrations than with normal air.

The effect of different oxygen concentrations on the seeds of field bindweed, leafy spurge, hoary cress, and horsenettle was determined. Oxygen concentrations of 5% produced little to no germination of hoary cress, horsenettle, and leafy spurge. At 10% oxygen and below, these three weeds germinated at a rate far below normal, whereas bindweed was reduced somewhat. The highest percentage germination was found at about 21% oxygen (normal air) for leafy spurge and hoary cress; but the best germination for horsenettle was at 36% oxygen and for field bindweed 53%.

Wild oat and charlock germination can be greatly suppressed by reducing the oxygen supply by soil compaction. Cultivation increased sixfold the number of wild oat that germinated and the number of charlock twofold, compared with the compacted plots. Cultivation increased soil aeration and thus increased the content of oxygen in the soil atmosphere.

Excess water in the soil cuts down seed germination of most plant species. Researchers believe that lower germination is related to a smaller supply of oxygen in waterlogged soils, rather than merely excess water.

Many small-seeded weed seeds germinate only in the upper 1–2 in. of soil, mostly in the upper 1 in. A limited number, however, germinate below 2 in. In sandy soils, seeds germinate deeper than in clay soils as a result of better aeration or better oxygen supply in the sands. Some seeds buried deep in the soil do not germinate but lie dormant for many years. When brought to the surface, they germinate promptly. Aeration, involving increased oxygen supply, is probably responsible.

By using herbicides, successive crops of weed seedlings may be killed without disturbing the soil. Few to no viable weed seeds may remain in the upper soil layer. The soil surface may then remain relatively free of weeds. Repeated treatments to kill annual weeds before they produce seed is especially useful in areas where the soil is not disurbed by cultivation, such as in some perennial crops, permanent sod, or turf areas.

Light

Some kinds of seed germinate best in light, others in darkness, and others germinate readily in either light or darkness. Among several hundred species in which the role of light has been investigated, about half require light for maximum germination. The length of day and quality (color) of light also have influence. Here is a brief review of the electromagnetic spectrum (color of wavelengths).

Name	Wavelength (Ångströms)
X ray	10–150
Ultraviolet	Below 4000
Visible spectrum	
Violet	4000–4240
Blue	4240–4912
Green	4912–5750
Yellow	5750–5850
Orange	5850–6470
Red	6470–7000
Infrared	Greater than 7000

Germination of lettuce seed was promoted by radiation at 6600 Å (red light) and inhibited at 7300 Å (infrared light). A later study with lettuce showed that the inhibitory effect of infrared light could be reversed by red light. Regardless of the number of alternating periods of red and infrared light to which the seeds were exposed, the final type of light determined the percentage germination. For example, when the final light was infrared, germination was about 50%; but when the final light was red, germination reached almost 100%. Without light, germination fell to about 8%.

Here are a few examples of species and the light requirement of their seeds for germination.

Germination Favored by Light		Germination Favored by Darkness	Germination in Either Light or Darkness	
Bluegrass	Dock	Onion	Salsify	Wheat
Tobacco	Primrose	Lily	Bean	Rush
Mullein	Buttercup	Jimsonweed	Clover	Toadflax

These effects vary from species to species. In some seeds, the light requirement can be replaced by after-ripening in dry storage, alternating the daily temperature, higher temperatures, and treatment in potassium nitrate or gibberellic acid solutions.

Allelopathic Effect

Many plants produce and release substances to the soil that inhibit other plants. This is now known as the allelopathic effect. (see page 28). Quackgrass, bermudagrass, johnsongrass, wild oat, and walnut trees are but a few examples (see Figures 1-3 and 3-2). Herbicides may chemically inhibit the germination of a seed, a bud of a plant, or a tuber. For example, trifluralin will inhibit the sprouting of a johnsongrass or bermudagrass bud on the rhizome, and EPTC may inhibit the sprouting of a nutsedge or a potato tuber.

Seed Coat Impermeable to Water, Oxygen

Seed coats may be waterproof and may prevent the seed from absorbing water. Such seed will not germinate even if soil moisture is plentiful.

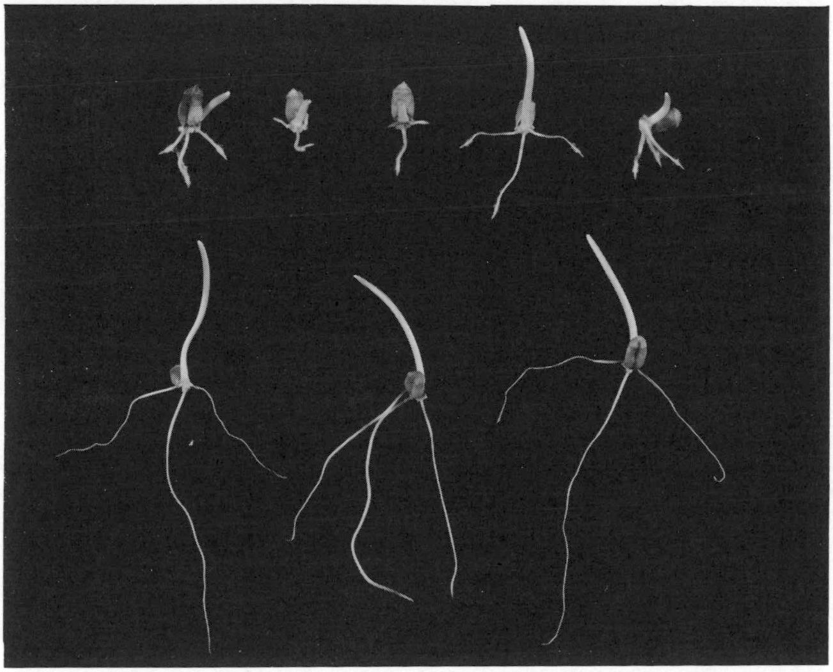

Figure 3-2. Allelopathic effects. *Upper row*: Stunted wheat seedlings caused by a water leacheate from quackgrass roots and rhizomes. *Lower row*: Normal wheat seedlings grown in tap water. (T. Kommedahl, University of Minnesota.)

Seed that fails to germinate because of a waterproof seed coat is called *hard seed.* Hard seed is common in annual morningglory, lespedeza, clovers, alfalfa, and vetch. Researchers believe that many weed species have hard seeds.

Some seed coats are impermeable to oxygen but not water. Cocklebur has two seeds per fruit, one set slightly below the other. The lower seed usually germinates during the first spring and the upper remains dormant until the next year. Both can be made to germinate immediately by breaking the seed coats, or by simply increasing the oxygen supply (see Figure 3-1). Seeds of ragweed, several grasses, and lettuce also show this type of dormancy.

As with waterproof seed coats (hard seeds), anything that breaks the seed coat—scarification, acids, soil microorganisms—will break this type of dormancy. Under laboratory conditions, oxygen dormancy can usually be broken by increasing the oxygen supply. Cultivation of the soil often has a similar effect by increasing the oxygen level in the upper layer of soil.

Mechanically Resistant Seed Coats

A tough seed coat may forcibly enclose the embryo and prevent germination. While the seed absorbs oxygen and water, it builds pressures in excess of 1000 psi. The seed will quickly germinate if the seed coat is removed. Pigweed, wild mustard, shepherdspurse, and pepperweed have this type of dormancy.

As long as the seed coat remains moist, it remains tough and leathery—for as long as 50 years. Any factor that weakens the seed coat will help break dormancy. Drying at temperatures of 110°F or mechanical or chemical injury to the seed coat may break this type of dormancy.

Immature Embryos

The outside of the seed may appear fully developed, but it may have an immature embryo that needs more growth before the seed can germinate. Therefore, the seed appears dormant, although the embryo is slowly growing and developing. Seeds of orchids, holly, smartweed, and bulrush show this type of dormancy.

After-Ripening

In some species the embryos appear completely developed, but the seed will not germinate although the seed coat has been carefully removed to permit easy absorption of water and oxygen. Light and darkness have no effect. In this case germination occurs normally after a period of *after-ripening.* Occasionally cool temperatures for several months will end this type of dormancy. After-ripening is especially common in the grass, mustard, smartweed, rose, and pink families.

After-ripening is a physiological change of a complex physiochemical nature. Although the exact processes are not completely understood, they may be associated with changes in the storage materials present, substances promoting germination may appear, or substances inhibiting germination may disappear (see Figure 3-2).

GERMINATION AS AFFECTED BY CERTAIN CONDITIONS

Burning

Burning fields after weed seeds have matured gives only partial and erratic destruction of seeds. Seeds that lie on the ground may readily escape, whereas those held on the plants often burn completely. The degree of weed-seed destruction depends largely on the intensity of the heat, and this in turn on dryness and the amount of litter and debris to be burned.

The after-effects of burning are usually pronounced, especially after forest fires. Weed species absent, for the most part, from the area for many years may suddenly appear after the fire and dominate other vegetation.

Data and experience clearly show that moderate heat may end seed dormancy. Five other factors that normally follow a fire may also terminate dormancy; these factors are (1) greater alternation of temperature in the upper soil layers between day and night, (2) more light is reaching the surface soil, (3) removal of litter, (4) removal of competition from other plants, and (5) removal of plants previously living in the area that had soil-inhibiting substances that prevented seed germination. With removal of those plants, those substances are no longer present. This is probably an alleopathic effect. See page 28.

Cutting (Stage of Maturity)

Cutting weeds to prevent seed production is a common recommendation. The practice is important in agricultural croplands, in turf, and in hay crops.

To determine the viability of seed cut before maturity, weeds were cut in various stages of maturity and allowed to dry in the sun. Table3-5 shows the results. Four of seven weed species showed high seed germination although cut when in flower (Gill, 1938).

Table 3-5. Germination of Weed Seeds Cut at Various Stages of Maturity.

	Percent Germinated		
Weed Seeds	Cut When Dead Ripe	Cut When in Flower	Cut in Bud Stage
Groundsel, ragwort	72	80	0
Sowthistle, common	100	100	0
Groundsel, common	90	35	0
Sea aster	90	86	0
Dandelion	91	0	0
Catsear, spotted	90	0	0
Canada thistle	38	0	0

In South Dakota, Canada thistle and perennial sowthistle heads were removed from the plant and dried at daily intervals after the flowers opened. The experiment was continued for three years. Perennial sowthistle heads harvested three days after blooming had 0.0% viable seed; six days after blooming they had an average of 6% viable seed; and eight days after blooming, 65% viable seed.

Canada thistle harvested 6 days after blooming had an average of 0.03% viable seed; 8 days after blooming, 6.7% viable seed; and 11 days after blooming, 73% viable seed.

Another study compared removal of the heads as described above with the effect of cutting the entire plant and leaving the heads on the plant during drying. Results of the two methods were similar; thus little seed development takes place after either type of cutting.

In summary, either mowing in or before the bud stage prevents viable seed production. If seeds reach medium ripeness, probably a large percentage of viable seeds will be produced. In the case of many species, cutting after that time does little or no good in preventing viable seed production. With other species, mowing is not effective because some of the heads are short and missed by the mower, and these produce seed.

Storage in Silage

Many seeds lose their viability in silage. Many weed seeds lose their germinating power 10–20 days after being placed in silage. Other reports indicate, however, that some seeds will germinate after being in the silo for periods up to four years. These variations are possibly a result of differences in the silage as to moisture content, temperature, and amount of organic acids produced (Tildesley, 1936–1937).

Table 3-6. Effect of length of Time in Cow Manure Upon the Viability of Various Weed Seeds (Harmon and Keim, 1934)

Kind of Seeds	Percent Viability Before Burial	Percent Viability After Storage: 1 Month	2 Months	3 Months	4 Months
Velvetleaf	52.5	2.0	0.0	0.0	0.0
Field bindweed	84.0	4.0	22.0	1.0	0.0
Sweetclover	68.0	22.0	4.0	0.0	0.0
Pepperweed	34.5	0.0	0.0	0.0	0.0
Smooth dock	86.0	0.0	0.0	0.0	0.0
Smartweed	0.5	0.0	0.0	0.0	0.0
Cocklebur	60.0	0.0	0.0	0.0	0.0
Puncturevine	52.0	0.0	0.0	0.0	0.0

Storage in Manure

When manure is spread fresh from the stable, viable weed seeds are usually spread with it. But if manure is stored, heating and decomposition will begin, and, in time, the weed seeds are destroyed.

In Table 3-6 only three weeds showed any viability after a one-month storage. All seeds were destroyed at the end of four months.

Stoker, et al. (1934) found complete destruction of hoary cress and Russian knapweed seeds in moist, compacted chicken manure at the end of one month. However, seed of field bindweed was still viable at the end of four months.

These tests were conducted during the summer in cow or chicken manure. Horse and mule manure tends to heat, whereas cow or chicken manure does not. Decomposition of the weed seed would be more rapid at the higher temperature. If the manure is frozen or cold, the seed would live longer.

If the edges or outside of the manure pile dry out, decomposition slows down and viable weed seeds would likely persist. Therefore, manure should be turned occasionally to kill all seeds. Well-rotted manure is free of viable seeds.

STORAGE CONDITIONS FAVORING LONG LIFE OF SEEDS

The conditions during storage are more important in maintaining viability than the length of storage. Under farm storage conditions, many crop seeds remain viable for only one to three years. Actually, many crop seeds are killed by warm, moist storage for one to three months.

Soybean seeds in humid regions usually are not considered good for seed if more than nine months old. However, when stored at about 9% moisture and about 32°F, they have shown good germination when over 10 years old. Vegetable seeds–carrot, tomato, eggplant, lettuce, onion, and pepper–were stored at about 5% moisture and at 25°F for 17 years with little or no loss in percentage of germination.

For most seeds, three primary factors lengthen the time that seeds can be stored and remain viable. These include storage at a low moisture content (5% for many seeds), low temperature (below 40°F), and a low oxygen supply. To produce a low oxygen supply, an inert gas such as nitrogen may be substituted.

DISSEMINATION BY RHIZOMES, STOLONS, TUBERS, ROOTS, BULBS, AND BULBLETS

Although weeds reproduce and spread most widely by means of seeds, they also multiply by vegetative, or asexual, methods. Rhizomes, stolons, tubers, roots, bulbs, and bulblets are all vegetative, or asexual, methods of reproduction. Many perennial weeds classed as "serious" reproduce vegetatively, and most of these also reproduce by means of seeds. Vegetative organs occasionally have a short period of dormancy.

Most plants spread slowly by vegetative means alone. Without help from man and his cultivation equipment, weeds such as quackgrass, field bindweed, johnsongrass, bermudagrass, nutsedge, and Canada thistle would spread vegetatively less than 10 ft/year. However, the rhizomes, stolons, roots, and tubers are dragged about the field with soil-tillage equipment. Wherever these plant pieces drop, a new infestation is likely. Disc-type cultivation equipment is less likely to drag the plant parts than are shovels, sweeps, and plows.

Repeated tillage will kill most plants possessing rhizomes, stolons, roots, and tubers. If cut off and in *dry soils,* the vegetative parts may quickly dry and die, preventing new growth.

In most soils, the cut vegetative parts quickly take root and establish new plants. Under such conditions, repeated tillage may exhaust the underground food reserves. Most such weeds are killed through *carbohydrate starvation* by repeated tillage for one to two years (see Figure 2-3).

Some chemicals mixed into the soil will retard the development of new roots, especially after the plant has been cut off. With a combination treatment of an effective herbicide plus repeated cultivation, many serious perennial weeds may be controlled in a short time. For example, johnsongrass can be controlled by trifluralin plus repeated cultivation.

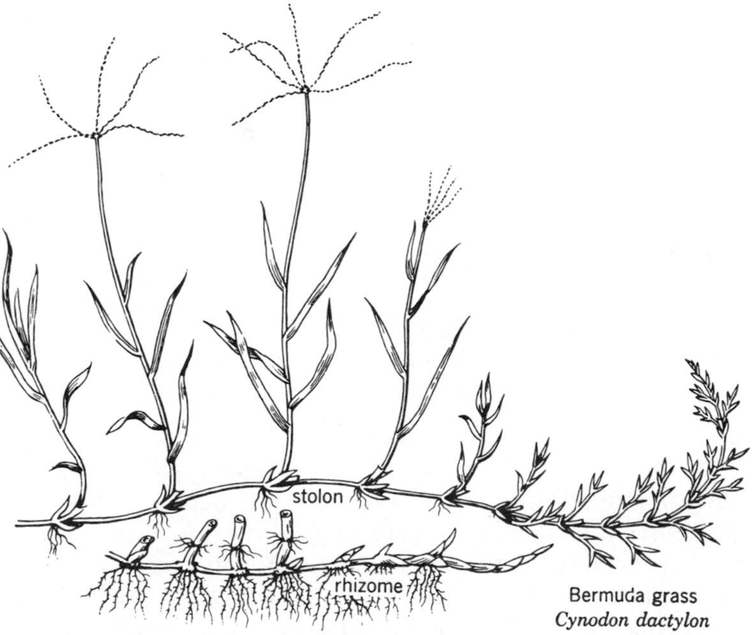

Figure 3-3. Bermudagrass reproduces by seeds, stolons, and rhizomes. (North Carolina State University.)

SUGGESTED ADDITIONAL READING

Bruns, V. F., and L. W. Rasmussen, 1958, *Weeds* **6**, 42.

Furrer, J. D., 1954, *North Central Weed Cont. Conf. Proc.* **11**, 26.

Gill, N. T., 1938, *Ann. Appl. Biol.* **25**, 447.

Harmon, G. W., and F. D. Keim, 1934, *J. Amer. Soc. Agron.* **26**, 762.

Mayer, A. M., and A. Poljakoff-Mayber, 1975, *The Germination of Seeds* p. 236, Macmillan, New York.

Stevens, O. A., 1932, *Amer. J. Bot.* **19**, 784.

Stoker, G. L., D. C. Tingey, and R. J. Evans, 1934, *J. Amer. Soc. Agron.* **26**, 600.

Tildesley, W. T., 1936–1937, *Sci. Agric.* **17**, 492.

Toole, E. H., and E. Brown, 1946, *J. Agr. Res.* **72**, 201.

4 Herbicides and the Plant

When a herbicide comes in contact with a plant, its action is influenced by the morphology and anatomy of the plant as well as numerous physiological and biochemical processes that occur within the plant. These processes include (1) absorption, (2) translocation, (3) molecular fate of the herbicide in the plant, and (4) effect of the herbicide on metabolism. The interaction of these plant factors with the herbicide determines the effect of a specific herbicide on a given plant species. When one plant species is more tolerant to the chemical than another plant species, then the chemical is considered to be *selective*.

We'll expand on the first four points in the first half of this chapter, and discuss selectivity in the last half.

The life processes of plants are many and varied; they are complex and delicately balanced. Disturb one, even slightly, and a chain of events may be set off that will cause major changes in the plant's metabolism.

Even a minor change in the plant's environment may result in a major change in its life processes. For example, many perennial plants lie dormant belowground all winter. If the soil temperature is raised but a few degrees, a complex series of reactions is set in motion that may continue through the growing season.

Developing an understanding of the *mode of action* of herbicides is difficult because the problem is complex. However, considerable progress has been made in the last two decades because of the thousands of studies made by many hundreds of research workers. Usually these studies have failed to show the mechanism as a single response, but rather as a series of biochemical responses. You can find further information in books by Ashton and Crafts (1981), Audus (1976), and Kearney and Kaufman (1975, 1976).

To understand the physiological and biochemical processes at work in the use of herbicides, we first need to define several terms that will be used repeatedly.

A *herbicide* is a chemical used to kill or inhibit growth of plants.

Mode of action refers to the *entire* chain of events from the first contact of the herbicide with the plant to its final effect, which could be death of the plant.

Mechanism of action is the biochemical or biophysical reaction(s) that brings about the ultimate herbicidal effect.

The *symplast* (*sym* means "together") comprises the sum total of living protoplasm of a plant. It is continuous throughout the plant; there are no islands of living cells. The phloem is a major component of the symplast. Phloem translocation is by way of the symplast.

The *apoplast* (*apo* means "apart," "separate" or "detached") comprises the total *nonliving* cell-wall continuum of the plant. The xylem is a major component of the apoplast. Xylem translocation is by way of the apoplast.

ABSORPTION

To be effective, herbicides must enter the plant. Some plant surfaces absorb the herbicide quickly, but other plant surfaces absorb the chemical slowly, if at all. The chemical nature of the herbicide is also involved. Therefore, differential absorption or selective absorption may account for differences in plant responses.

The two most common points of entry in plants are the leaves and the roots. In addition, some chemicals are effectively absorbed through the stems, including coleoptiles or young shoots as they grow through treated soil. Also, seeds may absorb the herbicide.

Leaf Absorption

Initial leaf penetration may take place either through the leaf surface or through the stomates. The volatile fumes of some herbicides and some solutions enter through the stomates. Probably of far greater importance, however, is the direct penetration of the leaf surface. Here the herbicide must first penetrate the cuticle. The cuticle is not homogeneous in composition (see Figure 4-1). Externally it is primarily composed of wax; internally, it is primarily composed of cutin; there is a continuous gradation of one into the other. At the cuticle–cell-wall interface, the cutin is in contact with the pectin, and the pectin is in contact with the cellulose of the cell wall.

There is a gradual transition in the polar nature of the cuticle–cell-wall complex from the surface wax to the cellulose. The cuticular wax is the most nonpolar (hydrophobic), followed by cutin, pectin, and cellulose, respectively. Cellulose is polar (hydrophilic). Therefore, polar compounds have considerable difficulty entering the cuticular wax, but

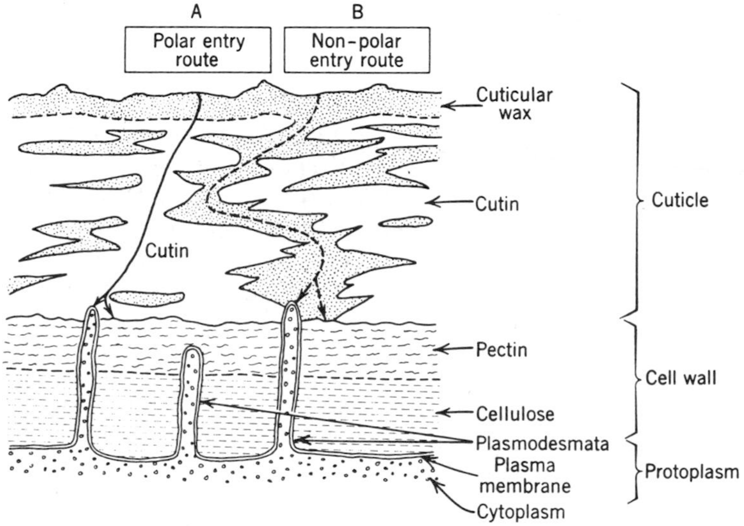

Figure 4-1. Hypothetical diagram representing the foliar absorption aspects of cuticle–cell-wall–protoplasm structure. Hypothetical routes of entry of (*a*) polar and (*b*) nonpolar herbicides. (Adapted in part from Orgell, 1954.)

once they pass this barrier they enter each succeeding phase more readily. In contrast, nonpolar compounds readily enter the cuticular wax but have increasing difficulty passing into each succeeding phase. Thus the polar nature of the herbicide may have considerable influence on its rate of absorption.

Figure 4-1 shows the hypothetical routes of entry of both polar (*A*) and nonpolar (*B*) herbicides and their ultimate entry into the symplastic system via the plasmodesmata or the apoplastic system via the cell wall. The exact molecular requirements for entry into these two systems are unknown. However, atrazine, monuron, and chlorpropham enter primarily the apoplastic system, whereas 2,4-D, amiben, and fenac enter primarily the symplastic system. Several herbicides, such as amitrole, dalapon, and picloram, readily enter both systems.

Any substance that will bring the herbicide into more intimate contact with the leaf surface should aid absorption. Surfactants increase leaf absorption by (1) reducing interfacial tension to give better "wetting" and/or (2) modifying the wax-like and oil-like substances of the cuticle. The amount of amitrole absorbed 24 hr after application was 13.1% without surfactant but 77.8% with surfactant (Freed and Montgomery, 1958).

The addition of a surfactant tends to equalize foliar herbicide absorption in all types of plants. Therefore, a surfactant may reduce the

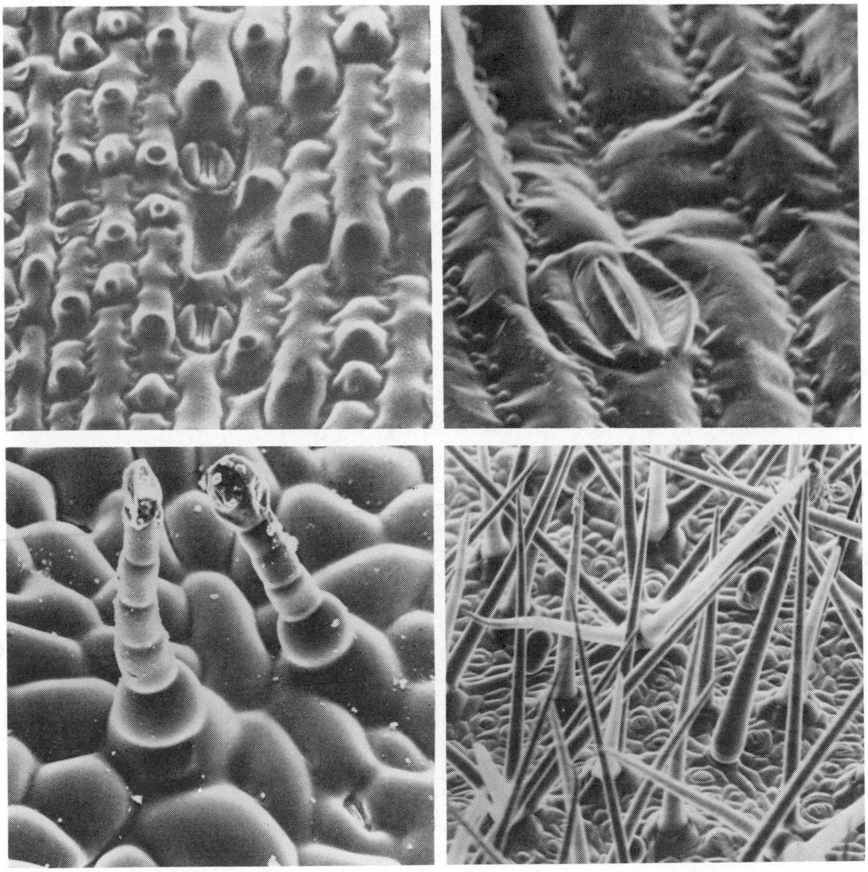

Figure 4-2. Leaf surfaces as they appear on a scanning electron microscope. *Upper left*: Bermudagrass (450×). *Upper right*: Nutsedge (1050×). *Lower left*: Redroot pigweed (350×). *Lower right*: Velvet leaf (170×). (D. E. Bayer and F. D. Hess, University of California, Davis.)

selectivity of the herbicide if that selectivity is dependent on selective foliar absorption.

Increases in temperature may cause more-rapid absorption. In many respects, the absorption process is a chemical process. Within biological limits, the *rate* of a chemical process tends to double with each increase of 10°C (17°F). Therefore, selectivity of postemergence herbicides dependent on selective foliar absorption may be reduced by excessive temperatures after treatment. This is true for postemergence treatments of dinoseb (Meggitt et al., 1956). Within prescribed temperature limits, selectivity can usually be maintained by using lower rates of dinoseb at higher temperatures.

Root Absorption

Roots absorb many herbicides from the soil. Comparative studies have shown that roots absorb some herbicides (monuron, simazine) very rapidly and others (dalapon, amitrole) more slowly (Crafts and Yamaguchi, 1964). Some herbicides are absorbed passively (monuron), whereas others (2,4-D) appear to be absorbed actively, requiring an expenditure of energy (Donaldson et al., 1973).

Herbicides appear to enter roots by three routes: apoplast, symplast, and apoplast-symplast (see Figure 4-3). The apoplast route involves movement exclusively in cell walls to the xylem. This route appears to require that the herbicide pass through the casparian strip and enter the xylem. The casparian strip is a watertight barrier in the cell walls of the endodermis separating the cortex and the stele.

The symplast route involves initial entry into cell walls and then into the protoplasm of the cells of the epidermis, cortex, or both. The herbicide remains in the protoplasm and sequentially passes into the endodermis, stele, and phloem by means of the interconnecting protoplasmic strands (plasmodesmata).

The apoplast-symplast route is identical to the symplast route but for the fact that the herbicide may reenter the cell walls after bypassing the casparian strip and may then enter the xylem.

Although certain herbicides may be restricted to one route of entry, others may enter by more than one route. The chemical and physical

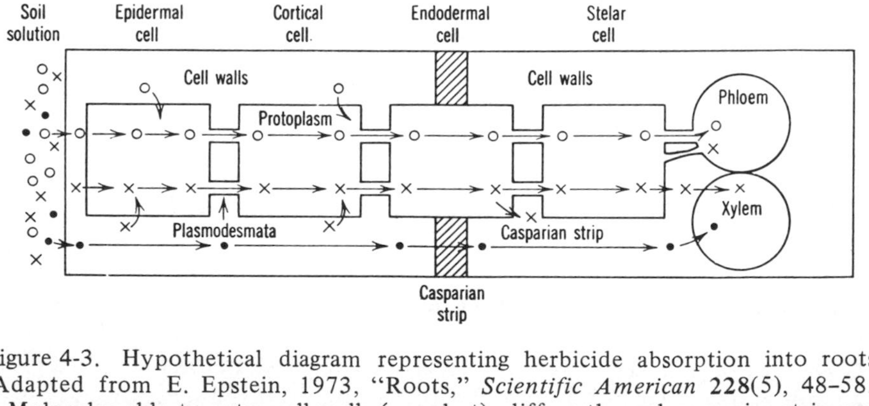

Figure 4-3. Hypothetical diagram representing herbicide absorption into roots. (Adapted from E. Epstein, 1973, "Roots," *Scientific American* **228**(5), 48–58.)

- • Molecules able to enter cell walls (apoplast), diffuse through casparian strip, and enter xylem.
- o Molecules able to enter protoplasm (symplast), pass from cell to cell through the plasmodesmata, and enter the phloem.
- X Molecules able to enter both cell walls (apoplast) and protoplasm (symplast), and enter both xylem and phloem.

properties of each herbicide primarily determine which route is followed.

Under most conditions, there is rapid translocation upward from roots in the xylem (transpiration stream), but only limited upward transport in the phloem. Therefore, in roots, herbicide entry into the xylem is more important than entry into the phloem.

Shoot Absorption

Absorption by leaves and roots has been considered so far. In addition some compounds may be absorbed from soil by coleoptiles and young shoots as they develop and push upward through the soil following germination of seeds. For example, the young shoot of barnyardgrass is the main site of EPTC uptake and also the prime site of injury (Dawson, 1963).

Stem Absorption

Direct applications of herbicides to stems of plants is not a common practice except for control of woody plants. However, the chemical may inadvertently come in contact with the stem in a foliar application or directed spray. Generally, foliar absorption is much more important than stem absorption.

In stem treatment of woody plants the herbicide may be applied to (1) the bark, usually for small tree or brush control, (2) cuts through the bark (cut-surface method) for the control of large trees, and (3) stumps of cut trees or brush to prevent sprouting (see Chapter 27 for details of woody-plant control).

TRANSLOCATION

Translocation of herbicides is of major importance in control of weeds. It is particularly important for those plants with belowground reproductive organs. Herbicides are translocated within the plant through the symplastic system, the apoplastic system, and intercellular translocation. Some herbicides are primarily translocated in the symplastic system, others in the apoplastic system, and still others in both systems (see Table 4-1).

Herbicides move within the plant much the same as other solutes. Figure 4-4 shows the movement of herbicides through the symplastic and apoplastic systems.

Table 4-1. Relative Mobility and Primary Translocation Pathway(s) of Herbicides[1] (Ashton and Crafts, 1981)

Free mobility			Limited mobility			Little or No Mobility
In Apoplast	In Symplast	In Both	In Apoplast	In Symplast	In Both	
Cacodylic acid	Glyphosate	Amitrole	Barban	Phenoxys[3]	Chloramben[3]	Bensulide
Chlorpropham	Maleic hydrazide	Asulam	Bipyridyliums		Endothall[3]	DCPA
Chloroacetamides		Dalapon	Chloroxuron		Fenac[3]	Dinoseb
Desmedipham		Dicamba	Fluridone[3]		Naptalam	Dinitroanilines
Diphenamid[3]		DSMA	Perfluidone		Nitriles	Diphenyl ethers
Methazole		MAA			Phenoxys[3]	
Napropamide[3]		MSMA			Propanil	
Norflurazon[3]		Picloram				
Phenmedipham		2,3,6-TBA				
Pronamide						
Thiocarbamates						
Triazines						
Uracils						
Ureas[2]						

[1] Herbicides may also move from the symplast to the apoplast and vice versa. The rate of translocation may vary between plant species and under different environmental conditions.
[2] Except chloroxuron, limited apoplast.
[3] Translocation rate varies widely between species.

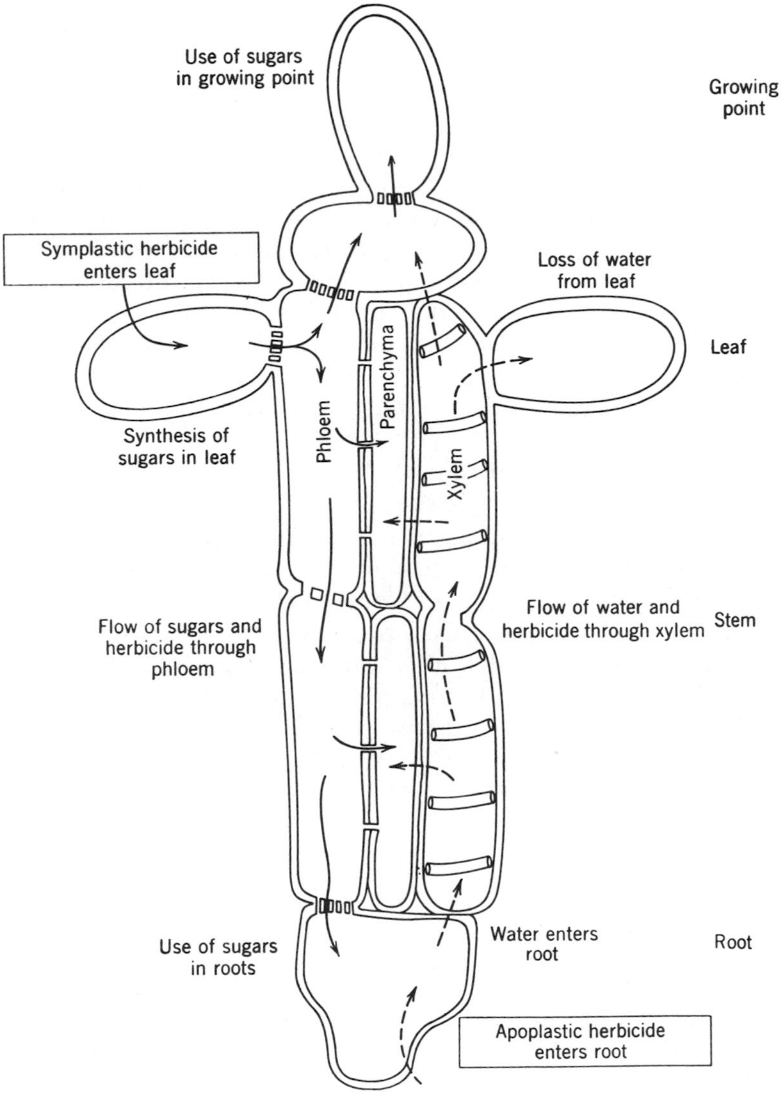

Figure 4-4. Diagram representing routes of translocation of herbicides in plants. (Adapted from *Principles of Plant Physiology* by James Bonner and Arthur W. Glaston, 1952, W. H. Freeman and Company.)

Symplastic Translocation

When applied to leaves, symplastic mobile herbicides follow the same pathway as sugar formed there by photosynthesis. Such herbicides move from cell to cell in the leaf via the interconnecting protoplasmic strands (plasmodesmata) until they enter the phloem. Then they move out of the leaf and move both up and down the stem via the phloem, accumulating in those areas where sugar is being used for growth. Growth is most rapid in the apical growing point, expanding young leaves, rapidly elongating stems, developing fruits and seeds, and root tips.

When a perennial plant with underground reproductive organs is being treated, translocation of the chemical to belowground parts is most rapid and most effective when large amounts of food reserves are being moved toward the roots. This usually occurs after full leaf development.

Effective use of the phenoxy herbicides to kill the underground parts of perennial weeds largely depends on the maintenance of live phloem cells. But some farmers have figured–wrongly–that they would raise the dosage and get better results. This is not so. The excessive rate quickly immobilizes or kills the phloem cells, stopping translocation into the underground parts, and the tops quickly die. This method has much the same effect as cutting off the tops. The plant, however, may quickly resprout from belowground buds.

Translocation through the phloem appears to involve a *mass flow* of solution. One explanation of this driving force is the difference in turgor pressure between the photosynthetic cells and the cells utilizing the products of photosynthesis, primarily sugar. Photosynthetic cells (high-pressure) are often called the *source* and using cells (low-pressure) the *sink*. The plasmodesmata and phloem are living; therefore, herbicides with great acute toxic properties kill these, stopping symplastic translocation. See Crafts and Crisp (1971) for detailed coverage of phloem transports in plants.

Apoplastic Translocation

Apoplasticly mobile herbicides that are absorbed by roots follow the same pathway as water. They enter the xylem and are swept upward with the transpiration stream of water and soil nutrients (see Figure 4-4). The driving force for this movement is the removal of water from the leaves by transpiration.

The xylem and cell walls are the principal components of the apoplastic system. They are considered nonliving. Therefore, all types of herbicides, including very toxic or poisonous chemicals, can be absorbed from the soil and quickly translocated to all parts of the plant. This movement does not injure the xylem because the xylem is nonliving. Absorption and translocation may continue for a time even though acutely toxic herbicides have killed the root.

Actually, acutely toxic herbicides may move downward through the xylem tissue under special conditions. Plants growing under conditions of low soil moisture and high transpiration rates develop water deficits within the plant. When strongly acidic or basic solutions are sprayed on living leaves and stems, the cells of these structures are made permeable. If there is a moisture deficit, the toxic substances are drawn into the plant and may be drawn downward rapidly through the xylem (Crafts, 1933; Kennedy and Crafts, 1927).

Interaction of Apoplastic and Symplastic Translocation

Translocation of most herbicides is not restricted to either the symplast or apoplast, but may involve both. However, many herbicides appear to be limited primarily to one or the other system. For example, chloramben is translocated mainly in the symplast whereas monuron is translocated mainly in the apoplast. Amitrole is readily translocated in both systems and actually appears to circulate in the plant.

As the herbicide travels through long transport channels, xylem or phloem, some may move into adjacent cells. This may occur by simple diffusion or active uptake. From these adjacent cells, the herbicide may then move into the other long transport channel, phloem or xylem, and be translocated in it.

MOLECULAR FATE

The *molecular fate* of a herbicide concerns the changes in the chemical structure of the herbicide molecule within the plant. Most of these changes reduce the phytotoxicity of the molecule (*inactivation*), for example, simazine to hydroxysimazine. However, with some changes the phytotoxicity is increased (*activation*), for example, 2,4-DB to 2,4-D. Since different species of plants have varying degrees of ability to modify the chemical structure of a herbicide, this difference often determines their tolerance to a given herbicide. Some plants can inactivate a herbicide so rapidly that they are not injured, whereas other plants are killed (e.g., atrazine on corn).

Molecular fate has also been referred to as degradation; however, many herbicides are not only degraded (from the Latin: *de,* "down" + *gradus,* "a step") but may also form conjugates, molecules containing the herbicide molecules combined with normal plant constituents (e.g., sugars, amino acids).

Higher plants have been shown to alter the molecular configuration of herbicides by a wide variety of chemical reactions. Most of these are probably catalyzed by specific enzymes; however some appear to be nonenzymatic. In most cases the specific enzyme or enzymes have not been isolated and characterized. The following reactions have been

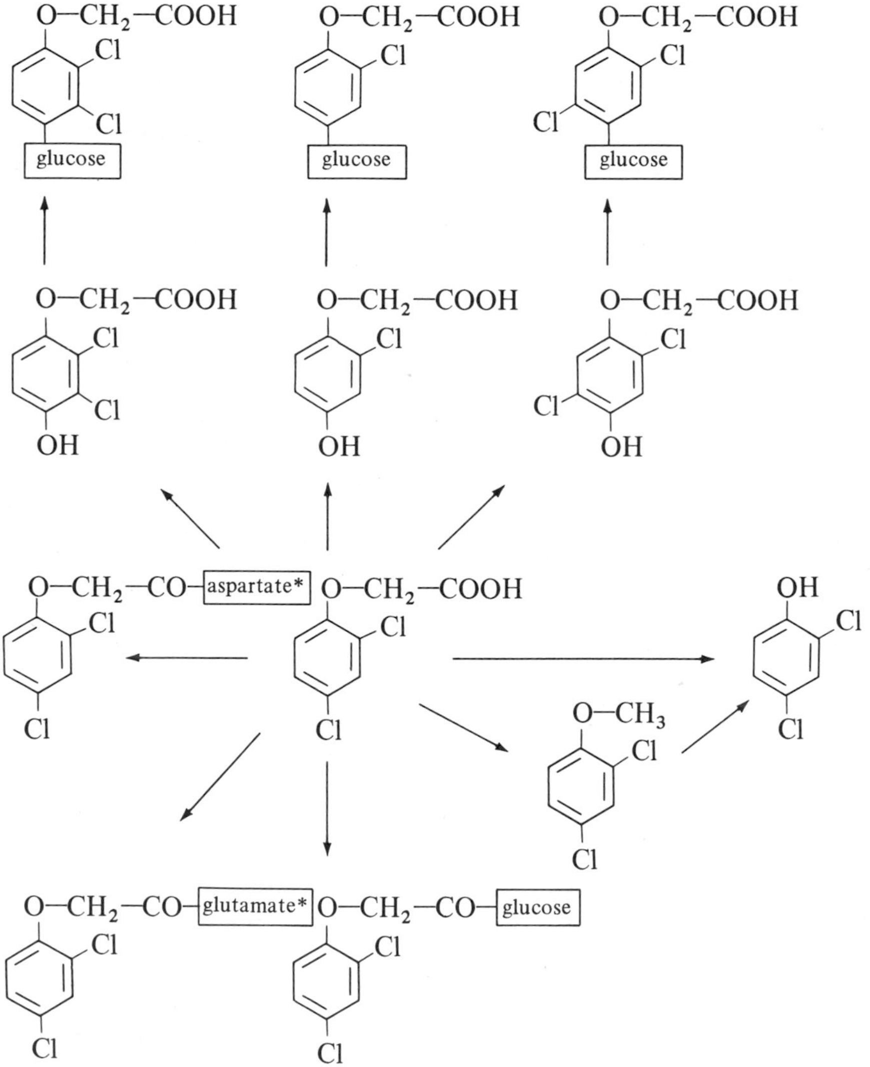

Figure 4-5. Metabolism of 2,4-D in higher plants. *Conjugates of 2,4-D with other amino acids (namely, alanine, valine, leucine, phenylalanine, and tryptophan) have also been reported in soybean callus tissue. (Ashton and Crafts, 1981.)

shown to be involved in herbicide degradation in higher plants: oxidation, reduction, hydrolysis, decarboxylation, deamination, dehalogenation, dethiolation, dealkylation, dealkyoxylation, dealkylthiolation, hydroxylation, and conjugation. An example of the molecular fate of a herbicide, 2,4-D, is given in Figure 4-5. For more information on degradation, see Ashton and Crafts (1981) and Kearney and Kaufman (1975, 1976).

PLANT METABOLISM

Plant metabolism includes the numerous biochemical reactions that occur in the protoplasm of living plant cells. Although most of these take place in all cells, for example, respiration, some occur only in specific cells, for example, photosynthesis in cells containing chlorophyll. A given herbicide may initially interfere with a single biochemical reaction (e.g., monuron rapidly inhibits the oxygen-evolving step of photosynthesis), or a herbicide may be relatively nonspecific and interfere with several reactions simultaneously (e.g., dalapon). Biochemical reactions are closely coupled, and often when one reaction is altered by a herbicide, others are soon affected.

Biochemical reactions that may be affected by herbicides are photosynthesis, respiration, carbohydrate metabolism, lipid metabolism, protein metabolism, and nucleic acid metabolism (Ashton and Crafts, 1981). By disrupting any of these biochemical reactions, herbicides may upset plant metabolism and thus injure or kill the plant.

GROWTH AND PLANT STRUCTURE

Growth and plant structure are the results of previous biochemical reactions. For example, seeds will not germinate and grow unless a source of energy is available; this energy comes from the metabolism of storage materials (carbohydrates, fats, or proteins) in the seeds produced by the parent plant. A plant containing chlorophyll and exposed to light produces its own energy by photosynthesis.

Herbicides may induce abnormal plant growth through morphological, anatomical, and cytological effects. However, these effects are generally specific for a given herbicide on a particular plant species. Some of these abnormal responses are (1) seed germination failure (Figure 4-6), (2) leaf chlorosis, (3) leaf form (Figure 4-7), (4) stem swelling, (5) root elongation (Figure 4-6), (6) cell division, and (7) chloroplasts (Figure 4-8).

So far we've examined four major processes within the plant: (1) absorption, (2) translocation, (3) molecular fate of the herbicide inside the plant, and (4) effect of the herbicide on plant metabolism. These processes profoundly affect herbicide–plant interactions. But there is a fifth major action—*selectivity*. Why and how do some herbicides harm or kill a weed without hurting the crop?

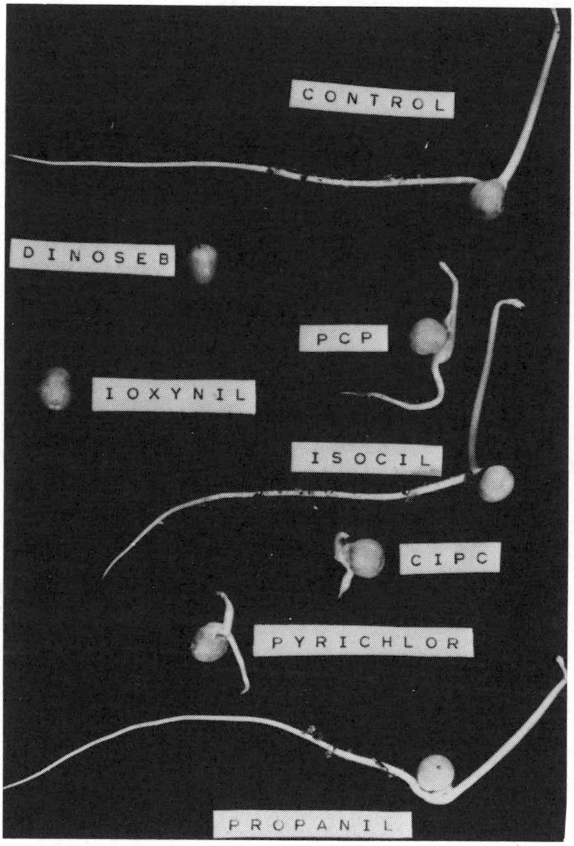

Figure 4-6. Effect of several herbicides on seed germination and root elongation in peas. (van Hoogstraten, 1972.)

SELECTIVITY

A selective herbicide retards growth or kills one plant species (the weed), whereas another plant species (the crop) is tolerant to the same treatment. Ideally, the weed is killed, but sometimes it is only necessary to retard its growth long enough for the crop to become dominant. A herbicide is selective to a particular crop only within certain limits. These limits are determined by a complex interaction between *plants,* the *herbicide,* and the *environment* (see Ashton and Harvey, 1971).

Role of the Plant

Seven factors affect plant response (both weeds and crops) to a chemical: age, growth rate, morphology, physiology, biophysical processes, biochemical processes, and inheritance.

Figure 4-7. The grape leaves on the right were exposed to 2,4-D; those on the left were not exposed to 2,4-D.

Age

The younger the plant, the higher the percentage of the plant that is meristematic (growing) tissue, thus creating higher total biological activity of the plant. Thus the age of a plant often determines its response to a particular herbicide; young plants are less tolerant than older ones. Preemergence treatments that kill germinating weed seeds or seedlings commonly have little or no effect on established weeds.

Growth Rate

Probably for this same reason, the growth rate of plants has a pronounced effect on their reaction to some herbicides. In general, fast-growing plants are more susceptible to treatment than are slow-growing plants.

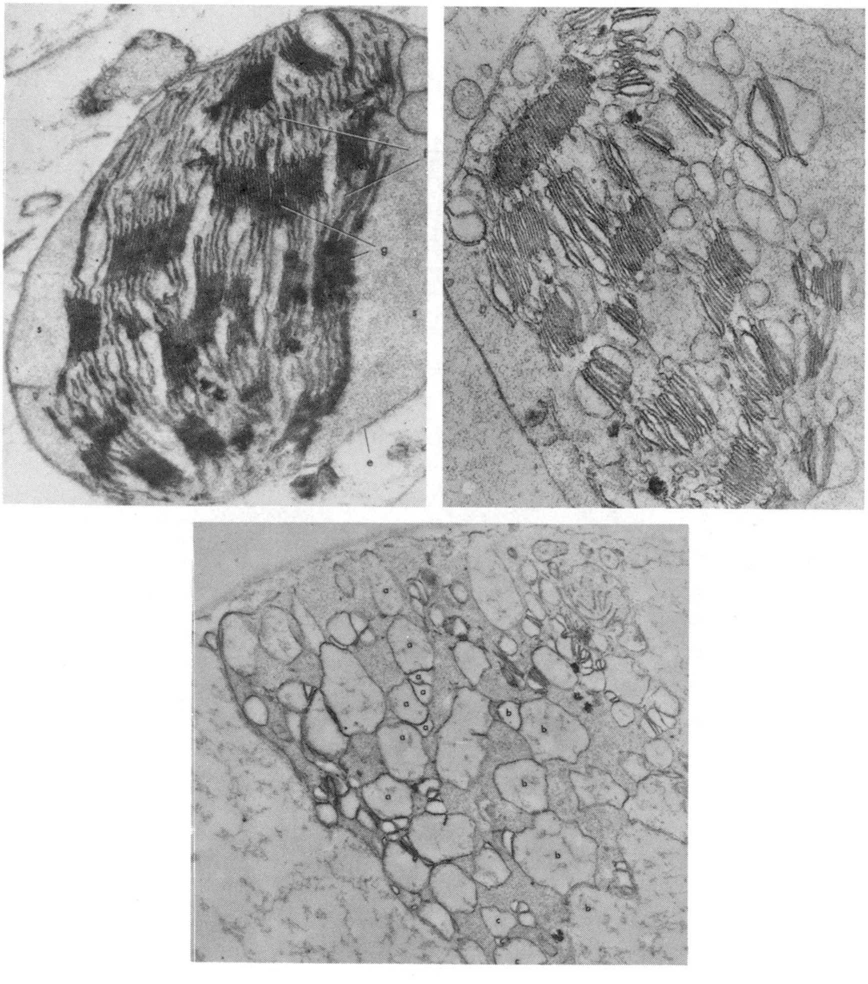

Figure 4-8. Effect of atrazine on the ultrastructure of bean chloroplasts. *Upper left*: Control. *Upper right*: Moderate injury. *Lower*: Severe injury.

Morphology

The morphology of a plant can determine whether or not it can be killed by a specific herbicide. Morphological differences are found in three places–(1) root systems, (2) location of growing points, and (3) leaf properties.

Root systems. Annual weeds in a perennial crop can be controlled because most annual weeds have shallow root systems, whereas perennial crops (e.g., alfalfa) have deep, extensive root systems. Such deep root systems may escape injury from a herbicide that remains near the soil surface through depth-protection, whereas the shallow roots may be killed. This way, it is possible to control many annual weeds among deep-rooted plants such as fruit trees and woody perennial ornamentals.

Location of growing points. The growing point of grasses is located at the base of the sheathed internode; thus it is protected from overtop-contact herbicides (sprayed from above). Therefore, a contact spray may injure the leaves of cereals but not the growing point.

Most broadleaf plants have exposed growing points at the tips of the shoots and in leaf axils. These growing points are directly exposed to chemical sprays. If all the growing points are killed, the plant dies.

Leaf properties. Certain leaf properties protect crops treated by selective herbicides. Liquid spray droplets can stick to only a small part of the surface of narrow, upright leaves (as in cereals and onions), or waxy leaf surfaces, or leaves that are corrugated or formed of small ridges. When sprays hit such leaves, the droplets have a tendency to bounce off or wet the surfaces only in small spots, thus reducing the effect of the herbicide.

Broadleaf plants usually have wide, smooth leaf surfaces, extending horizontally from the plant stem. Such leaves intercept and hold more of the spray; droplets are less apt to bounce off. Therefore, when broadleaf weeds such as lambsquarters, wild radish, pigweed, or wild mustard are sprayed with contact herbicides, the spray solution tends to spread as a thin film, or the droplets spread so as to wet a large portion of the leaf, thus killing the weed. The same sprays on cereals or onions rebound and leave the plants uninjured.

Physiology

The physiology of the plant also determines the amount of herbicide taken up by the plant (absorption) and how it moves within the plant (translocation). Generally, plants that absorb and translocate the greatest amount of herbicide will be killed.

Biophysical Processes

Two biophysical differences may determine whether or not a plant is killed: (1) adsorption and (2) membrane stability.

Adsorption. Adsorption of herbicides by plant-cell constituents inactivates herbicidal material, probably by a physical rather than biochemical process. Radioactive tracer studies have shown that movements of herbicides are slowed down by surrounding plant tissues. In extreme cases a herbicide may be bound so tightly to some plant constituent that it is not readily translocated from the point of application to the site of action; or it may even be held so tightly that it is unavailable for herbicidal action.

Lack of adsorption of a herbicide at its biochemical site of action results in resistance of certain weed biotypes. For example, different biotypes of groundsel are susceptible or resistant to control by atrazine, depending whether or not atrazine is adsorbed to a specific biochemical site in the chloroplast (Pfister et al., 1979).

Membrane stability. Tolerance to oils, existing in carrots and other crops of the carrot family, is one of the oldest examples of biophysical selectivity. Selective oils used for weed control in carrots kill weeds by damaging the cell membranes and allowing the cell sap to flow into the intercellular spaces (causing the leaves to appear water-soaked); this causes the cells to die and tissues to dry out later. Because their cellular membranes are resistant to this damage, the carrots are not killed.

Biochemical Processes

Biochemical reactions in various plants may protect them from injury by certain herbicides. Their reactions include herbicide (1) inactivation or (2) activation.

Herbicide inactivation. Inactivation of a herbicide into a nontoxic form usually involves its degradation to a different molecular form. See "Molecular Fate" earlier in this chapter for specific reactions. When degradation occurs rapidly in a plant, the plant is not injured (atrazine in corn); however, if degradation occurs slowly, toxic levels of the herbicide accumulate, and the plant is injured (atrazine in pigweed).

Herbicide activation. Activation of a harmless chemical into a herbicide sometimes can be used for selective weed control. For example, the relatively harmless compound 2,4-DB is changed in some sensitive plants into the weed killer 2,4-D. In resistant plants (e.g., alfalfa) this reaction takes place very slowly. In the resistant alfalfa, lethal concentration of 2,4-D does not accumulate from the 2,4-DB treatment.

Inheritance

The genetic complement of a plant determines the extent to which it responds to its environment. Many of these responses are morphological, physiological, biophysical, or biochemical. These responses vary from genus to genus, but within a genus, plant reactions to a given herbicide tend to be similar. However, there are exceptions—tolerance to a herbicide may vary considerably from species to species within a genus or even from variety to variety within one species. Thus it is potentially possible to develop and select crop varieties that are tolerant to a specific herbicide.

Three broad forces interact to determine selectivity of any herbicide. One is the plant, just discussed. The other two are the herbicide and the environment.

Role of the Herbicide

Molecular Configuration

Variation in molecular configuration of a herbicide changes its properties, which in turn modifies its effects on plants. This is illustrated by Figure 4-9, which shows the herbicides trifluralin and benefin. The only difference is that a methyl ($-CH_2-$) group is moved from one side of the molecule to the other. Trifluralin kills lettuce at rates required for weed control, but benefin at recommended rates will control weeds without harming lettuce.

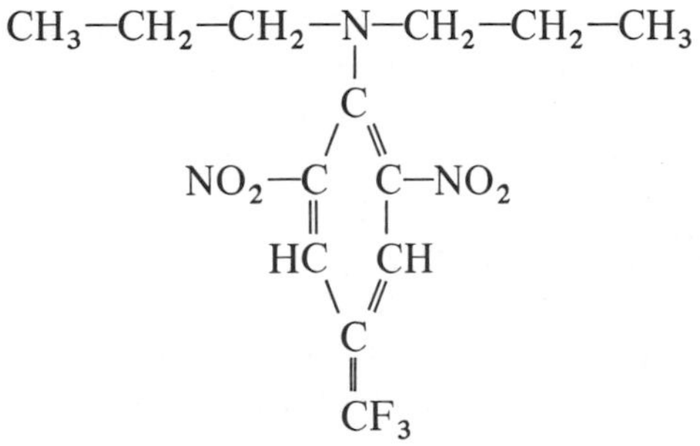

trifluralin

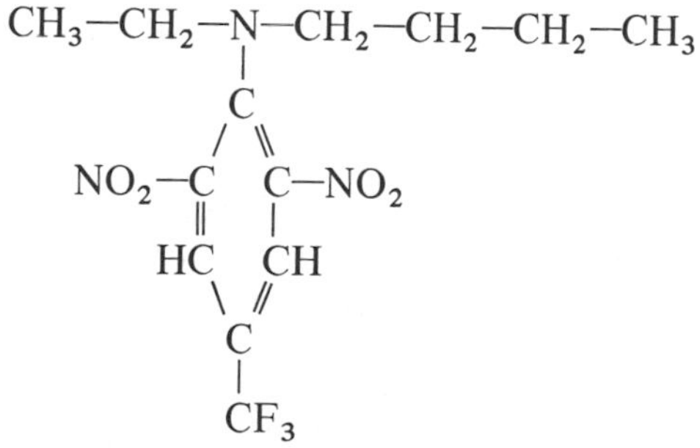

benefin

Figure 4-9. Chemical structures of trifluralin and benefin are similar; however, there are differences in plant selectivity.

Concentration of the Herbicide

Concentration may determine whether the herbicide inhibits or stimulates the plants. It has been clearly established that in low concentrations, dinitrophenol may stimulate respiration, whereas stronger concentrations directly inhibit respiration (Bonner, 1949a, 1949b; Kelly and Avery, 1949). Under many conditions 2,4-D may speed up the rates of respiration and cell division, but in excessive quantities it may immediately slow down the rates.

The concentration of the herbicide *at a vital location* in the plant *at any one given time* may determine the herbicide effectiveness. The same amount of herbicide over an extended length of time may have little or no effect. If the structure of the herbicide must be altered within the plant to an active form, or if absorption or translocation is slow, the concentration of the chemical at any one moment may be less than the lethal dosage.

Formulation

The formulation of a herbicide is vital in determining whether it is selective or not with regard to a given species. Perhaps the most striking example is the solid, granular form that permits the herbicide to "bounce off" the crop and fall to the soil. Other substances known as adjuvants and surface-active agents (surfactants) are often added to improve the application properties of a liquid formulation; these additives may increase or decrease toxicity. The addition of nonphytotoxic oils or surfactants to atrazine or diuron induces foliar contact activity in these normally soil-active herbicides. These herbicides have little foliar activity in their usual forms.

How the Herbicide Is Used

A herbicide can be applied so that most of it covers the weed but little of it contacts the crop. This can be done by using (1) shielded or (2) directed sprays.

Shielded spray. Shields prevent the herbicidal spray from touching the crop while the weeds are covered by the spray. The spray nozzles are simply placed under a hood or the crop is covered with a shield (see Figure 4-10).

Shielded sprays may not be needed if you choose nozzles that have little spray drift, if nozzle height and direction are carefully controlled, and if the crop is taller than the weeds (see Figures 2-1 and 4-10).

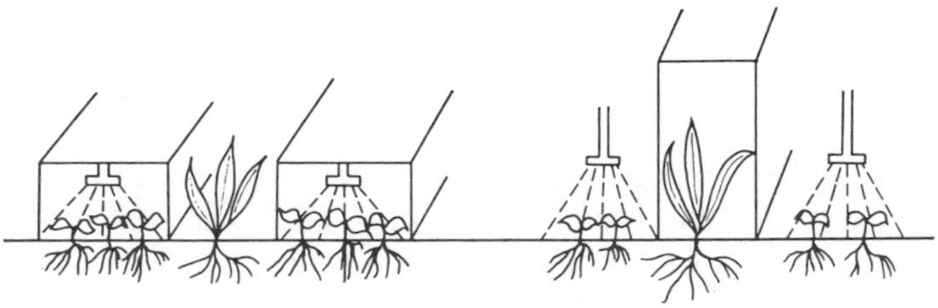

Figure 4-10. Shielded sprays protect crops from being sprayed with herbicides. Confine the spray within shields (*left*) or cover the crop by shields (*right*).

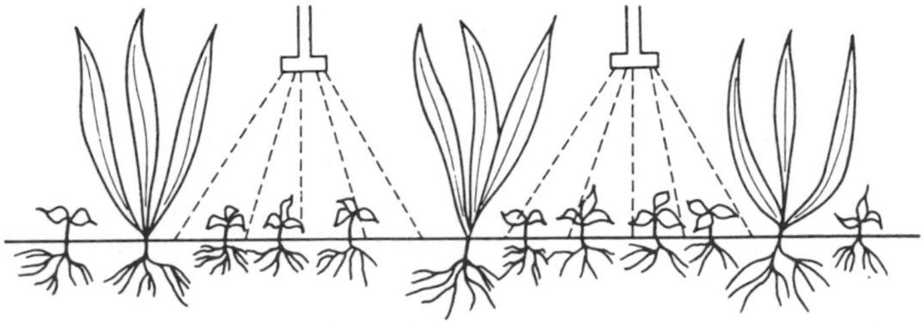

Figure 4-11. Directed sprays are aimed toward the base of the crop plant, favoring minimum coverage of crop and maximum coverage of weeds.

Directed spray. This method is usually used where the crop is grown in rows and where the crop is higher than the weeds. Drop nozzles are used to spray weeds between the crop rows, and little herbicide contacts the crop.

Role of the Environment

Dominant environmental factors that affect selectivity are soil type, rainfall or overhead irrigation, and temperature. See Chapter 5 for details of soil–herbicide interactions.

In general, soil type and the amount of rainfall determine the actual location of a specific herbicide in the soil. Among factors affecting movement of herbicides in soil are adsorption by soil particles, water solubility of the herbicide, amount of rainfall, and soil type. The intensity with which the molecule is held by the soil particle will strongly affect movement or lack of movement in the soil. Also, high water solubility of the herbicide, high amounts of rainfall, and light soil types favor a deep penetration of the herbicide. Some herbicides are extremely resistant to leaching, whereas others readily move with the water. This movement is normally downward, but the herbicide may move up as water evaporates from the surface.

Some herbicides not inherently selective may be made to function as such by their location in the soil. Such selectivity depends on different rooting habits of crop and weed. If you want to remove deep-rooted weeds while leaving shallow-rooted crops intact, you must use a herbicide that readily moves beyond the rooting zone of the crop into the rooting zone of the weed. But if you want to remove shallow-rooted weeds from a deep-rooted crop, you must choose a herbicide that remains near the soil surface (see Figure 4-12).

The temperature of the environment in which a plant is growing has considerable influence on the rates of its physiological and biochemical reactions. For example, the optimum temperature for germination of seeds of different species varies greatly (e.g., spinach at 41°F and canta-

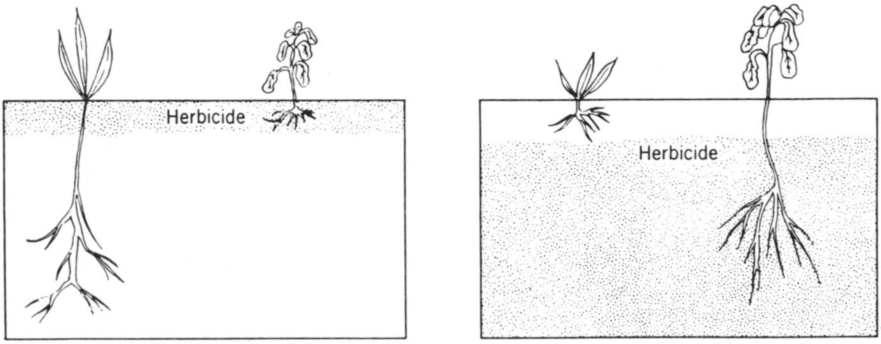

Figure 4-12. Herbicides differ in their tendency to leach. These diagrams show the selectivity differences that may result from various locations of the herbicide in the soil. *Left*: Deep-rooted crop (*left*) is not affected by herbicide that remains near soil surface. Shallow-rooted weed is killed by herbicide that stays near surface. *Right*: Shallow-rooted crop (left) remains alive if herbicide moves beyond its rooting zone. Deep-rooted weed is killed when the herbicide is leached into the deeper zones of the soil.

loupe at 77°F). The selectivity of various plants to herbicides also varies as the temperature differentially affects their processes.

The effect of temperature on these processes is often expressed as the temperature coefficient (Q_{10}). If the rate is doubled with a 10°C increase in temperature, the Q_{10} is 2. Most chemical reactions have a Q_{10} between 2 and 3, whereas physical processes often have a Q_{10} only slightly greater than 1. A herbicide is a chemical and most of the reactions which the herbicide influences in the plant are chemical in nature. Thus, a change from 80 to 97°F may result in doubling or tripling the activity of the herbicide.

SUGGESTED ADDITIONAL READING

Ashton, F. M., and A. S. Crafts, 1981, *Mode of Action of Herbicides,* Wiley, New York.

Ashton, F. M., and W. A. Harvey, 1971, Circular 558, California Agricultural Experiment Station and Extension Service, Berkeley.

Audus, L. J., 1976, *Herbicides,* Vols. 1 and 2, Academic, New York.

Bonner, J., 1949a, *Amer. J. Bot.* **36**, 323.

Bonner, J., 1949b, *Amer. J. Bot.* **36**, 429.

Bonner, J., and A. W. Galston, 1952, *Principles of Plant Physiology,* Freeman, San Francisco.

Crafts, A. S., 1933, *Hilgardia* **7**, 361.

Crafts, A. S., and C. E. Crisp, 1971, *Phloem Transport in Plants,* Freeman, San Francisco.

Crafts, A. S., and S. Yamaguchi, 1964, *The Autoradiography of Plant Materials*, Division of Agricultural Science, University of California, Berkeley.

Dawson, J. H., 1963, *Weeds* **11**, 60.

Donaldson, T. W., D. F. Bayer, and O. A. Leonard, 1973, *Plant Physiol.* **52**, 638.

Freed, V. H., and M. Montgomery, 1958, *Weeds*, **6**, 386.

Kelly, S., and G. S. Avery, Jr., 1949, *Amer. J. Bot.* **36**, 421.

Kearney, P. C., and D. D. Kaufman, 1975, 1976, *Herbicides*, Vols. 1 and 2, Dekker, New York.

Kennedy, P. B., and A. S. Crafts, 1927, *Plant Physiol.* **2**, 503.

Meggitt, W. F., R. J. Aldrich, and W. C. Shaw, 1956, *Weeds* **4**, 131.

Orgell, W. H., 1954, Ph. D. dissertation, University of California, Davis.

Pfister, K., S. R. Radosevich, and C. J. Arntzen, 1979, *Plant Physiol.* **64**, 995.

van Hoogstraten, S. D., 1972, Ph. D. dissertation, University of California, Davis.

5 Herbicides and the Soil

Numerous soil factors, many different kinds of herbicides, and large numbers of plant species and climatic variations make the study of herbicides in soils very complex and diverse. There are at least 10 different soil variables of major importance, 150 different herbicides, and hundreds of different plant species involved. Thus the complexity of herbicide–soil–weather–plant interactions is enormous.

Herbicides applied to the soil are directly affected by soil characteristics; however, herbicides applied to the foliage of plants are *not* directly affected by soil differences even though in foliar applications some of the herbicide may fall directly on the soil and usually minute amounts reach the soil later as the plants die and disintegrate.

Herbicides are applied directly to the soil as (1) preplanting treatments, (2) preemergence treatments, or (3) postemergence treatments. The time of application may refer to the crop or to the weed.

Some preplanting treatments are mechanically mixed into the soil, whereas others are applied to the surface. When mechanically incorporated into the soil, the chemical is usually immediately effective on seeds germinating in the area—with no added moisture. When applied to the soil surface, most herbicides must be moved into the soil by water to be effective.

The success of an incorporated preplanting treatment or a preemergence treatment depends largely on the presence of a high concentration of the herbicide in the upper 2 in. of soil. This is where most annual-weed seeds germinate. Also, there must be a relatively low concentration of the herbicide in the zone where the crop seeds germinate, unless the crop seed is tolerant to the chemical (see Figure 5-1).

If the viable weed seeds in the surface soil are killed, the area may remain weed-free long after the chemical has disappeared. This happens because most weed seeds will not germinate if buried deeply in the soil.

For effective soil sterilization, the chemical must remain active in the rooting zone to kill both germinating seeds and growing plants.

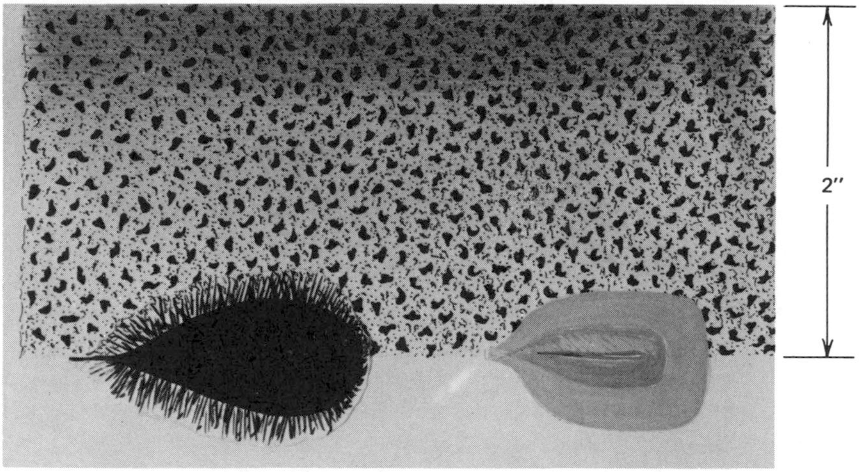

Figure 5-1. Small-seeded weeds (black specs) germinate principally in the upper 2 in. of soil. Preemergence herbicides depend upon water to leach the herbicide into this area. With soil incorporation the herbicide is mechanically mixed with the soil. Many weed seeds below the 2-in. level remain dormant for many years. (North Carolina State University.)

PERSISTENCE IN THE SOIL

The length of time that a herbicide remains active or persists in the soil is extremely important because it determines the length of time that weed control can be expected. Therefore, a certain amount of persistency is usually desirable. Residual toxicity is also important because it relates to phytotoxic aftereffects that may prove injurious to succeeding crops or plantings. In such cases, excessive persistence may restrict crop rotation options available to the farmer.

Herbicides may disappear faster with large amounts of water, which provides leaching, and with repeated cultivation or mixing of the soil. In some cases, fertilizers can be added to reduce the injurious aftereffects. For example, nitrate nitrogen reduces the phytotoxic effects of sodium chlorate.

Seven factors affect the persistence of a herbicide in the soil: (1) microbial decomposition, (2) chemical decomposition, (3) adsorption on soil colloids, (4) leaching, (5) volatility, (6) photo-decomposition, and (7) removal by higher plants when harvested. Each factor is discussed later in this chapter (see Figure 5-2).

Table 5-1 gives approximate lengths of persistence for various herbicides under a given set of conditions. This table is based on experimental work and general observations. In general, the conditions stated are favorable to rapid herbicide decomposition or a short period of herbi-

Table 5-1. Persistence of Biological Activity at the Usual Rate of Herbicide Application in Moist-Fertile Soils, Summer Temperatures, in a Temperate Climate[1]

1 Month or Less (Temporary Effects)	1–3 Months (Early Season Control)	3–12 Months[3] (Full Season Control)	3–12 Months[3] (continued)	Over 12 Months[4] (Total Vegetation Control)
Acrolein	Acifluorfen	Alachlor	Oryzalin	Arsenic
Amitrole	Bentazon	Ametryn	Oxyflurfen	Borate
AMS	Bifenox	Atrazine	Pendimethalin	Bromacil
Barban	Butachlor	Benefin	Perfluidone	Chlorate
Benzadox	Butylate	Bensulide	Pronamide	Fenac
Cacodylic acid	CDAA	Bromoxynil	Propazine	Fluridone (in soil)
Chloroxuron	CDEC	Buthidazole	Prosulfalin	Hexaflurate
Dalapon	Chloramben	Chlorobromuron	Secbumeton	Karbutilate
2,4-D	Chlorpropham	Cyanazine	Simazine	Picloram
2,4-DB	Cycloate	Cyprazine	Terbutol	Prometon
Dinoseb (DNBP)	Desmedipham	DCPA	Trifluralin	Tebuthiuron
Diquat[2]	Diallate	Dicamba		Terbacil
DSMA	Diphenamid	Dichlobenil		2,3,6-TBA
Endothall	Dipropetryn	Difenzoquat		
Fluorodifen	EPTC	Dinitramine		
Glyphosate	Mecoprop	Diuron		

Ioxynil	Methazole	Ethalfluralin
Metham	Naptalam	Fenuron
Methyl Bromide	Norea	Fluchloralin
MCPA	Pebulate	Fluometuron
MCPB	Prometryn	Fluridone (in water)
MH	Propachlor	Isopropalin
Molinate	Propham	Linuron
MSMA	Pyrazon	Metobromuron
Nitrofen	Siduron	Metolachlor
Paraquat[2]	Silvex	Metribuzin
Phenmedipham	TCA	Monolinuron
Propanil	Terbutryn	Monuron
Propham	Thiobencarb	Napropamide
	Triallate	Norflurazone
	2,4,5-T	Norea
	Vernolate	

[1] These are approximate values and will vary as discussed in the text.

[2] Although diquat and paraquat molecules may remain unchanged in soils, they are adsorbed so tightly that they become biologically inactive.

[3] At higher rates of application, some of these chemicals may persist at biologically active levels for more than 12 months.

[4] At lower rates of application, some of these chemicals may persist at biologically active levels for less than 12 months.

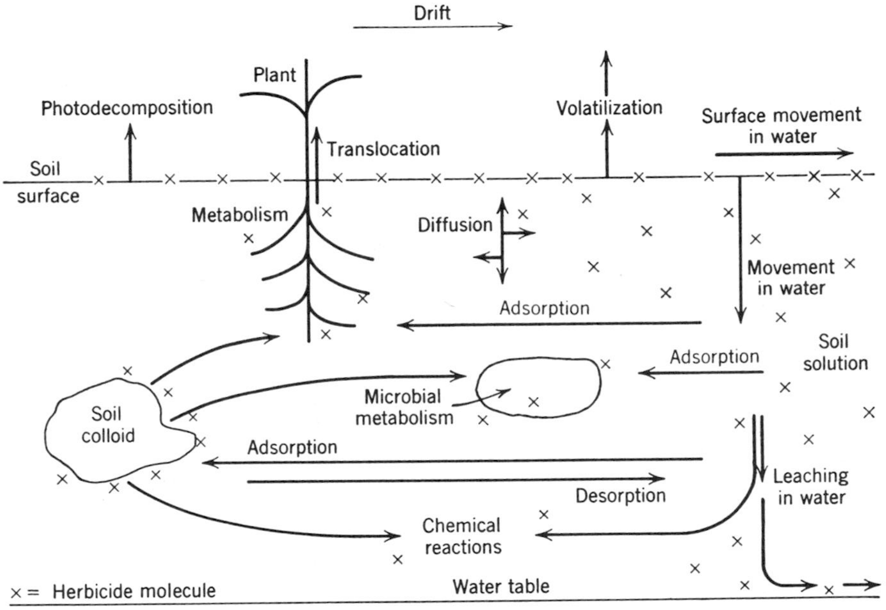

Figure 5-2. Diagram of the interrelations of the processes that lead to detoxication, degradation, and disappearance of herbicides. (Sheets and Kaufman, 1960.)

cide toxicity. Those herbicides persisting 1 month or less may be used to control weeds present at the time of treatment. Those persisting 1–3 months will protect the crop only during a short period early in the growing season. Those providing 3 to 12 months of control may provide protection to the crop for the entire growing season. Those providing more than 12 months of control are used primarily for total vegetation control or where persistence is desirable. If this group is used on crop-

Figure 5-3. Total-vegetation-control herbicides may provide annual-weed control up to two years. There is no mowing or fire problem around these posts and under the steel cable. (E.I. du Pont de Nemours and Company.)

Figure 5-4. Simazine, at rates of 4–20 lb/acre, may provide annual-weed control up to two years. There is no mowing or fire problem in this gas-storage area. (Ciba-Geigy Chemical Corporation.)

land, it is at low rates of application, with rotations of two or more years' duration, and where crops in the rotation are known to be tolerant (see Figures 5-3 and 5-4).

Kearney and Kaufman (1975, 1976) prepared reviews on the *degradation of herbicides* and they are suggested for additional reading.

Microbial Decomposition

The principal microorganisms in the soil are algae, fungi, actinomyces, and bacteria. They must have food for energy and growth. Organic compounds of the soil provide this food supply, the exception being a very small group of organisms that feed on inorganic sources.

Microorganisms use all types of organic matter, including organic herbicides. Some chemicals are easily decomposed (easily utilized by the microorganisms) whereas others resist decomposition.

The microorganisms use carbonaceous organic matter in their bodies primarily through aerobic respiration. In the process, oxygen is absorbed and CO_2 is released.

If an organic substance, that is, a herbicide, is applied to the soil, microorganisms immediately attack it. Those that can utilize the new food supply will likely flourish and increase in number. In effect, this hastens decomposition of that organic substance. When the substance has decomposed, the organisms decrease in number because their new food supply is gone.

Other factors beside food supply may quickly affect the growth and rate of multiplication of microorganisms. These factors are temperature, water, oxygen, and mineral nutrient supply. Most soil microorganisms are nearly dormant at 40° F, 75–90° F being the most favorable. Without water most organisms become dormant or die. Aerobic organisms are very sensitive to an adequate oxygen supply, and deficiency of nutrients, such as nitrogen, phosphorus, or potash, may reduce microorganism growth.

Thus the herbicide may remain toxic in the soil for considerable time if the soil is cold, dry, poorly aerated, or if other conditions are unfavorable to the microorganisms. If the organisms are destroyed by soil sterilization (steam or chemical methods), decomposition of the herbicide may temporarily stop.

Soil pH also influences the microorganisms. In general, the bacteria and actinomyces are favored by soils having a medium to high pH, and their activity is seriously reduced below pH 5.5. Fungi tolerate all normal soil pH values. In normal soils, therefore, fungi predominate at pH 5.5 and below.

Above pH 5.5, the fungi are reduced through competition with bacteria and actinomyces.

Thus a warm, moist, well-aerated, fertile soil with optimal pH is most favorable to microorganisms. Under these ideal conditions, the organisms can most quickly decompose organic herbicides. Microbial decomposition of many herbicides follows typical growth curves for bacterial populations (see Figure 5-5).

At the usual rate of herbicide application on farmlands, the total number of organisms is seldom changed to any great extent because the herbicide may benefit one group of organisms and injure another group. When the herbicides are decomposed, the microorganism population returns to "normal." The biological activity of most herbicides applied

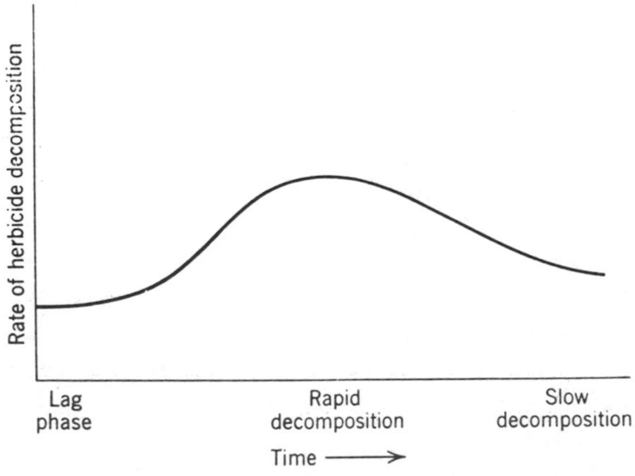

Figure 5-5. Rate of herbicide decomposition by microorganisms. (A. D. Worsham, North Carolina State University.)

at rates recommended for cultivated crops disappears in less than 12 months (Table 5-1). Therefore, no longtime effect on the microorganism population of the soil is expected.

Chemical Decomposition

Chemical decomposition destroys some herbicides and activates others. Chemical decomposition may involve reactions such as oxidation, reduction, and hydrolysis. For example, potassium cyanate and dalapon will slowly hydrolyze in the presence of water, rendering both ineffective as herbicides.

In soil saturated with water, oxygen will likely become a limiting factor. Under such conditions, anaerobic degradation of organic compounds can be expected. It has not been established whether this is chemical or microbial, but it is probable that both are involved. Using trifluralin, and under standing water, degradation was complete in seven days at 76°F in nonautoclaved soils, whereas only 20% had degraded at 38°F. For further details see page 180.

Adsorption on Soil Colloids

Colloids (from the Greek word *kolla* meaning "glue") have unusual adsorptive capacities. As a soils term, colloid refers to the microscopic (1 micron or less in diameter) inorganic and organic particles in the soil. These particles have an extremely large surface area in proportion to a given volume. It has been calculated that 1 cubic in. of colloidal clay may have 200–500 square ft of particle surface area.

Many clay particles resemble the negative radical of a weak acid, such as COO^- in acetic acid. Thus the negative clay particle attracts to its surface positive ions (cations) such as hydrogen, calcium, magnesium, sodium, and ammonium. These cations are rather easily displaced, or exchanged (from the clay particle) for other cations. They are known as *exchangeable ions.* This replacement is called *ionic exchange* or *base exchange.*

The base-exchange capacity of a soil is expressed as milliequivalents (meq) of hydrogen per 100 g of dry soil. A soil with a base-exchange capacity of 1 meq can adsorb and hold 1 mg of hydrogen (or its equivalent) for every 100 g of soil. This is equivalent to 10 ppm or 0.001% hydrogen.

Adsorptive capacity, and thus exchange capacity, is closely associated with inorganic and organic colloids of the soil. Inorganic colloids are principally clay. There are two principal groups of clay: *kaolinite* and *montmorillonite.* Because of their chemical nature and lattice structure, the kaolinite colloids have a relatively low adsorptive or base-exchange capacity. Kaolinite clays have a tendency to predominate in areas of high rainfall and warm-to-hot temperatures. Thus clay soils of the tropics and the Southeastern United States are principally kaolinite. Aluminum and iron oxides are important constituents of kaolinic soils.

Montmorillonite clays have an adsorptive capacity of perhaps three to seven times that of kaolin clays. The montmorillonite clays predominate in areas of moderate rainfall and moderate temperatures. They are typical of Corn Belt soils.

Organic colloids involve the humus of the soil. Organic colloids have a very high adsorptive capacity—about 4 times the base-exchange capacity of a montmorillonite clay, and perhaps 20 times that of kaolinite on a weight basis.

Much evidence supports the fact that organic matter and clay (especially montmorillonite) play an important role in determining phytotoxicity and residual persistence through adsorption, leaching, volatilization, and biodegradation. Adsorption of herbicides with basic properties is regulated to a great extent by the pH of the soil. Basic herbicides such as the *s*-triazines and tebuthiuron are more effective herbicides on low-pH (acid) soils than on high-pH (basic) soils. This happens because a higher percentage of the herbicide molecules are ionized in acid soil.

Scientists have studied the exchange of negatively charged particles (anions) much less than the exchange of cations. Anion exchange does occur. Anions such as OH^-, $PO_4^{\equiv}$, $HPO_4^{=}$, $H_2PO_4^-$, NO_3^-, HCO_3^-, and $CO_3^{=}$ may be involved. Anion exchange is probably more important in humus and in kaolinite clays than in montmorillonite clays. Because the active portion of many herbicide molecules is anionic, research designed to aid in the understanding of anionic adsorption should prove helpful.

Observations in research work as well as in the field have shown the following:

1. Soils high in organic matter require relatively large amounts of most soil-applied herbicides for weed control.
2. Soils high in clay content require more soil-applied herbicide than sandy soils for weed control.
3. Soils high in organic matter and clay content have a tendency to hold the herbicides for a longer time than sand. The adsorbed herbicide may be released so slowly that the chemical is not effective as a herbicide.

The adsorption of monuron was studied in various soils; some of the results are shown in Table 5-2. These findings may help to clarify the principles of monuron adsorption.

Table 5-2. Monuron Adsorption Correlated with Certain Soil Properties

Soil Property	Correlation Coefficient
Organic matter	0.991
Clay	0.209
Silt	0.358

The percentage of organic matter had a high correlation with the amount of monuron adsorbed. The type of clay (see the previous discussion) may explain the relatively low adsorption of the clay in this study.

In another study, the toxicity of diuron to cotton and ryegrass in 12 different soils was measured. Organic-matter content, cation exchange capacity, and total exchangeable bases gave a multiple regression coefficient (R) of 0.968 for cotton and 0.956 for Italian ryegrass (Upchurch, 1958).

Scientists have attempted to develop equations for predicting safe, effective rates of herbicides for various soil types. Of the several chemical and physical properties of soils that were measured, organic matter gave the best prediction of performance. The use of other properties with organic matter in the equations did not greatly improve predictability. Such prediction equations are useful for many soils; however, the theoretical value of a given soil may be considerably different than that observed in practice. Therefore, such values should be used as guidelines but not as absolute recommendations.

"Activated" carbon is one of the most effective adsorptive materials known. It has been used to protect plants from herbicides. Roots of strawberry plants have been coated with "activated" carbon prior to setting the plants in herbicide-treated soil. Also, bands of "activated" carbon have been placed over previously seeded rows soon after planting and before a preemergence herbicide treatment. Well-decomposed organic matter could presumably have properties similar to those of "activated" carbon.

These facts indicate that a certain amount of herbicide is required to saturate the adsorptive capacity of a soil. Above this "threshold level," heavier rates will greatly increase the amount of herbicide in the soil solution, and thus increase the herbicide toxicity to plants.

Therefore, the nature and strength of the "adsorbed linkage" or "bonding" is of considerable importance for both cations and anions. Apparently the nature and characteristics of the colloidal organic matter, as well as the clay, may affect the tenacity of this bonding (see Figure 5-6).

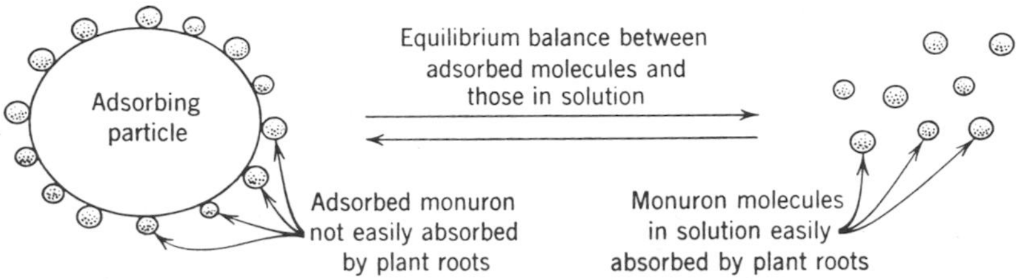

Figure 5-6. Plants absorb monuron in solution more easily than monuron that is adsorbed on the soil colloids.

In summary, the various soils show large differences in their adsorptive capacities. In practice, however, the *range* of herbicidal rates of application is much less than might be predicted from the very wide ranges in adsorptive capacity of the soils.

Leaching

Leaching is the downward movement of a substance by water through the soil. The movement of a herbicide by leaching may determine its effectiveness as a herbicide, may explain selectivity or lack of it, or may account for its removal from the soil. Preemergence herbicides are frequently applied to the soil surface. Rain leaches the chemical into the upper soil layers. Weed seeds germinating in the presence of the herbicide are killed. Large-seeded crops such as corn, cotton, and peanuts planted below the area of high herbicidal concentration may not be injured (see Figure 5-1). In addition to the protection offered by depth, crop tolerance to the herbicide is desirable.

Some herbicides can be removed from the soil by leaching (see Figure 5-7). For example, sodium chlorate can be applied to shallow-rooted turf and the chemical leached to kill deep-rooted plants without injuring the turf (See Figure 4-12).

The extent to which a herbicide is leached is determined principally by:

1. Adsorptive relationships between the herbicide and the soil.
2. Solubility of the herbicide in water.
3. Amount of water passing downward through a soil.

Solubility is sometimes cited as the principal factor affecting the leaching of a herbicide. Simple calculation of the amount of water in the usual rainfall disproves this assumption. A 4-in. rainfall weighs nearly 1,000,000 lb/acre. If you apply 1 lb of herbicide per acre, this equals 1 ppm of the herbicide in water; thus if the herbicide is soluble to the extent of 1 ppm, you might expect a 4-in. rain to remove essentially all of the herbicide from the surface inch of soil.

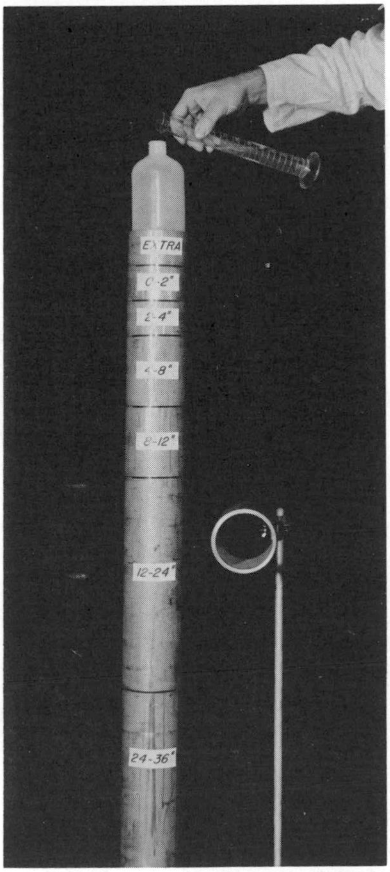

Figure 5-7. Laboratory study of herbicide leaching through a soil column. Simulated rainfall is being added at the top. The depth of leaching is determined by disassembling the steel columns and analyzing the soil enclosed. (R. P. Upchurch, formerly North Carolina State University.)

An example is given to illustrate the point. Monuron at 25°C is soluble in water to a concentration of 230 ppm. Monuron was applied at the rate of 2 lb/acre to a fine sandy soil, and 16 in. of water failed to remove the monuron from the surface inch of soil. This amount of water is capable of dissolving 400 times the amount of herbicide applied (Ogle and Warren, 1954).

The interrelationship between binding of herbicides to the soil, and water solubility can be demonstrated with 2,4-D. The salts of 2,4-D are water-soluble and readily leach through porous, sandy soils. Soils with high organic-matter content adsorb the 2,4-D, reducing the tendency to leach. The ester formulations of 2,4-D have a low water solubility. Their tendency to leach is reduced by both low solubility and by adsorption by the soil.

To restate the point, the strength of "adsorption bonds" is considered more important than water solubility in determining the leaching of herbicides. Organic-matter content in the soil is the most important single factor determining the adsorptive capacity of the soil. The second most important is the clay fraction.

Herbicides have been known to move upward in the soil. If water evaporates from the soil surface, water may move slowly upward and may carry with it soluble herbicides. As the water evaporates, the herbicide is deposited on the soil surface.

In arid areas where furrow irrigation is practiced, lateral movement of herbicides in soils also occurs.

Volatilization

All chemicals, both liquids and solids, have a vapor pressure. Water is an example of a liquid that will vaporize and naphthalene (moth balls) is an example of a solid that will vaporize. At a given pressure, vaporization of both liquids and solids increases as the temperature rises (page 109).

Herbicides may vaporize and be lost to the atmosphere as gases. The gases may or may not be toxic to plants, and the volatile gases may drift to susceptible plants. The *ester* forms of 2,4-D are volatile, and the vapors or fumes can cause injury to susceptible crops such as cotton or tomatoes (see Figure 6-8).

The herbicide may move into a porous soil as a gas. EPTC is thought to move in this way. Adsorbed by the soil, the EPTC may effectively kill germinating seeds.

The importance of volatilization and the loss of a herbicide from the soil surface is often underestimated. In volatility studies, EPTC volatilized from a free-liquid surface at the rate of 57 $\mu g/cm^2$ (about 5 lb/acre) of surface area per hour at 86°F (Ashton and Sheets, 1959). This high rate of vaporization could easily explain the loss of the herbicide. EPTC, trifluralin, and other volatile soil-applied herbicides are usually mechanically mixed into the soil soon after application to reduce loss.

Codistillation with water evaporating from the soil surface (steam distillation) is another means by which a volatile herbicide may be lost. This process has been studied little but may be of considerable importance in view of the immense amount of water lost from the soil surface through evaporation.

Herbicides such as atrazine with very low vapor pressures, however, may be lost from a surface over an extended period of time, especially if exposed to high temperatures. Soil-surface temperatures have been measured as high at 180°F.

Rain or irrigation water applied to a dry or moderately dry soil will usually leach the herbicide into the soil, or aid in its adsorption by the soil. Once adsorbed by the soil, the loss by volatility is usually reduced.

Photodecomposition

Photodecomposition, or decomposition by light, has been reported for many herbicides. This process begins when the herbicide molecule absorbs light energy; this causes excitation of the electrons and may result in breakage or formation of chemical bonds.

Most herbicides absorb radiation in the ultraviolet region (150–4000 mμ). Most herbicides are white, or nearly so, and have peak light absorption in the ultraviolet range (220–324 mμ), whereas yellow compounds such as the dinitroanilines and dinoseb have absorption peaks around 376 mμ. Solar energy below 295 mμ reaching the surface of the earth is considered to be negligible. The "sensitization" process may also be involved; light energy is absorbed by an intermediate molecule and transferred to the herbicide molecule by collision. Thus the effective wavelength of light could be outside the absorption spectra of the herbicide.

Some products of photodecomposition are similar to those produced by chemical or biological means.

An 88.3 ppm monuron solution in distilled water was sealed in quartz tubes and exposed to sunlight for 48 days. There was an 83% loss of the monuron compared with the control.

Chemicals applied to soil surfaces are frequently lost, especially if they remain for an extended period without rain. It is entirely possible that photodecomposition has taken place. However, other factors that may account for the loss should not be overlooked. Volatilization accentuated by high soil-surface temperatures, biological and chemical deactivation, and adsorption are a few of the factors that should be considered.

Removal by Higher Plants

Higher plants may absorb herbicides from the soil in which they are growing. The absorbed herbicide may then be removed when the crop is harvested. This may not be a major factor in persistence of herbicides under most conditions; however it has been used to help remove persistent herbicides from soils where they were applied as soil sterilants and the planting of ornamentals was desired (e.g., corn for simazine or atrazine removal).

Figure 5-2 shows the interrelations of the processes that lead to detoxication, degradation, and disappearance of herbicides in the environment.

SUGGESTED ADDITIONAL READING

Weed Science published by WSSA, 309 West Clark St., Champaign, IL 61820. See index at end of each volume.

Weeds Today. Same availability as above.

Ashton, F. M., and T. J. Sheets, 1959, *Weeds* **7**, 88.

Kearney, P. C., and D. D. Kaufman, Eds., 1975, 1976, *Herbicides* Vols. 1 and 2, Marcel Dekker, New York.

Ogle, R. E., and G. F. Warren, 1954, *Weeds* **3**, 257.

Sheets, T. J., and D. D. Kaufman, 1960, USDA, ARS-20-9.

Upchurch, R. P., F. L. Selman, D. D. Mason, and E. G. Kamprath, 1966, *Weeds* **14**, 42.

Weber, J. B., 1980, *Weed Science* **28**(5), 467.

Weber, J. B., 1980, *Weed Science* **28**(5), 478.

For chemical use, see the manufacturer's label and follow the directions. Also see the Preface.

6 Formulations, Drift, and Calculations

FORMULATIONS

Herbicides are formulated to make their application easier and to improve their effectiveness under field conditions. The chemist can change the formulation of a chemical to affect its solubility, volatility, specific gravity, toxicity to plants, and numerous other characteristics. He can add other substances, such as surface-active agents, which can considerably alter the effectiveness of a herbicide. He may dissolve the active ingredient in a suitable solvent for a liquid formulation or mix it with an inert material such as clay for a wettable powder formulation, add it to a coarse particle such as vermiculite for a granular formulation, or combine the ingredients and, with pressure, produce a pellet.

For effective weed control, the quantity of chemical (active) may be as low as $\frac{1}{8}$ lb/acre (only 2 oz/acre), whereas less-active compounds may require several pounds per acre. The formulated herbicide is usually diluted with water for field applications and applied at 1–10 gal/acre by aircraft, or 5–20 gal/acre by ground sprayers. Uniform, even applications are required so that germinating seeds or tiny seedlings, perhaps only $\frac{1}{8}$ in. across, are killed. Hence, uniform application is extremely important.

Herbicides are formulated to be applied as (1) solutions of water or oil, (2) emulsions, (3) wettable powder suspensions, and (4) granules. Detailed information on pesticide formulations is presented in a book by Yates and Akesson (1973).

Solutions

A *solution is a physically homogeneous mixture of two or more substances.* The substances may be solids, liquids, or gases. However, when we think of a solution, we usually have in mind solids or liquids dissolved in a liquid. Every part of the solution is like every other part.

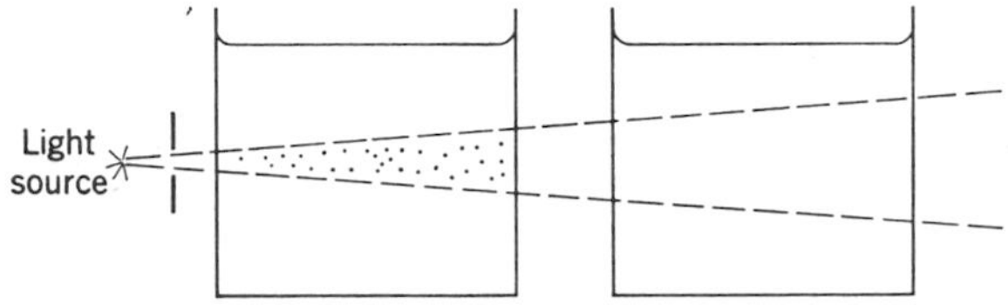

Figure 6-1. Reflective properties of true solutions and colloidal suspensions, the Tyndall effect. Light is not reflected when it passes through a true solution (*Right*). But light is reflected from colloidal suspensions (*Left*) of either suspended liquid particles (emulsion) or suspended solid particles (wettable powder).

The constituents cannot be seen separately, nor can they be separated by mechanical means. The solution is clear in appearance, even though it may have color. If darkly colored, the transparency of the solution may be lost (see Figure 6-1).

Solutions are formed when sugar or salt (solute) dissolves in water (solvent), when alcohol dissolves in water, or when kerosene dissolves in gasoline. The dissolved constitutent is the *solute,* and the substance in which solution takes place is the *solvent.*

The salts of most herbicides are soluble in water. A few examples are sodium and amine salts of 2,4-D, 2,4,5-T, MCPA, and silvex; amine salts of dinoseb; sodium salt of pentachlorophenol; sodium salt of TCA; and sodium salt of dalapon. These can be dissolved in convenient amounts of water and sprayed efficiently.

The "parent acid" formulations of some of these are oil-soluble. For example, DNBP and pentachlorophenol are soluble in oil. They are often used to increase the toxicity of oil sprays, or to *fortify* the oil. Ester formulations of 2,4-D and related products are also soluble in oil.

In a solution, a certain percentage of solute molecules is dissociated into ions. These ions are then free to combine with other ions of the solution. For example, "hard water" may have a high calcium or magnesium content. If a salt form of 2,4-D is added to hard water, it too becomes ionized. The 2,4-D ions are free to react to form calcium-2,4-D or magnesium-2,4-D salts. The calcium salt of 2,4-D is soluble in water to the extent of 2.5 g/liter and the magnesium salt to 17.4 g/liter at 68°F. Thus the calcium and magnesium 2,4-D molecules form precipitates if the quantity exceeds these amounts. These precipitates may clog filters and nozzles on the sprayer.

Emulsions

An emulsion is one liquid dispersed in another liquid, each maintaining its original identity. Without agitation the liquids may separate. Once mixed, some emulsions require very little agitation to prevent separation; others, however, require constant heavy agitation to prevent

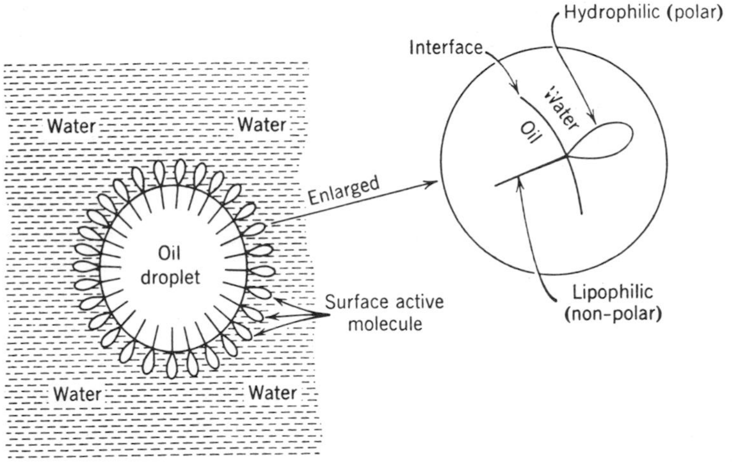

Figure 6-2. The surface-active molecule tends to bind the oil–water surfaces together, reducing interfacial tension.

separation. Remixing following separation will usually reform the emulsion.

Oil-soluble herbicides are often formulated for mixture with water as an emulsion. For example, formulations of CIPC, CDEC, CDAA, trifluralin, and ester formulations of 2,4-D, 2,4,5-T, MCPA, and silvex are usually mixed with water to form emulsions, which appear milky. Emulsions of the *oil-in-water* (O/W) type are easily applied as sprays. The *water-in-oil* (W/O) type of emulsion varies in viscosity; some may be too viscous to be applied as a spray.

As stated above, an emulsion is a liquid dispersed in another liquid. The emulsifying agent binds and also has a tendency to "insulate" the surfaces of the two liquids. There is, however, no direct contact be-between the two liquids (see Figure 6-2). Therefore, little or no opportunity exists for chemical reaction between the two. The emulsion offers the opportunity, at least for a short period, of mixing reactive chemicals without the formation of precipitates. The emulsifying agent must be stable in the presence of both. For example, ester formulations of 2,4-D are oil-like and form emulsions when mixed with water; even when mixed with "hard water," no calcium or magnesium precipitates are formed. The dispersed, oil-like droplets are suspended in the water but insulated by the emulsifying agent.

Emulsions are discussed more fully on page 104.

Wettable Powder Suspensions

A wettable powder suspension consists of finely divided, solid particles dispersed in a liquid. It usually consists of a mixture of the pesticide with clay, a dispersant, and a wetting agent. At the time of application it is suspended in water and sprayed.

Some herbicides are nearly insoluble in both water and oil-like substances. Therefore, concentrated solutions or emulsions cannot be prepared. In such cases, a wettable powder suspension may be the most practical formulation.

The suspension will usually give a "clouded" appearance to the liquid. It will also give reflection to a light cone (Tyndall effect) similar to an emulsion (see Figure 6-1). The addition of a surface-active agent is needed to maintain the suspension in water.

Some gel-like substances are made by chemically or physically treating special clays. Various wettable powder pesticides can be added to this gel-like substance. At the time of application the gel-like formulation is added to water and applied as a spray. Most "flowable" and "liquid" formulations are of this type. With long periods of storage, the wettable powder may settle to the bottom of the container.

Most wettable powder suspensions require agitation to prevent settling of the solid particles. The smaller the solid particles, the slower the rate of separation. Similar density of the solid particle and the liquid will also slow the rate of separation.

Herbicides sold as wettable powders include simazine, atrazine, monuron, diuron, DCPA, nitralin, oryzalin, diphenamid, and tebuthiuron. They are marketed either as dry wettable powders or as "liquid suspensions."

Granules

Some chemicals are applied at rates high enough so that crystals of the chemical itself can be uniformly applied. This is true with many of the borate and sodium chlorate compounds. In other cases, the herbicide is applied by mixing with a "carrier" to provide sufficient bulk for even distribution. Pellets are granules that are extruded under pressure to yield a particular size and shape. Many carriers are used, including clay, sand, vermiculite, and finely ground plant parts (ground corncobs, tobacco-leaf wastes, etc.). Granules are made in many different sizes. Granular materials can be spread by hand, or by mechanical spreaders resembling seed or fertilizer spreaders.

Granular materials have three advantages over sprays: (1) water is not needed to apply the material, (2) costlier spray equipment is not required, and (3) large granules have a tendency to fall off the leaves of valuable plants without injuring them.

However, granules also have disadvantages: (1) they are heavier and bulkier, and shipping charges may considerably increase their cost; (2)

some granular materials are easily moved by wind or water; and (3) application is seldom as uniform as it is with sprays.

Dust

Dusts are a popular method of application for insecticides and fungicides. However, *herbicides* are not applied as dusts because of the drift hazard.

SURFACE-ACTIVE AGENTS

Surface-active agents (surfactants) include wetting agents, emulsifiers, detergents, spreaders, sticking agents, and dispersing agents. One or more of these types of surfactants are used in essentially all herbicide formulations. We are interested in the properties of the different surfactants because these properties may explain why some sprays adhere to plant surfaces and others bounce or run off. A wetting agent may increase the toxicity of a herbicide in one case and decrease it in another.

Since certain surfactants may increase the phytotoxicity of specific herbicides and others decrease the phytotoxicity, the applicator should not add surfactant to a commercial formulation unless this is recommended by the manufacturer, or unless independent research has shown that a specific surfactant–herbicide combination increases herbicidal effectiveness.

Incidentally, the "mechanism of action" of surface-active agents used with herbicides is closely related to that of chemical detergents in the home laundry, textile dyeing and finishing, electrochemical plating, mining-ore flotation, paper toweling, and even fire fighting.

Basic Concepts

Water is not compatible with many chemicals used as herbicides, or with many plant surfaces (see Figure 6-3). For example, water does not mix with oil or oil-like chemicals: one usually repels the other. By adding a surfactant, in this case an emulsifying agent to the oil, the herbicide can be mixed with water to form an emulsion; as such, it can easily be sprayed through the sprayer.

Water is also repelled by the waxlike cuticle found on plant surfaces. By adding a surfactant, in the form of a wetting agent, the effectiveness of a herbicide may be completely changed, as shown in Figure 6-3. For example, an effective surfactant added to amitrole (in water spray) in-

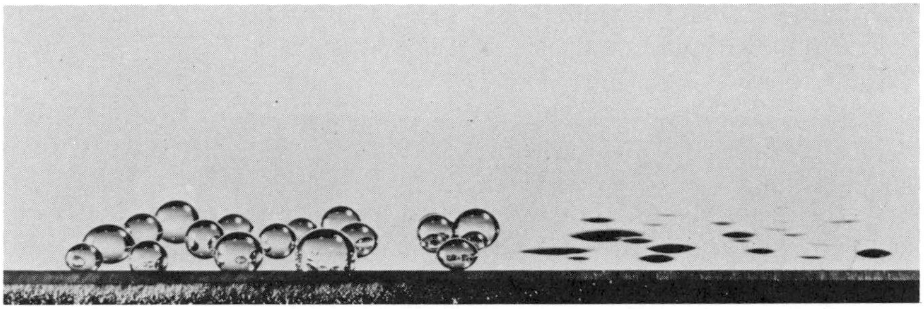

Figure 6-3. Water droplets on a waxed pane of glass. *Left*: Pure water with high surface tension. *Right*: Water droplets with a wetting agent added spread over the waxed surface as a thin layer. (Atlas Powder Company, Wilmington, DE.)

creased the amount of amitrole absorbed by bean leaves by six times over a 24-hr period.

Surface Relationships

Here are the four surface relationships, followed by examples:

1. *Liquid to liquid*—oil dispersed in water by agitation to form an *emulsion.*

2. *Solid to liquid*—clay suspended in water, or a wettable powder herbicide suspended in water.

3. *Solid to air*—carbon in the air to form smoke, or pesticides applied as *dusts.* A spray droplet that dries in the air to leave a small solid particle suspended in the air is another example.

4. *Liquid to air*—tiny water droplets suspended in the air to form fog, or spray droplets in the air.

In emulsions and suspensions, the water has a tendency to be repelled by the other liquid, or by the solid forming the suspension. This produces surface tensions or interfacial tensions. To "bind" the two surfaces, a substance is needed that has an affinity for both water and the other substance. Such a substance will be a molecule that orients itself between the two surfaces so they are "bonded" in a more intimate contact. Thus *a surface-active agent modifies the surface forces (interfaces) by orienting itself between the interfaces, providing a more intimate coupling.*

A molecule or ion will usually possess surface activity if it contains a strongly polar group that is attracted to water (hydrophilic) and a nonpolar group that is attracted to nonaqueous materials (lipophilic or hydrophobic). This nonpolar group is usually attracted to oils, fats, and waxes. Lipid means fat. Fats and oils are chemically similar, with the major difference the temperature of the melting point. Lipoid means fatlike, thus the term oil-loving or lipophilic.

You might visualize surfactant molecules as tadpole shaped, with the head portion being hydrophilic, and the tail portion lipophilic. When added to a mixture of oil and water, the lipophilic portion orients itself in the oil and the hydrophilic portion in the water (see Figure 6-2).

The distinct head–tail relationship is lacking in some of the newer synthetic detergents. However, their molecules do have hydrophilic and lipophilic regions that permit effective orientation between the two surfaces.

Surface Tension

The molecules of a liquid strongly attract other molecules that are close to them. Therefore, the surface molecules are attracted toward the center of a liquid body. Thus a "free" water droplet appears as if held by a tense, elastic membrane. This can be illustrated by the use of a waxed or oiled surface (nonpolar) that has little attraction for the water molecules (polar). Because the water molecules are mutually attracted, causing surface tension, a nearly spherical drop is formed (see Figure 6-3). In summary, we can define surface tension as *the tendency of the surface molecules of a liquid to be attracted toward the center of the liquid body.*

A small steel sewing needle will float on the surface of pure water; the needle is not heavy enough to break the surface tension. But the needle will not float if a wetting agent is added. Surface tension is not strong enough to support the needle with the wetting agent added to the water.

Hydrophile–Lipophile Balance (HLB)

The HLB of a surfactant molecule or a mixture of two or more surfactant molecules is a quantitative value of their polarity. This concept is particularly useful when surfactants are used as emulsifiers. It uses a scale of 0 to 20; the higher the value, the greater the hydrophilic prop-

Table 6-1. Approximation of HLB by Water Solubility (Becher, 1973)

Behavior When Added to Water	HLB Range
No dispersibility in water	1–4
Poor dispersion	3–6
Milky dispersion after vigorous agitation	6–8
Stable milky dispersion (upper end almost translucent)	8–10
From translucent to clear	10–13
Clear solution	13+

erties. The HLB value is calculated from the nature of the molecule or molecules, or is determined experimentally. A rough approximation of the HLB can be obtained by observing the water solubility (see Table 6-1).

Although HLB values are valuable in the chemical formulation of herbicides, little information is available on the correlation of the HLB of a herbicide formulation with herbicidal effectiveness. However, in one study certain foliar-applied herbicides reached maximum phytotoxicity at an HLB value of about 14 (Jansen, 1964; Grover, 1976).

Types of Surfactants

Surface-active agents can be classed as *nonionic* and *ionic,* depending on their ionization or dissociation in water. Nonionic surface-active agents have no particle charge, whereas ionic agents show either a positive or negative charge.

The *nonionic* surfactants ionize little or not at all in water. They are classed as nonelectrolytes and are usually chemically inactive in the presence of the usual salts. Thus they can be mixed with most herbicides and still remain chemically inert. Many emulsifying agents are of this type, and they are usually liquids.

Ionic surfactants ionize when in an aqueous medium, some being anionic (–) and others cationic (+). *Anionic* agents are those in which

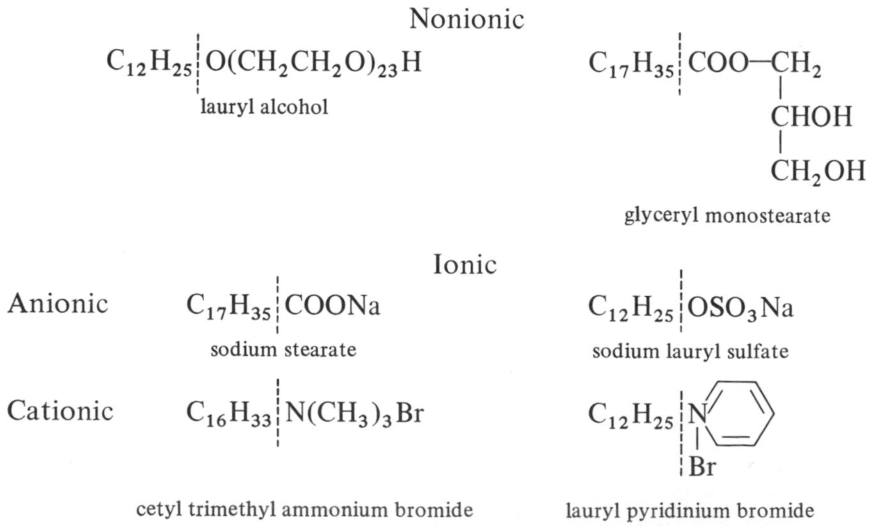

Figure 6-4. Structural formulas of the various types of surface-active agents. The portion of the molecule to the left of the dashed line is lipophilic; that to the right, hydrophilic. (Becher, 1973.)

the anion part of the molecule exerts the predominant influence. Wetting agents, detergents, and some emulsifiers fall in this group. *Cationic* agents (also known as reversed soaps) become ionized in water with the cation part of the molecule exerting the predominant influence. They have powerful bactericidal action, but they are expensive and have had only limited use in agriculture.

Most emulsifiers in commercial usage are blends of anionic and nonionic substances. Specific blends can give a specific result. In general, however, a blend with a high proportion of the anionic agents will improve performance in cold water and work best in soft water. Usually a blend with a predominance of nonionic types will perform better than other types in warm water and in hard water. See Figure 6-4 for the chemical structure of various surfactants.

Surfactants Classed According to Use

Wetting Agents

A wetting agent increases a liquid's ability to moisten a solid. It lowers the interfacial tension, bringing the liquid into intimate contact with the solid (see Figure 6-5). The degree of effectiveness of a wetting agent is shown by the increase in spread of the liquid over a surface area. The spread over the surface area determines the "contact angle" of the liquid with the surface.

Wetting agents may increase or decrease the effectiveness of herbicide sprays. For example, the effectiveness of contact herbicides depends largely on both complete and uniform wetting of the plant. A contact chemical applied in a spray at the rate of 10 gal/acre without a wetting agent will not "wet" the vegetation. Lacking a wetting agent, the chemical remains as droplets, burning small "pinholes" in the leaf. The plant is not killed. The addition of a wetting agent causes the spray droplets to spread, and this action helps to cover the plant uniformly. At low gallonage, the wetting agent may be expected to increase the effectiveness of the spray.

Now, let us suppose that the same *chemical* is applied at the *same rate*

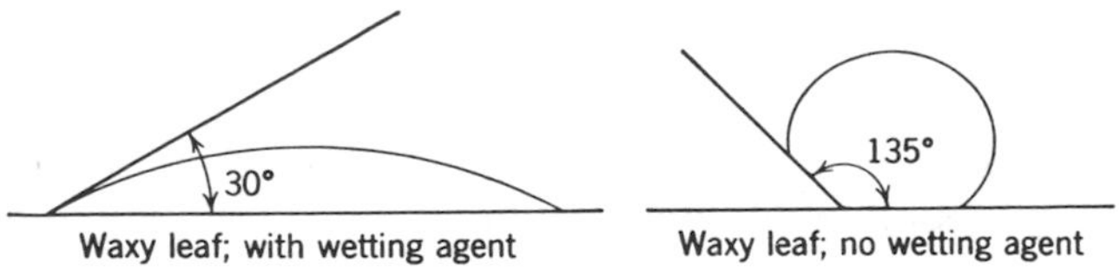

Figure 6-5. *Left*: Water droplets containing a wetting agent will spread in a thin layer over a waxed surface. *Right*: Pure water will stand as a droplet, with a small area of contact with the waxed surface.

per acre, but it is diluted in 200 gal of spray. Without a wetting agent, this rate heavily "wets" the foliage, with no runoff. If a wetting agent is added, the spray forms a thin layer over the plant. The low surface tension permits approximately two-thirds of the spray to run off the plants to the ground. The chemical is lost from the surface of the plant and the herbicide loses some of its effect. Thus with high-gallonage sprays, a wetting agent may actually decrease the effectiveness of contact herbicides. This is seldom a problem, however, because excessive spray quantities are not often applied.

Herbicidal *selectivity* may be lost by the addition of a wetting agent. This is true if selectivity depends on selective wetting or selective absorption.

Emulsions (see discussion on page 96)

If oil is added to water and shaken vigorously, the oil is momentarily suspended as small droplets in the water, forming an emulsion. The water is a continuous body and is therefore referred to as the *continuous phase*; however, the oil is dispersed and is therefore referred to as the *discontinuous phase.*

If the mixture is allowed to stand, the oil and water will separate. A third material may be added to decrease the tendency to separate or to increase the stability of the emulsion. This is called the *emulsifying agent* or *emulsifier.*

Emulsification is principally a "surface-action" phenomenon. The emulsifier molecule orients itself between the oil–water interface, coupling the two (see Figure 6-2). This prevents the dispersed droplets from coalescing to form large droplets that quickly separate.

For example, milk is an emulsion of butterfat (dispersed) and water (continuous), with casein acting as the emulsifying agent. This type of emulsion is referred to as an *oil-in-water* (O/W) emulsion. Most herbicidal emulsions are of the O/W type. When mixed with water they have approximately the same viscosity as water, with a "milky appearance."

Butter and mayonnaise are emulsions with water the dispersed phase and fat or oil the continuous phase. This type of emulsion is called a *water-in-oil* (W/O) emulsion. Water-in-oil emulsions are usually viscous, too thick to pass through the usual spray equipment; they resemble whipped cream. The W/O emulsion may be referred to as an *invert.* Invert 2,4-D and 2,4,5-T emulsions have been prepared and have been used particularly for brush control.

Three factors largely determine the stability of an emulsion. These are (1) size of dispersed particles, (2) relative density of the two liquids, and (3) viscosity of the emulsion. Size of particles is especially important. For example, in homogenized milk the fat globules have been broken into very tiny droplets. These separate much more slowly than large butterfat droplets in plain milk.

There are hundreds of detergents and emulsifying agents. McCutcheon (1949) listed more than 700 detergents and gave the manufacturers, chemical formulas, main uses, concentrations, and types. Bennett (1947) named about 600 emulsifying agents, giving the common name as well as the chemical name for many of the products. Cupples (1940) prepared a list of wetting, dispersing, and emulsifying agents, and gave the properties and uses of more than 300 compounds. Gast and Early (1956) rated the toxicity of 77 emulsifiers and 66 organic solvents when applied to beans, corn, cotton, cucumber, tobacco, and tomato plants. In general, the emulsifiers were more toxic than the organic solvents tested.

Detergents

Detergency indicates the cleaning power or the ability of a chemical to remove soil or grime. All detergents are surfactants. Many of the common detergent chemicals have been used with herbicides as wetting agents, spreaders, and emulsifiers; hence the word *detergent* has appeared in herbicidal literature.

Spreaders

Spreaders and wetting agents, although sometimes considered independently, are closely related. When the wetting agent reduces surface tension, spreading naturally follows. Therefore, when used in connection with herbicides, spreaders and wetting agents should be considered together.

Adhesive or Sticking Agents

Adhesive or sticking agents, as the name implies, are substances that cause the herbicide to adhere to the sprayed surface. Many of the surfactants discussed above may also act as sticking agents.

Dispersing Agents

A dispersing agent is a substance that reduces the cohesion between like particles. It promotes separation of like particles and in some cases its reaction may be closely related to deflocculation. Some dispersing agents are good wetting agents, but others have little or no effect on the surface tension. Some wetting agents and dispersing agents are not compatible and have a tendency to interfere with each other if used together.

Effects on Plants

Surfactants generally intensify the action of a foliar-applied herbicide. In some cases selectivity of the herbicide is lost. Surfactants probably have five important effects on the plant:

1. They favor uniform spreading of the spray, or uniform wetting of the plant.
2. Spray droplets usually stick to the plant—there is less bounce-off.
3. The chemical spray is brought into intimate contact with the plant surface. Droplets do not remain suspended on hairs, scales, or other projections.
4. The surface-active agent may alter nonpolar plant substances. Substances such as the waxy cuticle or the lipoidal portion of the cell wall may be altered so the plant readily absorbs the chemical. Also, the cell membranes may be modified, thus permitting the cell sap to leak into the intercellular spaces. When this happens the surfactant may affect the plant as oils do (see Figure 6-6 and Chapter 4).

Figure 6-6. A wetting agent was added to an ammonium nitrate solution. The spray was directed to hit the Pennsylvania smartweed and corn stems without hitting the corn leaves. The wetting agent nearly doubled the burning and killing effect on the weeds. (North Carolina State University.)

5. Detergents are believed to have many effects on proteins. These may include protein precipitation and denaturation, and inactivation of enzymes, viruses, and toxins.

CHEMICAL DRIFT THROUGH THE AIR

Chemicals may drift through the air from the area of application and cause considerable injury if they contact susceptible plants. Movement through the air may result from *spray drift* or *volatility*.

Spray Drift

Spray drift is the movement of airborne spray particles from the target area. The amount of spray drift depends upon (1) size of the droplets, (2) amount of wind, and (3) height above the ground that the spray is released.

The size of the droplet depends primarily on pressure, design of the nozzle, and surface tension of the spray solution. In general, low pressures have a tendency to produce large droplets, and high pressures, small droplets. Different nozzle designs produce different droplet sizes. Small nozzles usually produce small droplets, and solutions possessing low surface tension tend to produce small droplets.

Table 6-2 illustrates the importance of large droplets in reducing spray drift.

The 400 μm droplets (the size of fine rain) will drift only 9 ft–if released 10 ft above the ground in a 3-mph breeze. However, under the

Table 6-2. Spray Droplet Size and Its Effect on Spray Drift (Adapted from Brooks, 1947; Yates and Akesson, 1973)

Droplet Diameter (μm)	Type of Droplet	Number of Droplets/in.^{2} From 1 gal of Spray/Acre	Time Required to to Fall 10 ft in Still Air	Distance Droplet Will Travel in Falling 10 ft With a 3-mph Breeze
0.5	Brownian max.	–	6750 min	388 miles
5	Fog	9,000,000	66 min	3 miles
100	Mist	1,164	10 sec	440 ft
200	Drizzle	195	3.8 sec	17 ft
400	Fine rain	28	2.0 sec	9 ft
500 ($\frac{1}{50}$ in.)	Rain	9	1.5 sec	7 ft
1000 ($\frac{1}{25}$ in.)	Heavy rain	1.1	1 sec	4.4 ft

same conditions, 100 μm droplets (like mist) will drift 440 ft. The 5 μm droplets (fog) will drift 3 miles.

One gallon of spray per acre will apply *per square inch:* 28 droplets of 400μm size (fine rain), 195 droplets of 200μm, and 1164 droplets of 100μm (mist). These are more than ample droplets–especially for soil-applied herbicides from only *1 gal/acre.* Obviously, few herbicide spray operations require 10–50 gal/acre. As a practical range, droplets of size 300μm–500μm are suggested.

Many farm sprayer nozzles produce a dangerously high percentage of fog- and mist-size droplets. These nozzles continue to be used even though, at least theoretically, many sprays can be applied at rates of 1 gal/acre *without* danger of drift (droplets 300μm or larger). But in trying to reach low gallonages per acre, one needs to attain (1) uniform application, (2) adequate gallonage to mix with the chemical, (3) no nozzle stoppage, (4) no loss of chemical efficiency, and of course (5) no spray drift.

As droplets fall, evaporation may occur, leaving extremely small particles. Such particles may be nearly as small as smoke particles; the latter may be small enough to approximate the Brownian maximum. Such particles may drift as much as 388 miles under the conditions described in Table 6-2. If drift of the chemical is a hazard, the carrier liquid should have a low vapor pressure to slow the rate of evaporation, especially if applied by airplane.

Periods of strong wind velocities vary with locality. In general, winds are least turbulent just before sunrise, and another low period occurs just after sunset and throughout the night. As an average, winds are most gusty and turbulent between 2 and 4 p.m.

The height above ground that the spray is released is important for two reasons. First of all, the distance, and thus the time required for the droplets to reach the ground, is directly affected. Second, wind velocities are usually much lower close to the ground than at higher elevations.

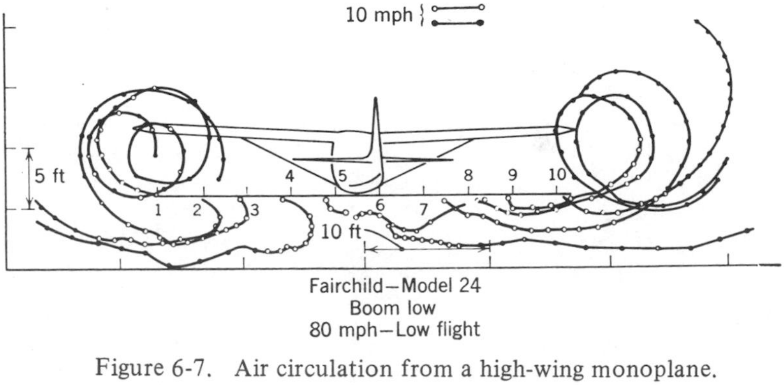

Figure 6-7. Air circulation from a high-wing monoplane.

Therefore, spray applications from aircraft present greater drift hazards than ground sprayers. Air currents produced by an aircraft have a major effect on the trajectory of fine particles released from it. Basically, any aircraft, rotary (helicopter) or fixed wing, has updrafts produced by wing-tip vortices and downdrafts under the middle of the aircraft (see Figure 6-7).

Nozzles that produce uniform droplets between 300 μ and 500 μm in diameter will deliver a spray with essentially no spray-drift hazard. Such sizes will permit adequate coverage for most herbicidal spray operations at relatively low gallonages per acre. Nozzle design is discussed further in Chapter 7.

Volatility

Volatility refers to the tendency of the chemical to vaporize or give off fumes. The amount of fumes or vapors emitted is related to the vapor pressure of the chemical.

Vapor drift is the movement of vapors or fumes. Vapor drift may damage susceptible crops, or may simply reduce, through loss, the effectiveness of the herbicide treatment.

The volatility of 2,4-D has perhaps received more attention than that of any other chemical. Figures 6-8 and 6-9 indicate differences found between the salts and esters. The amine and sodium salts of 2,4-D have

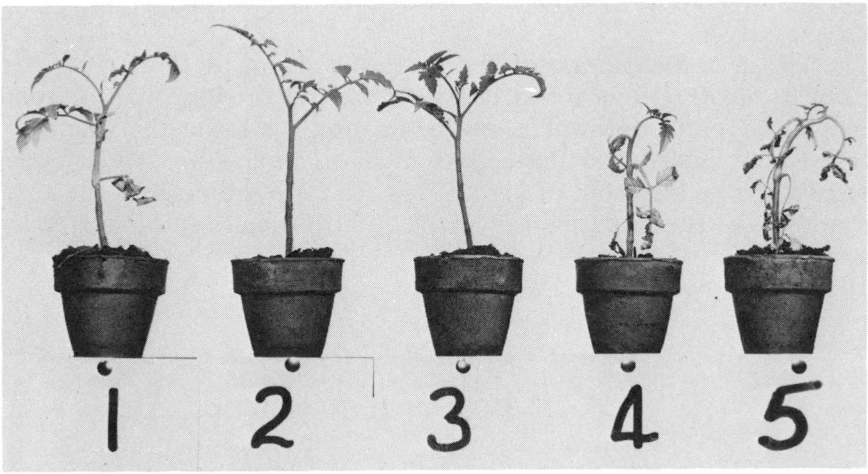

Figure 6-8. These tomato plants were kept under a jar that was exposed to the vapor or fumes from different 2,4-D formulations. *1*: Sodium salt of 2,4-D caused slight to no injury. *2*: Diethanolamine salt of 2,4-D caused no injury. *3*: Triethanolamine salt of 2,4-D caused no injury. *4* and *5*: Butyl and ethyl esters of 2,4-D killed the plants. (Klingman, 1947.)

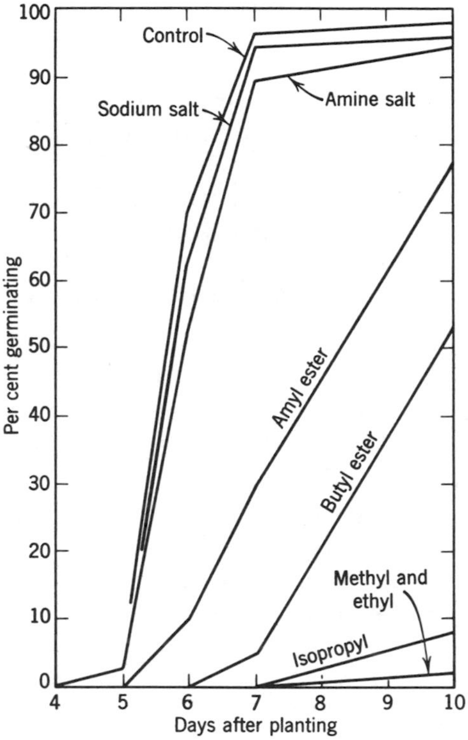

Figure 6-9. Percentage germination of pea seeds after being exposed to the vapors or fumes from different 2,4-D formulations. (Mullison and Hummer, 1949.)

little or no volatility hazard. The ester formulations vary from low to high volatility.

The length and structure of the alcohol portion of the 2,4-D ester molecule directly affect its volatility. In general the longer the carbon chain in the part contributed by the alcohol, the lower the volatility. Those esters made from five-carbon alcohols or less are usually considered volatile. Inclusion of an oxygen as an ether linkage in the alcohol portion of the molecule will also reduce the volatility of an ester of 2,4-D.

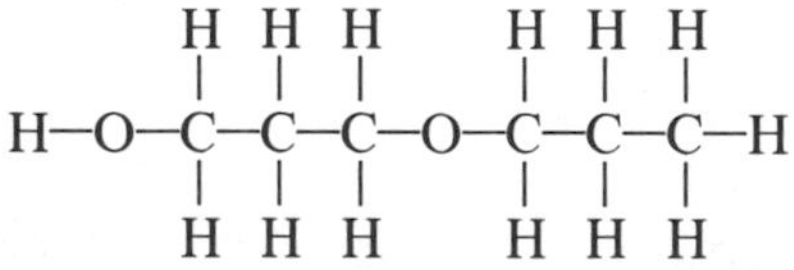

Grover (1976), comparing the relative volatility of various formulations, assigned 2,4-D amine a relative volatility of 1, low volatile esters of 2,4-D (propylene glycol butyl, butoxy-ethanol, and isooctyl) a relative volatility of 33, and high volatile ester of 2,4-D (butyl) a relative volatility of 440. Thus, where 2,4-D–susceptible plants are grown, the

2,4-D amine form should be chosen. Also, application methods that keep spray drift to a minimum should be used (see Chapter 7).

Soil surfaces exposed to direct sunlight often reach 180°F. At this temperature some chemicals quickly volatilize and may be carried away in the wind, and therefore are hazards to susceptible plants. Also, the herbicidal effect of the treatment may be lost.

CALCULATIONS

Active Ingredient

The active ingredient is that part of a chemical formulation that is directly responsible for the herbicidal effects. In some herbicides the entire molecule is considered to be the "active unit." Therefore, if the chemical is 99% pure, it is considered 99% active ingredient.

In others, the herbicide activity is more accurately calculated on an acid-equivalent basis.

Acid Equivalent

The acid equivalent refers to that part of a formulation that theoretically can be converted to the acid. In this case, the acid equivalent is given as the active ingredient.

The percentage of active ingredient or acid equivalent is given on the label. To calculate the weight of the commercial product required, the following formula may be convenient:

$$\frac{\text{Weight of chemical to be applied (active ingredient)}}{\text{Percentage expressed as a decimal (active ingredient)}} = \text{Weight of commercial material required}$$

For example, you buy a herbicide with 80% (0.80) active ingredient as diuron. You want to apply 1.0 lb of diuron per acre. Therefore,

$$\frac{1.0}{0.80} =$$ Thus $1\frac{1}{4}$ lb of the commercial product is needed to apply 1 lb of the active ingredient of diuron.

In this case, $1\frac{1}{4}$ lb of the commercial product are added to the amount of water required to cover 1 acre.

Liquid formulations may show on their labels both the *percentage active ingredient* or *acid equivalent,* and the *weight per gallon.* A label on a container of 2,4-D may read:

67% triethanolamine salt of 2,4-D by weight
40% 2,4-D acid equivalent by weight
4 lb of 2,4-D acid equivalent/gal

These statements can be on the same label and still describe the contents accurately. The 40% and 4 lb represent the acid equivalent in this case; 1 gal of the herbicide would obviously weigh 10 lb. The 67% represents the total chemical formulation—27% triethanolamine and 40% 2,4-D acid equivalent. The other 33% may be additives such as wetting agents, but is principally inert ingredients.

Liquid formulations can be weighed in the same manner as dry forms, but it is usually easier to use liquid measure. For example, the above 2,4-D product contained 4 lb of 2,4-D acid equivalent per gallon. Therefore, 1 qt will contain 1 lb of the 2,4-D acid equivalent, and if 1 lb of the 2,4-D is to be applied per acre, then 1 qt of the liquid should be included in the amount of spray required to cover 1 acre. Conversion factors that are useful in calculation are given in the appendix.

For applying to field conditions, the amount of the herbicide is usually given as pounds of active ingredient per given unit of volume. For example if a farm sprayer applies 10 gal of spray per acre, the farmer then adds the amount of herbicide per acre needed to each 10 gal of spray. Or in making a mixture for basal spraying of brush he may add 12 lb of 2,4-D per 100 gal of oil spray. If there are 4 lb of 2,4-D (active ingredient) per gal, this would mean adding 3 gal of the commercial 2,4-D to 97 gal of oil, to give a total of 100 gal of mixture. Each gallon of the mixture has 0.12 lb of 2,4-D (acid equivalent). In this case, 3.9 liquid oz of the 2,4-D concentrate would be added to enough oil to make 1 gal of the mixture.

Parts per Million

Parts per million (ppm) refers to the number of parts by weight or volume of a constituent in 1,000,000 parts of the final mixture, by weight or volume. The stated concentration should tell whether ppm is measured by weight or volume. For herbicidal purposes it usually refers to a given weight of a chemical in a given volume of the spray.

If we take the gram and the milliliter (1 ml of water = 1 g) as the units, then 1000 ppm of 2,4-D means that 1000 g of 2,4-D are dissolved in sufficient water to make 1,000,000 ml of solution; or 1 g is dissolved in enough water to total 1000 ml of solution. If 1 qt of 2,4-D contains the equivalent of 1 lb (453.6 g) of 2,4-D acid, then to make up a spray of 1000 ppm, the 1 qt is added to enough water to make up a spray weighing 453,600 g or nearly 120 gal.

Percent Concentration

Percent concentration is similar to ppm except that it is expressed as a percentage. For example, 1000 ppm is equivalent to 0.1%, and 5000 ppm to 0.5%. A 1.0% solution (10,000 ppm) contains 1 g of 2,4-D (acid equivalent) in 100 ml of spray. The conversion of percent to ppm, and vice versa, can be done according to the following equation:

$$\% = \frac{\text{ppm}}{10{,}000} \qquad \text{ppm} = \% \times 10{,}000$$

PRACTICE PROBLEMS. See page 429 for conversion factors, also see page 125.

1. You want to apply 1 lb of 2,4-D (acid equivalent) per acre. You buy a product that is 67% 2,4-D amine, 40% 2,4-D acid equivalent, and has 4 lb 2,4-D acid equivalent per gal. How much will you apply per acre in terms of —— pints; or —— liquid oz; or —— cc or ml?
2. You then see a product that has 3.34 lb/gal of 2,4-D acid equivalent. One pound of 2,4-D acid will be present in —— pints; or —— liquid oz; or —— cc.
3. You see a granular material that is 10% active CDEC. You want to apply 8 lb/acre of active CDEC. How many pounds per acre will you need?
4. You want to apply $\frac{1}{2}$ lb/acre active ingredient of diuron to cotton. The label indicates that it is 80% active. You will need to apply —— lb; or —— g of the commercial material per acre.
5. You want to apply $\frac{3}{4}$ lb/acre of the active ingredient trifluralin. The commercial product has 4 lb/gal of the active chemical. You will apply —— pints; or —— liquid oz; or —— cc/acre of the commercial material.
6. You have a sprayer that applies 7 gal of spray per acre. Based on problem 5, you will add —— pints; or —— liquid oz; or —— cc to each gallon of spray.
7. You have a sprayer that applies 25 gal of spray per acre. Using the rate given in problem 4, you will add —— lb; or —— g of commercial material to each gallon of spray.
8. From problem 6, your sprayer tank holds 55 gal. You need to refill the tank. You measure the amount of spray and find that 6 gal of spray remains. You will add —— pints; or —— liquid oz; or —— cc of Treflan® to the tank prior to filling with water.
9. From problem 7, your sprayer tank holds 55 gal. You need to refill the sprayer, but 13 gal of spray remains. You will add —— lb; or —— g of the commercial diuron prior to refilling the tank.
10. Using the product described in problem 1, you will need —— liquid oz; or

—— cc/gal when mixed with water to give a concentration of 2000 ppm. This concentration is the same as —— %.

11. You have a nozzle that delivers approximately 3.8 gal/hr. The nozzle will treat a $3\frac{1}{2}$-ft middle in corn. Walking 3 mph, $1\frac{1}{4}$ acres are treated per hr; —— gal of spray are applied per acre. If there are 4 lb of 2,4-D per gal, you will need —— liquid oz of 2,4-D per gal of spray to apply $\frac{1}{2}$ lb of 2,4-D per acre.
12. You have a sprayer that sprays a swath $16\frac{1}{2}$ ft wide. Under regular operating conditions, it sprays 80 gal of water per hr. At a continuous 8 mph, —— acres will be treated per hr; —— gal of spray are applied per acre.

SUGGESTED ADDITIONAL READING

Weed Science published by WSSA, 309 West Clark St., Champaign, IL 61820. See index at end of each volume.

Weeds Today. Same availability as above.

Becher, P., 1973, "The Emulsifier," p. 65, in W. van Valkenburg, Ed., *Pesticide Formulations,* Marcel Dekker, New York.

Bennett, H., 1947, *Practical Emulsions,* Chemical Publishing Co., New York.

Brooks, F. A., 1947, *Agr. Eng.* **28**, 233.

Cupples, H. L., 1940, USDA, Bureau of Entomology Quarantine Publication E504.

Gast, R., and J. Early, 1956, *Agr. Chem.* **11**, 42.

Grover, R., 1976, *Weed Science* **24**(1), 26.

Jansen, L. L., 1964, *J. Agr. Food Chem.* **12**, 223.

Klingman, G. C., 1947, *North Carolina Research and Farming* **4**, 3.

McCutcheon, J. W., 1949, *Soap Sanit. Chem.* **25**(8), 33; (9), 42; (10), 40.

Mullison, W. R., and R. W. Hummer, 1949, *Bot. Gaz.* **11**, 77.

Yates, W. E., and N. B. Akesson, 1973, "Reducing pesticide chemical drift," pp. 275, in W. van Valkenburg, Ed., *Pesticide Formulations,* Marcel Dekker, New York.

For chemical use, see the manufacturer's label and follow the directions. Also see the preface.

7 Application Equipment

Herbicides are formulated as solutions, emulsions, wettable powders, and granular materials (see Chapter 6). The herbicide may be applied broadcast, in narrow bands, as individual spot treatments, or to a particular part of the plant. The quantity of the chemical may vary from a few ounces to several thousand pounds per acre. Obviously, to provide uniform coverage, special equipment is needed.

Solutions, emulsions, and wettable powders are usually applied as sprays. Water is the usual diluent or carrier, but oils are frequently used, both as herbicides and as carriers. Granular materials are spread by special mechanical spreaders similar to those used for broadcasting seed or fertilizer.

In addition, herbicides are applied by (1) *mechanical incorporation* into the top 2–6 in. of soil, (2) *subsurface layering* (a horizontal layer a few inches below the soil surface), and (3) *injection* into (a) soil, (b) irrigation water and drainage canals, and (c) lakes and reservoirs.

SPRAYING

Spraying is the most common method of applying herbicides. Sprays have many advantages over granular forms. For one thing, sprays can be applied more uniformly because extremely small quantities of a herbicide can be sufficiently diluted to permit even coverage. The amount of total spray can be varied from 1 gal/acre to perhaps 500 gal/acre to suit the needs of the treatment. With properly designed and properly operated equipment, spray drift can be reduced to a minimum (see Chapter 6). Sprays can be accurately directed to a given area, either on the soil or on the plant.

Sprayers are classified as low-volume or high-volume sprayers. Low-volume sprayers usually apply less than 30 gal of total spray per acre. As shown in Chapter 6, it is theoretically possible to apply as little as 1 gal of spray per acre, with 28 drops/in.^{2}. These drops are large enough so that there is essentially no spray-drift hazard.

Systemic herbicides and *growth regulators* can be effectively applied in low-volume sprays. The entire plant need not be "wet", because the herbicide is translocated. However, with low-volume sprays, small droplets are usually more effective than large ones, probably a result of the spray's sticking to the plant (less bounce-off) and better coverage, especially of small weeds. Reducing drop size in half increases the number of drops eightfold.

Contact herbicides usually require thorough "wetting" of the plant. The chemical is translocated little, if at all; therefore, complete coverage is important. With complete wetting of the plant, droplet size of the spray does not influence effectiveness of the herbicide.

Uniform application is extremely important with both low and high volumes.

Sprays can be applied from a sprinkler can or a power or hand-pump-type sprayer. The power sprayer may be a small portable unit, a tractor-mounted sprayer, a jeep- or truck-mounted unit, or an airplane-type sprayer.

THE SPRAYER

Most spray units have some type of nozzle or nozzles, container (tank) to hold the spray, and pump to force the spray through the nozzles. There are also usually one or more filters or strainers, pressure gauges, pressure regulators, shut-off valves, and connecting hoses.

Nozzles

The nozzle is the most important part of the sprayer. Other parts exist only to help the nozzle operate properly. Uniformity of application, rate of application, and spray drift as influenced by droplet size are all largely determined by nozzle design and conditions of its operation.

Uneven sprays, largely due to faulty nozzles or faulty operation of the nozzles, often cause a 10-fold difference in rates of actual chemical applied. This unevenness causes crop injury for high rates of application and no weed control for low rates. These spotty variations may show up as narrow strips 2–3 in. wide.

The nozzle converts the spray liquid into spray droplets. Nozzle design (see Figure 7-1) and the pressure of operation largely determine the size and uniformity in size of the droplets. The nozzle design and construction also determine the uniformity of the spray across the width of the spray pattern.

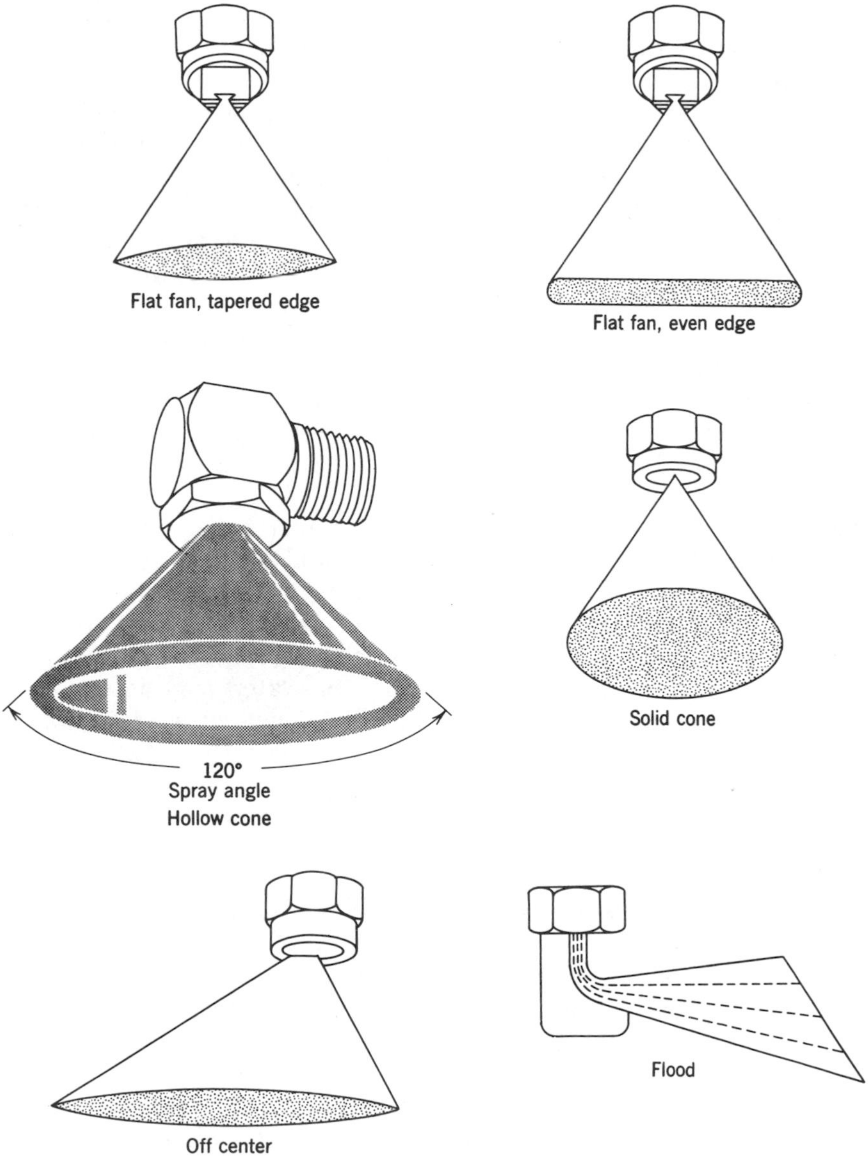

Figure 7-1. Nozzle designs used to apply herbicides.

At low pressures, the liquid escapes from the nozzle tip as a liquid film. This film can be seen with droplets forming at the outer edge. As the film expands, it forms ligaments, then droplets at the outer edge based on surface tension of the liquid. Liquids with low surface tension usually have small droplets.

As pressures increase, droplet formation occurs closer to the nozzle tip with the formation of small droplets. At high pressures, small droplets are formed directly from the nozzle tip as a result of escape from hydraulic force. With high pressures, droplets may be of fog and mist size, creating a drift hazard.

Most nozzles can be operated at a pressure that favors large-droplet formation. If nozzles are operated at this pressure, there will be a minimum amount of mist-size droplets to cause spray drift. At the selected pressure make sure that the nozzle is evenly applying the herbicide. For nozzles, this pressure may range from as low as 5 psi to as high as 30 psi (pounds per square inch). To reduce spray drift, use the lowest pressure that gives a suitable spray pattern.

Nozzle tips are made of brass, aluminum, stainless steel, nylon, rubber, plastic, and ceramic-type material, and choice between these materials depends largely on corrosive and abrasive effects of chemicals and on cost. Some chemicals are chemically corrosive whereas others are abrasive; for example, TCA is chemically corrosive on brass, whereas most wettable powders are abrasive.

Nozzles that depend on finely milled edges for proper operation are easily damaged by either abrasion or corrosion. Damage to the sharp edges will ruin the spray pattern. Such damage may occur after a few hours' use with sprays containing abrasive wettable powders. Also, such nozzles are easily damaged by cleaning with a wire, knife, or other hard object. Rinsing the clogged nozzle parts in water may loosen and remove the obstruction. A bristle brush, wooden peg, or wooden match will usually loosen lodged materials without damaging the nozzle.

Nozzles deliver many different spray patterns. No one pattern is best for all herbicide work. The nozzle must be chosen purely on the basis of satisfactory performance of a given job.

Spray patterns shown in Figure 7-1 are most commonly used for herbicide applications. *Flat-fan, tapered-edge* nozzles, and the *wide-angle hollow-cone* nozzles are used for broadcast applications, and may be arranged in a series to give spray swaths up to 50 ft. The hollow-cone nozzle (120°), when tipped (30–45°) to spray to the rear (or forward), delivers a tapered-edge spray pattern as needed for uniform broadcast spray application. Research conducted by the senior author indicated that this nozzle applied the spray more evenly than other nozzles tested, had less spray drift, would apply the chemical uniformly at low gallonages per acre, and can be designed to work effectively at 5 psi.

The flat-fan, even-edge, and hollow-cone nozzles (70–80° spray angle—not shown) are used for band application 8–14 in. wide. The herbicide is placed in a band over the crop row. Using this band-herbicide application, weeds between the rows are removed later by cultivation.

Table 7-1 shows comparative data for several nozzle types that farmers often use for broadcast spray. An important difference is the percentage of spray volume smaller than 100 μm (right-hand column). As explained

Table 7-1. Nozzle Performance at 40 psi and 5 mph

Type of Nozzle[1]	Flow (gpm)	Median Droplet Size (μm)	Spacing (in.)	GPA	% Spray Volume Under 100 μm
RA Raindrop® 2	0.20	330	40	5.9	1.0
RA Raindrop® 5	0.50	590	40	14.8	0.6
RD Raindrop® 1	0.18	310	20	10.7	1.0
RD Raindrop® 2	0.29	410	20	17.2	0.8
Flat fan Spray LF-2	0.20	190	20	11.8	16.5
Flat fan Spray LF-5	0.50	220	20	29.8	13.0
Flooding D-1	0.20	200	40	5.9	15.5
Flooding D-2.5	0.50	235	40	14.8	11.5

[1] Nozzle data provided by Delavan Corp., West Des Moines, IA 50265

in Chapter 6, the 100-μm droplet compares in size to *mist,* and such droplets will drift 440 ft in a 3-mph breeze while falling 10 ft. If the droplet dries, it will be even smaller, and the drift hazard will increase drastically. The 5-μm droplet will drift 3 miles under conditions listed above.

The small RA Raindrop® nozzle and the RD Raindrop® nozzles reduced spray drift 16.5 times compared with the flat-fan nozzle and 15.5 times compared with the D-1 flood nozzle. The RA Raindrop® nozzle has spray characteristics similar to those of the hollow-cone nozzle (120°) previously discussed (see Figure 7-2).

Figure 7-2. Raindrop® nozzle spraying at 40 psi. (Photo by Ann Hawthorne, Delavan Corporation.)

Pumps

Pumps are of two types—one delivers gas or air pressure and the other liquid pressure. Gas (or air) pressure is used mainly on research-plot sprayers. If gas pressure is used, the entire system is under pressure; the spray tank is pressurized to force the liquid from the tank through the lines, boom, and nozzles. Compressed or liquified gases such as air, nitrogen, or carbon dioxide can be released through regulating valves to provide gas pressure for hand equipment. The spray tank must be strong enough to withstand the pressures. Because the tank is more expensive, the gas-pressure system often costs more than the liquid-pressure type and is usually limited to small hand-operated equipment.

In liquid-pressure systems, pressure normally exists only from the pump through the lines and boom to the nozzles. The spray tank is not pressurized. Part of the liquid may be bypassed back to the tank for agitation and for pressure control.

There are many designs of pumps for liquids, each with certain advantages and disadvantages. Because most herbicides can be applied at pressures of 5–40 psi, high-pressure pumps may be of no advantage. The pump must resist corrosion by the chemicals to be applied, provide relatively long, trouble-free service, and be priced so as to attract the user. Because pumps vary widely in design, cost alone is not necessarily a good measure for choosing a pump for a particular job.

Pumps having positive displacement require a pressure-relief system if the sprayer has a cut-off valve. This will prevent broken hoses and possible damage to the pump. Pumps not having positive displacement may not need the pressure-relief system.

Many ingenious methods can be used to develop liquid pressure. The more common types of pumps include rotary-impeller, centrifugal or turbine, gear, diaphragm, and piston pumps.

Rotary-Impeller Pumps

Probably more rotary-impeller pumps, especially those using the roller-type impeller, are used for agricultural spray application than any other type of pump. Rotary-impeller pumps consist of a rotor set to one side within the pump housing. The rotor maintains contact with the outer pump housing through flexible rubber vanes or through rollers that adjust to the outer wall of the pump housing. The rollers are held to the outer wall by centrifugal force. Space between the rotor and the pump housing expands during half of each revolution and contracts during the other half. The expanding phase creates vacuum, and the contracting phase creates pressure.

The principle of operation is the same for the flexible-rubber impeller and for the roller-type impeller. The flexible-rubber impeller may be built as shown in Figure 7-3, or the rubber impeller may consist of a

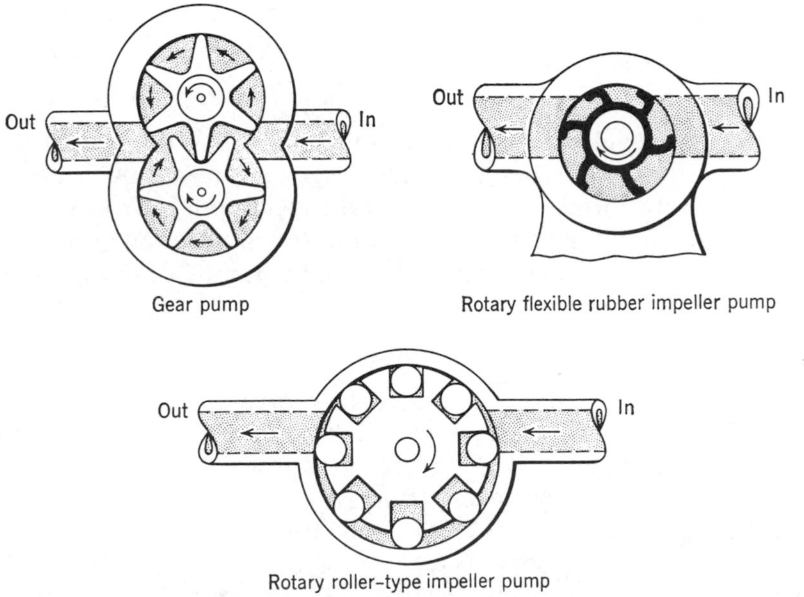

Figure 7-3. Diagrams illustrating the method of developing pressure in the gear, rotary flexible-rubber impeller, and rotary roller-type impeller pumps.

vaned rubber disc that rotates at an angle to a flat surface (not pictured). In the roller-type impeller pump, the rollers are usually made of nylon or hard rubber.

With good care, especially when in storage, the roller-type impeller pump is moderately long-lived. After use, it should be rinsed with clean water and the inside of the pump lubricated with oil.

Centrifugal or Turbine Pumps

Centrifugal or turbine pumps develop pressure as a result of centrifugal force. Blades, fins, discs, or similar structures rotate rapidly and produce liquid velocity. The velocity of the liquid, combined with its weight, gives it pressure. The liquid is usually released at the outer edge of the pump housing. These pumps will usually handle coarse and abrasive materials; naturally, they are *not* self-priming.

Gear Pumps

Gear pumps depend on the meshing of gear teeth to develop both vacuum and pressure. Figure 7-3 illustrates the principle of operation. Also, one gear may be made to rotate internally to an external gear, with the same type of action. The gear pump is usually not long-lived, especially with abrasive materials, and as with most other pumps, the higher the pressure, the greater the wear. Some can provide up to 5000 psi; however, those used for herbicides usually operate in the range of 30–60 psi.

Diaphragm Pumps

Diaphragm pumps work on the same principle as the diaphragm fuel pump on an automobile. They can handle sprays that are both chemically and abrasively corrosive. The chemical touches only the valves, the diaphragm, and pump housing. Therefore, the pump can give long, trouble-free service with minimum upkeep. The pump delivers a pulsating effect similar to a piston pump.

Piston-Type Pumps

Piston-type pumps were used on nearly all sprayers before 1945, and they are still used when pressures over 100 psi are necessary. Piston-type pumps can produce very high pressures; the only limits are the structural strength of the pump and the power supply. They are usually reliable and long-life pumps, but usually expensive as well.

The piston pump is also used as a metering pump. Because the pump delivers a given amount of liquid with each stroke of the pump, the amount of spray per acre remains constant regardless of speed (within practical limits). The pump speed may be governed by the ground speed through special power take-off arrangements or by a wheel-driven mechanism.

The piston pump produces a significant pulsating effect, which can be partially corrected by the use of an air chamber.

Power Supply for the Pump

The most common sources of power to drive the pump include (1) the tractor power take-off; (2) gasoline engines or electric motors as direct drive, belt drive, or gear box drive; (3) a ground wheel traction drive; (4) on airplanes, a small propeller to drive the pump, and (5) water pressure from the water hydrant to provide power for small sprayers for home use.

Pump Capacity

Pump capacity needed is determined by the total rate of discharge through the nozzles plus the bypass needed for agitation. From 3 to 5 gal/min bypass is enough agitation for most farm sprayers holding up to 100 gal. The bypass liquid should be released at the bottom of the tank, so that the liquid force directly agitates the liquids at the tank bottom. To calculate the rate of discharge from the nozzles, use the following equation:

$$\underset{\text{(gal/min)}}{\text{Pump capacity}} = \frac{\underset{\text{(gal/acre)}}{\text{Spray}} \times \underset{\text{(ft)}}{\text{Boom spray width}} \times \underset{\text{(mph)}}{\text{Speed}}}{495}$$

For example, if you want to apply 10 gal of spray per acre through a $16\frac{1}{2}$-ft boom at 6 mph, the pump capacity must be a minimum of 2 gal/min. Additional capacity is needed for bypass agitation.

Occasionally, a farmer may want to increase the liquid volume discharged by his sprayer. For example, his sprayer may apply 5 gal/acre, but he wishes to apply 30 gal/acre. He can accomplish this by using larger nozzles. If he does this, the pump, lines, and boom capacity must be large enough to accommodate the increased volume. If the area is small, he may wish to accomplish the same purpose by reducing the speed to one-sixth his normal speed.

Filters or Strainers

Filters or strainers may be built as a part of the nozzle, installed as a line filter, or placed on the intake in the spray tank. A coarse strainer or funnel strainer is also desirable to filter the liquid as the spray tank is filled. This is especially important if the water contains trash or dirt.

The size of the strainer openings is regulated by the size of the nozzle orifices (openings). If the strainer openings are half as large as the nozzle orifices, you will usually have little difficulty with clogged nozzles.

Wettable powders are seldom ground fine enough to pass through extremely fine filters. Thus coarse filters as well as large nozzle openings are usually needed for wettable powders. Most wettable powders will pass through 50-mesh, or coarser, screens.

Pressure Regulators

The pressure regulator reduces pressure to a relatively constant desired pressure.

The usual regulator on farm sprayers is a spring-loaded type that gives trouble-free service because of its simple construction. The regulator works only when there is enough liquid to require a bypass. The strength of the spring must be directly related to the chosen spray pressure. For example, with spray pressures of 15 psi, the regulator spring should be operative between the pressures of 10 and 20 psi.

Hoses

Hoses should be chosen to fit the pressure or vacuum expected, and hose materials should be picked that will withstand the chemicals to be used.

Good-grade garden hose, rubber or plastic, can be used for pressures up to 65 psi. Heavy hoses are available for higher pressures. If the sprayer is built so that a vacuum may develop between the spray tank and pump, a heavy-walled hose or metal pipe should be used in place of the hose. Hose is also made with a metal interior to prevent collapse. A vacuum may develop from clogged suction strainers in the hose, the inlet opening held to the wall of the tank, or from collapsed or twisted hose. Insufficient liquid flow will damage some pumps and will reduce the efficiency of all pumps.

Oil-resistant hoses made of neoprene, plastic, or other oil-resistant materials may be most satisfactory with oils and oil-like herbicides. In addition to a longer life span, the oil-resistant hoses will resist absorption of the chemical, making it less difficult to remove the herbicide.

Pressures in a hydraulic system are the same, regardless of hose size, minus any friction loss involved in movement. Therefore, hoses should be chosen that are large enough *not to restrict liquid flow* (see page 434).

Foaming

Herbicides often include surface-active agents as wetting agents, emulsifiers, and so on. Different surface-active agents have different foam-producing qualities. Agitation that tends to incorporate air may cause foaming. Returning the bypass solution *below* the surface of the liquid will reduce the amount of air incorporated and thus reduce the amount of foaming.

If foaming continues, it can usually be stopped by adding about 1 pint of kerosene or fuel oil per 100 gal of spray. The excess surface-active agent is used in emulsifying the oil.

Calibrating the Sprayer

Several methods can be used to determine the number of gallons of spray applied per acre. Four methods are listed and briefly discussed.

Prepared Tables

Prepared tables that give nozzle spacing, pressures, speed, and various nozzle sizes giving various gallons of spray per acre are available from some nozzle manufacturers; the proper nozzle size can be selected from these tables. The disadvantage to this method is that there is no assurance that speeds and pressures used in spraying are correct.

Special Measuring Devices

Special measuring devices and prepared charts or graphs can be used. The spray is usually collected for a prescribed period of time or distance. The amount of spray collected is then converted, through tables or charts, into gallons per acre. Glass jars with the table printed directly on the jar are available to catch the spray. This has the same disadvantages as those given for prepared tables.

Measurement of Gallons of Spray Delivered

Measurement of gallons of spray delivered per hour and calculations of acres treated per hour at a given speed can be used to determine gallons per acre as follows:

$$\text{Gal/acre} = \frac{\text{Gallons applied/hr}}{\text{Acres sprayed/hr}}$$

The rate of delivery of the spray can be determined by timing the rate of delivery of a known quantity of spray. From this timing, the gallons of spray per hour can be calculated.

The acres per hour can be calculated as follows:

$$\text{Acres/hr} = \frac{\text{mph} \times 5280\ (\text{ft/mile}) \times \text{Spray width (ft)}}{43{,}560\ (\text{ft}^2/\text{acre})}$$

For example, a sprayer traveling 6 mph, covering a swath $16\frac{1}{2}$ ft wide will spray 12 acres/hr of continuous operation. If the sprayer applies 60 gal of spray per hr, using the formula given first, the sprayer will apply 5 gal of spray per acre.

Spraying a Known Size Area and Measuring the Amount of Spray Applied

Perhaps the most accurate method is to actually spray an area of known size. By starting with a full tank and measuring the gallons required to refill the tank after spraying a specific area, the gallons per acre are quickly calculated. This method can be used with the aerial sprayer, the farm sprayer, or hand sprayer. The size of the area sprayed must be large enough to give an accurate measurement of the spray applied, and it is also imperative that all nozzles are the same size and that they operate properly. The airplane sprayer may need 5–10 acres, the tractor sprayer 1–2 acres, and the hand sprayer 1,000 ft^2 or 1/43.5 acres (43,560 ft^2 = 1 acre).

Cleaning the Sprayer

Cleaning the sprayer before storage, even for short periods of time, will usually prove beneficial. Emptying the spray tank and rinsing it with water may be sufficient for short-time storage. Rinsing the pump and tank (both inside and out) with fuel oil will protect most metal parts from corrosion, but oil may injure parts made of natural rubber.

By cleaning the sprayer of leftover chemicals, the possibility of injuring or killing sensitive plants in future sprayings is also reduced. 2,4-D and related products have caused many problems through the neglect of this operation.

Farmer experience has shown that the barrel or tank is by far the most important source of contamination. It is especially difficult to clean barrels lacking a bottom drain. Therefore, if contamination is expected, the barrel should be changed.

If the sprayer is to be cleaned, first rinse it with a material that acts as a solvent for the herbicide. Kerosene and fuel oils carry away herbicides known to be oil-soluble (chemicals that form emulsions when mixed with water are oil-soluble). Following the oil rinse, a rinse with a surfactant in water will help to remove the oil. The oil-soluble herbicides, such as 2,4-D esters, are usually the most difficult herbicides to remove. The 2,4-D salts are water-soluble and removed by thorough rinsing with water. Check the spray on susceptible plants, such as tomatoes, to make certain that the 2,4-D has been removed.

For wettable powder herbicides, examine the tank to see that *none of the wettable powder remains in the bottom*; otherwise, a thorough rinsing with water is usually sufficient.

RECIRCULATING SPRAYER, ROLLER APPLICATORS, AND ROPE WICKS

These three specialized methods are adapted to treating weeds that are taller than the crop. The recirculating sprayer directs the spray horizontally as a narrow stream above the crop canopy, hitting only the tops of tall weeds. Spray that is not intercepted by tall weeds is caught and recycled through the sprayer (see Figure 7-4).

Roller applicators use a rotating, carpeted roller to *wipe* herbicide on tall weeds.

Rope-wick applicators use a straight section of plastic pipe with rows of wicks exposed to wipe tall weeds (see Figure 7-5).

Although these applicators do an excellent job on tall weeds, their use must be delayed until the weeds are taller than the crop. Thus crop yields are reduced before this method is used. Glyphosate is the principal herbicide applied this way.

CONTROLLED-DROPLET SPRAYER

The possibility of applying spray at low rates, free of small droplets causing spray-drift problems, has been studied for many years. The

Figure 7-4. Recirculating sprayer equipped with solid-stream nozzles to apply herbicide to weeds that are taller than crop. Spray caught in the box is returned to be sprayed again. (USDA-SEA, Stoneville, MS.)

Figure 7-5. Rope-wick applicator, showing herbicide-soaked ropes attached to horizontally positioned pipe reservoir. (Southern Weed Science Lab. USDA, Stoneville, MS.)

spray must be nearly free of droplet sizes that are subject to spray drift. One proposed solution to the problem is the rotary disc or rotary atomizer. The rotating disc may have teeth on the edge to aid in uniform droplet formation. In general, the higher the flow rate, the larger the droplets; and the slower the disc turns, the larger the droplet size. Most whirling-disc applicators for herbicide application are aiming at uniform

Figure 7-6. The Micron Herbi® uses a spinning disc to produce uniform-size droplets at low gallonage per acre. Larger units are available for tractor and airplane application. (Micron Corporation, P.O. Box 19698, Houston, TX 77024.)

droplet sizes of about 250 μm. Usually the disc is electrically driven (see Figure 7-6).

GRANULAR MATERIALS

Granular-application equipment should apply the granular material uniformly as shown in Figure 7-7. Granular materials are spread in a manner similar to that used with seed or fertilizer. (For further details see Chapter 6.)

SOIL INCORPORATION

Herbicides are incorporated (mixed) into the soil to (1) reduce volatility losses of relatively volatile herbicides, (2) place the herbicide close to the germinating weed seed, and (3) improve dependability of weed control. The latter is particularly important in areas where rainfall is somewhat erratic, and where there may not be sufficient water to leach surface-applied herbicides into the soil. Most preemergence herbicides must be in contact with the germinating weed seeds to be effective.

Control of serious perennial weeds with rhizomes, such as johnsongrass, depends especially on mixing the chemical evenly into the soil. The large rhizome clumps must be broken up and the chemical brought into intimate contact with each rhizome. After choice of the proper

Figure 7-7. Granular materials being applied in band over the row at planting time. (Gandy Company.)

herbicide and its proper rate of application, even application is the first step in getting good weed control without injuring the crop. The 120° whirlchamber nozzle is especially well adapted for attachment to the disc harrow used to mix in the herbicide.

Virtually all types of cultivation equipment have been used to mix herbicides into the soil. However, all implements do not work equally well for every situation or herbicide. For overall applications, a disc harrow or ground-driven rolling cultivator can be used. For band applications, power-driven rotary tillers are usually used. Figure 7-8 shows the

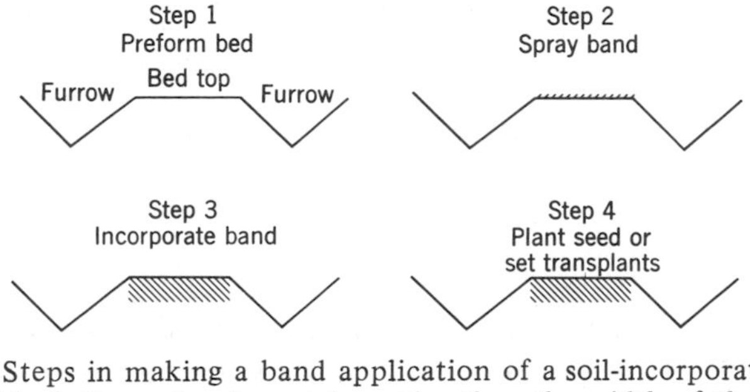

Figure 7-8. Steps in making a band application of a soil-incorporated herbicide to a preformed bed. Spray band must be wider than the width of the head of the incorporator. Selectivity may be influenced by the depth of seed placement within the treated soil. All of these steps can be performed in a single pass through the field by placing all the equipment required on one tractor.

procedure for incorporating a herbicide into a preformed bed. In addition to application rate, three major factors must be considered: depth of incorporation, soil conditions, and correct ground speed.

Depth of Incorporation

The depth of mixing is critical. If herbicides requiring shallow incorporation are placed too deep, they may lose some of their effect by dilution in a greater volume of soil. If volatile herbicides are mixed too shallow, some volatility loss may still occur. The more volatile herbicides (EPTC, pebulate, butylate, etc.) require deep mixing of 2–4 in. Less-volatile herbicides (alachlor, trifluralin) may need to be mixed moderately deep, –2 ins. As a rule of thumb, the chemical is usually placed half as deep as the depth of discing.

The disc harrow is perhaps the most common incorporation tool. The second discing is commonly at right angles to the first. Speeds must be fast enough to effectively mix the soil. The small disc harrow (finishing disc) is better adapted to incorporation than the large heavy disc. Disc blades should be spaced no more than 8 in. apart. Large discs recently available may incorporate the chemical too deep in the soil, especially on sandy and sandy loam soils. Depth-gauge wheels set for the disc to cut 4–5 in. deep may help. Following the disc with a heavy spike-tooth harrow may further help to evenly mix the herbicide in the soil. The manufacturer's label should be checked for the depth of incorporation and equipment recommended for each herbicide.

Soil Conditions

Weed control may be erratic when many large clods are present. A fine seedbed with clods less than $\frac{1}{2}$ in. in diameter is recommended. The herbicide does not readily mix with excessively moist soil, which can also increase volatility losses by decreasing the adsorption of herbicides on soil particles. The development of a plow sole, soil compaction, or loss of soil structure may occur by working soil with excessive moisture.

Ground Speed

Ground speeds of 5–6 mph are necessary to obtain adequate mixing of the herbicide into the soil with discs, harrows, and ground-driven rolling cultivators. Lower speeds (1–2 mph) can produce "streaking" and poor weed control. With power-driven rotary tillers, ground speeds of 1.5–3 mph are recommended. Good results have been obtained at a tiller speed of about 500 rev/min and 2-mph ground speed. The tillers must be close enough together to provide a "clean sweep" of the soil. This may require L-shaped knives.

For an overall flat preplant soil-incorporation treatment followed by bedding (listing and bed forming), the lister shovels should always be set to run 1–2 in. shallower than the depth of incorporation. This setting prevents placing untreated soil on top of the bed. For a band application of a preplant soil-incorporation treatment to preformed beds, care must be taken not to remove treated soil or place untreated soil on top of the beds by subsequent planting or cultivating operations (see Figure 7-8).

SUBSURFACE LAYERING

The spraying of a horizontal layer of herbicide a few inches below the soil surface (subsurface layering) has produced effective control of several hard-to-control perennial weeds. One example is field bindweed control with trifluralin or dichlobenil. This method is also called the *spray-blade method* or simply *blading* or *layering.*

The spray-blade method consists of a blade with backward-facing nozzles attached under the leading edge. Modified sweeps and V-shaped blades have been used. The effectiveness of the control also can be seen in Figure 7-9. The concentrated layer of herbicide acts as a "protective wall," preventing the weed shoots from growing through it; thus the deeper storage roots starve and the plant dies. Any disturbance of this layer by cultivation or natural cracking of the soil often permits the shoots to emerge, and control is therefore greatly reduced.

Figure 7-9. Bindweed control is clearly evident in this research plot. Treflan-blade-treated area is rectangle in foreground, with solid bindweed on all sides. (Elanco Products Company, a division of Eli Lilly and Company.)

INJECTION

Herbicides are injected into soil or water to control terrestrial and aquatic weeds. The most common chemical injected into soil is methyl bromide. This herbicide is usually applied by a commercial applicator because of its toxicity and because the treated area must be covered with a gas-tight plastic tarpaulin for 24–48 hr. Methyl bromide is applied with chisel-type applicators that inject the chemical 6–8 in. below the soil surface. The chisel-injection units are spaced not more than 12 in. apart.

HERBIGATION

The injection of herbicides into water requires precise metering with adequate mixing. This is particularly important when applying herbicides through a sprinkle or furrow irrigation system. The precision of herbigation application is no better than the accuracy of the water distribution and is usually less accurate than a spray application.

To control aquatic weeds, herbicides are injected into flowing waters, (irrigation and drainage canals) and injected from a boat into static waters (lakes and reservoirs). See Chapter 29 for details of aquatic-weed control.

AIRCRAFT

Aircraft sprayers are less commonly used to apply herbicides than other pesticides because of the drift hazards (see Chapter 6). In addition, the precision of application is somewhat less than that of ground sprayers. Despite these limitations, aircraft are especially adapted for spraying or applying granular formulations to areas not readily accessible to ground equipment, such as utility lines and firebreaks through remote woody areas, flooded rice fields, very large pasture or range areas, and large cereal grainfields.

Both fixed and rotary (helicopter) wing aircraft are used. In general, the components of sprayers for aircraft are similar to those of ground sprayers. Included are a tank, agitators, pump, boom, valves, screens, nozzles, pressure regulator, and pressure gauge. Because the design of aircraft sprayers requires special engineering knowledge, the details of design are not discussed here.

CALCULATIONS

Calculations involving rates of application and various mixtures, in addition to practice problems, are presented in Chapter 6. Conversion factors useful in calculations are given in the appendix.

SUGGESTED ADDITIONAL READING

Weed Science published by WSSA, 309 West Clark St., Champaign, IL 61820. See index at end of each volume.

Weeds Today. Same availability as above.

Akesson, N. B., and W. E. Yates, 1976, *Weeds Today* **7**(6), 8.

Bode, L. T., M. R. Gebhardt, and C. L. Day, 1968, *Trans. Amer. Soc. Agr. Eng.* **11**, 754.

Bouse, L. F., J. B. Carlson, and M. G. Merkle, 1976, *Weed Science* **24**, 361.

Brooks, F. A., 1947, *Agr. Eng.* **28**(6), 233.

Ennis, Jr., W. B. and R. E. Williamson, 1963, *Weeds* 11,67.

Hedden, O. K., 1960, *American Society of Agricultural Engineers Meeting Paper* 60-602.

Hess, F. D., D. E. Bayer, and R. H. Falk, 1981, *Weed Science* **29**(2), 224.

Holly, K., 1956, *Proc. British Weed Contr. Conf.* **1**, 605.

Klingman, G. C., 1964, *Weeds* **12**(1), 10.

McWhorter, C. G., and O. B. Wooten, 1961, *Weeds* **9**,42.

Ogg, A. G., 1980, *Weed Science* **28**(2), 201.

Tate, R. W., and L. F. Janssen, 1966, *Trans. Amer. Soc. Agr. Eng.* **9**,303.

Thompson, L., and S. W. Lowder, 1981, *Weeds Today* **12**(2), 19.

Ware, G. W., W. P. Cahill, and B. J. Esteson, 1975, *J. Econ. Entomol.* **68**, 320.

Welker, W. V., and T. Darlington, 1980, *Weed Science* **28**(6), 705.

Williford, J. R., O. B. Wooten, and W. L. Barrentine, 1968, *Weeds* **16**, 372.

Wilson, R. G., and F. N. Anderson, 1981, *Weed Science* **29**(1), 93.

Yates, W. E., N. B. Akesson, and D. E. Bayer, 1978, *Weed Science* **26**(6), 597.

See the manufacturer's instructions for use, and follow those instructions.

8 Aliphatics

The chemicals in this chapter are aliphatics. Aliphatic means no rings in the structural formula. The aliphatic herbicides are glyphosate, the chlorinated aliphatic acids (TCA and dalapon), the organic arsenicals (cacodylic acid, MAA, MSMA, DSMA, and MAMA), methyl bromide, and acrolein. Except for the aliphatics, most organic herbicides have chemical structures containing one or more rings.

GLYPHOSATE

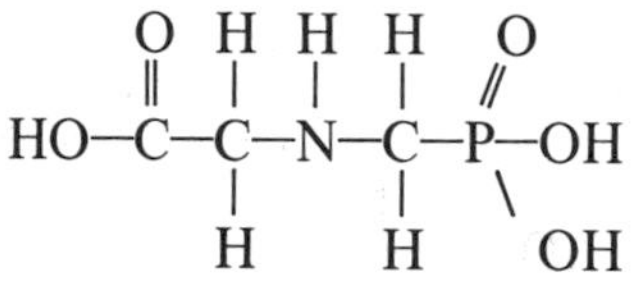

glyphosate

Glyphosate is the common name for *N*-(phosphonomethyl) glycine. The trade name is Roundup®. It is a white solid, soluble in water, about 1.2 g/100 g water at 25°C, formulated as the isopropylamine salt for an aqueous spray. Glyphosate is relatively nontoxic to mammals; the LD_{50} is 4320 mg/kg for rats.

Uses

Glyphosate, a nonselective broad-spectrum herbicide, is very effective on annual, biennial, and perennial herbaceous species of grasses, sedges, and broadleaf weeds. It also controls many woody brush and tree species. It is applied to the foliage and has little effect when applied to the soil. Glyphosate is used in noncrop areas and also in crops when contact with aboveground parts of the crop can be avoided. Directed sprays in tree crops and grapes are examples of the latter use. It has also been used for chemical fallow. Spray drift onto desirable plants must be avoided.

Spot treatments are also used to control small areas of serious weeds in certain crops. However serious crop damage usually occurs in the treated area.

Recirculating sprayers may be used for glyphosate applications when weeds are taller than the crop plants. These sprayers direct the spray solution horizontally onto the weeds but over the top of the crop. The unused solution is collected in a recovery chamber and returned to the spray solution tank for reuse (see Figure 7-4).

Glyphosate is active through the leaves, but has little effect when applied to the soil. Hence certain soil-active herbicides are commonly combined with glyphosate to control both emerged weeds and later-germinating weed seeds in the same treatment.

Mode of Action

The common symptom of glyphosate injury is foliar chlorosis followed by necrosis. Regrowth of perennial plants often shows malformed leaves as well as white spots and striations (Putnam, 1976; Fernandez and Bayer, 1977; Marriage and Kahn, 1978). Multiple shoots ("witches' broom") commonly develop from a single node (see Figure 8-1). Disruption of the chloroplast envelope and swelling of the endoplasmic

Figure 8-1. Development of multiple shoots ("witches'-broom") in bermudagrass 15 days after treatment with 0.5% glyphosate. (C. Fernandez.)

reticulum with subsequent formation of vesicles also occur (Campbell et al., 1976).

Glyphosate is readily absorbed by leaves and translocated throughout the plant. It moves via the symplastic system and probably later in the apoplastic system, as evidenced by its high mobility. Glyphosate appears to be degraded slowly in higher plants (Gottrup et al., 1976). Most research on the mechanism of action of glyphosate supports the findings of Jaworski (1972), who showed that glyphosate interferes with aromatic amino acid synthesis. However, some recent reports suggest that the action of glyphosate is more complex. Chlorosis and necrosis are consistent with reported increases in free ammonia and membrane damage. Some of the abnormal growth observed with glyphosate resembles ethylene responses. Glyphosate, especially at low rates of application, acts slowly, and herbicidal effects may not be seen for 7–10 days. See Ashton and Crafts (1981) for many references on glyphosate's mode of action.

CHLORINATED ALIPHATIC ACIDS

TCA and dalapon are chlorinated aliphatic acids. They are usually formulated as the sodium salt and are particularly effective on grasses. Although closely alike chemically, they show decided differences in phytotoxicity and translocation. Dalapon is about 10 times more phytotoxic than TCA. Dalapon is systemic from both root and leaf applications, but TCA moves in the plant only from root applications.

TCA

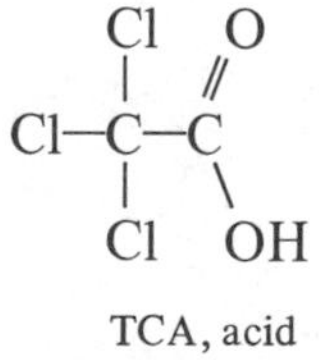

TCA, acid

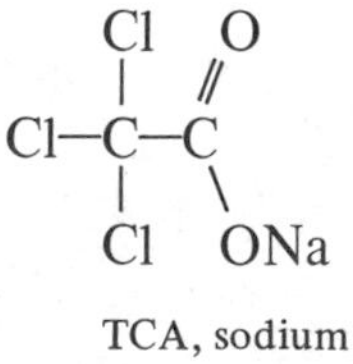

TCA, sodium salt

TCA is the common name for trichloroacetic acid. It is usually formulated as the sodium salt. Sodium TCA is a white solid, very soluble in water (83.3 g/100 g of water). It is hygroscopic. Exposed to 90–95% relative humidity at 70°F, the chemical will absorb its weight in water in 8–10 days. Thus the chemical must be stored in moistureproof containers. The acute oral LD_{50} is about 5000 mg/kg for rats.

Uses

Sodium TCA is primarily a grass killer. It has proven useful as a non-

Figure 8-2. Sodium TCA was applied to the plot $4\frac{1}{2}$ months before this picture was taken. (Texas Agricultural Experiment Station, College Station, TX.)

selective treatment on perennial weedy grasses such as johnsongrass, bermudagrass, and quackgrass (see Figure 8–2).

TCA is also used as a selective, preemergence treatment in sugarbeets to control annual grasses. In sugarcane, one preemergence or early postemergence treatment controls seedling grasses, including johnsongrass seedlings.

When handling and using sodium TCA, the user should be especially careful to keep it away from skin and eyes. Solutions of sodium TCA stronger than 10% may irritate and burn the skin and eyes unless rinsed off immediately. Ordinary glasses or sunglasses offer considerable eye protection from spray drift, and enclosed protective goggles provide more and better protection.

Mode of Action

When applied to foliage, TCA often causes rapid necrosis by contact action. It inhibits the growth of both shoots and roots, and causes leaf chlorosis and formative effects, especially to the shoot apex. At subherbicidal concentrations, however, growth may be stimulated (Mayer, 1957; Meyer and Buchholtz, 1963).

TCA is readily absorbed by leaves and roots (Blanchard, 1954). It is translocated throughout the plant from the roots, but only small amounts are translocated from leaves; therefore, it is primarily translocated via the apoplastic system. Perhaps its rapid contact action prevents symplastic movement. TCA is degraded slowly, if at all, by higher plants (Mayer, 1957).

Because TCA is used by the chemist to precipitate protein, this basis for its effectiveness as a herbicide has also been suggested. However, this has not been generally accepted. Perhaps it modifies sulfhydryl or amino groups of enzymes or induces conformational changes in enzymes.

At rates up to 100 lb/acre, TCA will usually disappear from a moist, loam soil with temperatures of 70–100°F in 50–90 days. It persists somewhat longer in soils with a high organic-matter content and in dry or cold soils. It leaches readily in most soils.

Dalapon

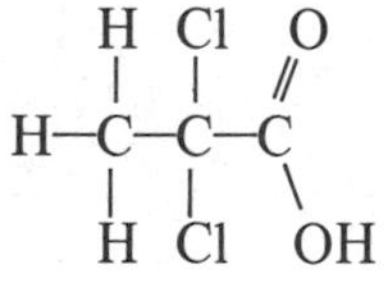

dalapon

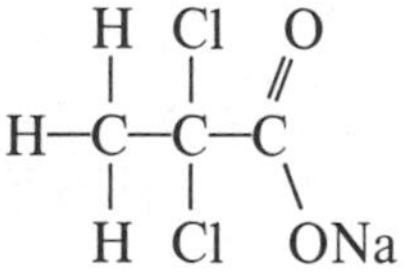

dalapon, sodium salt

Dalapon is the common name of 2,2-dichloropropionic acid. Its chemical structure is similar to that of TCA, except that one chlorine atom in TCA has been replaced by a methyl ($-CH_3$) group.

A trade name of the powdered sodium salt formulation of dalapon is Dowpon®. Other trade names are also used depending on the formulation or manufacturer. A formulation containing 72.5% of the sodium salt of dalapon and 12% of the magnesium salt of dalapon is less hygroscopic than the sodium salt of dalapon alone. This mixture of dalapon salts has superior handling properties and is less subject to caking and other characteristics described below for the sodium salt alone.

The sodium salt of dalapon is a white solid, quite soluble in water (50.2 g/100 g water at 25°C). It is hygroscopic; in 90% relative humidity, it will absorb enough water to dissolve itself. Exposed to 90% relative humidity at 70°F, the chemical absorbs its weight in water in 10–12 days. Therefore, storage requires moisture-proof containers.

When dalapon absorbs water, hydrolysis may occur, and the chemical may lose its herbicidal effects. At 25°C (77°F) this reaction is very slow, whereas at 50°C (122°F) conversion is much faster (Report of Research Committee, 1954). Therefore, spray solutions must be prepared and applied immediately, especially at high temperatures. Breakdown of dalapon is no problem if it is applied within 24 hr after mixing with water.

The sodium salt of dalapon is relatively nonvolatile and nonflammable. It is less toxic to man and animals than is sodium TCA. The main precaution is to keep dalapon away from the skin and eyes. Glasses worn when spraying will aid in keeping the spray drift from the eyes. The acute oral LD_{50} is 7570 for female rats and 9330 for male rats.

Uses

Dalapon is an effective grass killer—considerably more effective than TCA in foliage application. It is more easily dissolved in water, somewhat less toxic to the skin and eyes, less corrosive to metals, and has a slightly shorter period of residual toxicity in the soil than TCA.

Dalapon controls annual grasses and perennial grasses such as bermudagrass, quackgrass, and johnsongrass. Two or three applications spaced 5–20 days apart at 5–10 lb/acre per application have usually given better control than one heavy application. Dalapon is also effective on cattails. An autumn application is most effective, about three to four weeks before leaves lose their green color (see Figures 8-3 and 8-4).

Dalapon is registered for use in asparagus, citrus, cotton, flax, peas, potatoes, and sugarbeets as well as certain tree crops and grapes.

Figure 8-3. Taking part of the hard work out of spot spraying. The cotton is the correct size for the first application of dalapon. (Dow Chemical Company, Midland, MI.)

Figure 8-4. *Top*: Dalapon applied at the rate of 6 lb/acre, three weeks prior to planting the corn, provided control of quackgrass. *Bottom*: Not treated with dalapon. Corn yields will be seriously reduced, perhaps resulting in a crop failure. (K. P. Buchholtz, University of Wisconsin.)

Mode of Action

Dalapon inhibits the growth of both shoots and roots (Ingle and Rogers, 1961; Meyer, and Buchholtz, 1963). It also causes leaf chlorosis and formative effects, especially to the shoot apex, and interferes with cell division of root tips (Prasad and Blackman, 1964) and probably also shoot tips.

Dalapon may be absorbed through both foliage and roots. It enters the foliage through both cuticle and stomates. Surface-active agents aid movement into the stomates. The rate of dalapon absorption by plant leaves is affected by the rate of treatment, type of plant surface, humidity, temperature, and addition of surface-active agents. Absorption in the first 6 hr is by far the most important, although absorption usually continues for at least 48 hr.

Translocation downward is primarily through the phloem and upward movement is through the xylem (Foy, 1975). The amount of dalapon translocated from the leaves is in direct proportion to the rate of application, provided there is no burning or acute toxicity. With contact injury, translocation of dalapon is greatly diminished or prevented. Translocation from leaves is through living tissue.

Dalapon applied at high rates has acute plant toxicity. Quick killing of the leaves, whether resulting from frost, heat, or a contact chemical, does not favor translocation to other parts of the plant. Repeated application at low to moderate concentrations gives the best control of deep-rooted perennial grasses. Applied in this way, the chemical shows chronic toxicity and kills very slowly. The plant may show little to no effect after the first treatment. But the chemical and method are highly effective after the second or third treatment (Foy, 1975).

Dalapon is readily absorbed by the roots and translocated to all parts of the plant. It was absorbed through the roots of cotton and sorghum and swept upward in the transpiration stream, reaching all parts of the plant within 1 hr (Foy, 1975). Under most circumstances, however, the chemical is applied more efficiently to the leaves for foliar absorption. At 40 lb/acre, dalapon may remain toxic in moist, warm, loam soils for 20–60 days; cold or dry weather will lengthen the period of toxicity. Dalapon also leaches readily in most soil types.

Like TCA, dalapon probably acts by interfering with the activity of many enzymes by alkylating the sulfhydryl or amino groups (Foy, 1975), inducing conformational changes (Ashton and Crafts, 1981), or both. A rather specific effect appears to be an increase in ammonia. This is most apparent in susceptible species that are not able to form amides at a sufficient rate to prevent the accumulation of toxic levels of ammonia. It has been shown that dalapon interferes with carbohydrate, lipid, and nitrogen metabolism; however, these may be secondary effects induced by the previously discussed reactions.

ORGANIC ARSENICALS

The organic arsenical herbicides include two similar compounds, cacodylic acid and MAA (methanearsonic acid), and their salts.

Cacodylic Acid

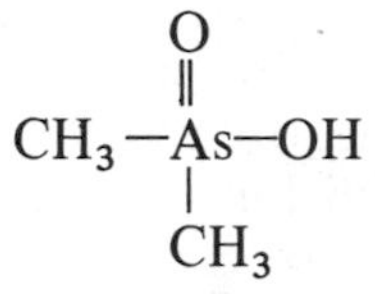

cacodylic acid

The common name of hydroxydimethylarsine oxide is cacodylic acid. It was one of the first organic arsenicals to be introduced. It is a colorless, crystalline solid, soluble in water, (66.7 g/100 ml of water). It is often formulated as the sodium salt. Like other organic arsenical herbicides it has a much lower toxicity than elemental arsenic. Its acute oral LD_{50} is 830 mg/kg for rats.

Cacodylic acid and its sodium salt are used as general contact sprays. They desiccate and defoliate a wide variety of plant species. They are used in noncrop areas (rights-of-ways, ditch banks, fences, industrial sites), forests, lawn-renovation areas, and citrus orchards. Directed sprays are required in citrus orchards. It is also used for cotton defoliation.

MAA

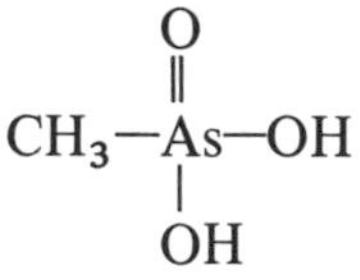

MAA

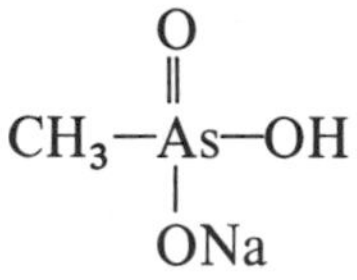

MSMA

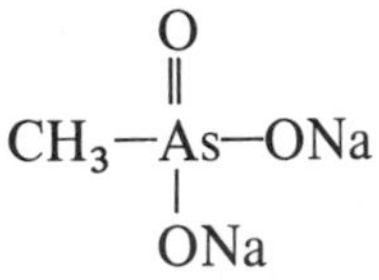

DSMA

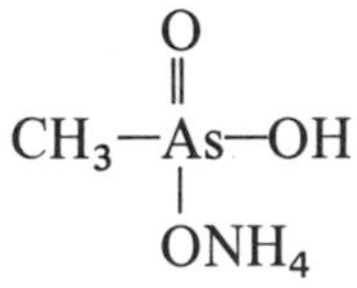

MAMA

MAA is the common name for methanearsonic acid. It is usually formulated as monosodium methanearsonate (MSMA), disodium methanearsonate (DSMA), or monoammonium methanearsonate (MAMA). DSMA, the more widely used formulation, is a white, crystalline solid. It is soluble in water (25.6 g/100 g of water). The acute oral LD_{50} of MAA is 1800 mg/kg for rats.

The first major use of these compounds was for postemergence control of crabgrass in turf. They are now used to control certain other annual weeds in tolerant lawn grasses. They also remove dallisgrass and johnsongrass from bermudagrass turf. In cotton, citrus, and noncrop areas, many annual weeds and perennials such as johnsongrass and nutsedges are controlled. Directed sprays are required for the crops. They are also combined with other herbicides to broaden the spectrum of weeds controlled (see pages 419 and 421).

Mode of Action—Organic Arsenicals

Cacodylic acid desiccates and defoliates many species of plants. MAA and its salts induce foliar chlorosis followed by gradual browning and finally necrosis. They also inhibit growth in general, inhibit the sprouting of rhizome and tuber buds, and cause aberrant cell division.

Cacodylic acid is primarily translocated in the apoplast. Extensive symplastic transport is probably prevented by its rapid contact action, which injures the phloem. However, the MAA derivatives are translocated in both the symplast and the apoplast.

These compounds appear to conjugate with sugars, organic acids, and amino acids. However, only the amino acid conjugates have been widely reported. The carbon-arsenic bond of these compounds appears to be very stable in higher plants.

The biochemical mechanism of action of these compounds is not known. However, it has been suggested that it is related to an increase in the amino acid concentration or an accelerated utilization of starch in the storage organs of perennial species. See Woolson (1976) for more information on the mode of action of organic arsenical herbicides.

METHYL BROMIDE

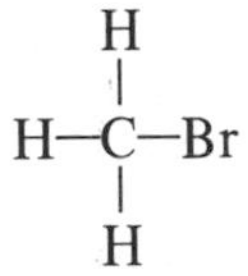

methyl bromide

Methyl bromide is a colorless, nearly odorless liquid or gas. At 1 atm of pressure and at 38°F (3.56°C), the liquid boils (turns to gas). The gas is 3.2 times heavier than air at 68°F. It is slightly soluble in water, very soluble in alcohol and ether, and is generally considered nonflammable and nonexplosive. However, mixtures containing between 13.5 and 14.5% of the gas in air may be exploded by a spark.

Methyl bromide gas is poisonous to humans and animals, and the effects of exposures are cumulative. Because methyl bromide is nearly odorless (or has a very slight, sweetish odor), traces of other gases such as chloropicrin are sometimes added as warning agents.

Uses

Methyl bromide is effective as a temporary soil sterilant. It kills living plant, stem, or root tissue, most seeds, nearly all insects, and most disease organisms. As such, it makes an excellent soil treatment for seedbeds of tobacco, flowers, vegetables, turf, and tree seedlings. It is also an effective treatment for propagating beds and for preplanting treatment of trees, shrubs, fruits, and gardens.

Methyl bromide treatments are expensive (~\$1,000/acre); therefore, they are primarily used in high-value crops for serious pest problems. Formulations of methyl bromide (45%) and chloropicrin (55%) give increased control of certain plant pathogens.

The chemical is applied under an airtight cover, usually plastic, at the rate of 1–2 lb/100 ft^2 of surface, or 100 ft^3 of soil. It may be injected into the soil by power-driven equipment for large areas, or merely released under the plastic tarpaulin for small areas (see Figure 8–5).

Methyl bromide may kill beneficial organisms in the soil as well as disease organisms, so that normal decomposition of organic matter into ammonia, then nitrite, and finally nitrate may be disturbed and toxic substances accumulate. This poisoning occurs if the organisms responsible for one or more of these processes are killed.

For example, in sterilized soils high in organic matter, ammonia or nitrite may build up to toxic levels. Such responses are seldom noted in sands or other low-organic-matter soils. Usually the beneficial microorganisms multiply sufficiently in one to three months in warm, moist soils so they restore a normal microorganism equilibrium and normal plant growth. That is why in preparing compost mixtures, it is usually desirable to sterilize the soil first and then add the peat, sawdust, or other organic matter.

Figure 8-5. Methyl bromide being released under a sealed plastic cover. (Dow Chemical Company, Midland, MI.)

ACROLEIN

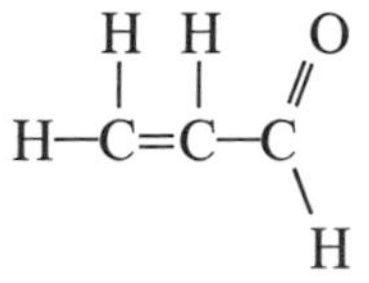

acrolein

Acrolein is the common name for acrylaldehyde or 2-propenal. The trade name is Aqualin®. It is a colorless liquid, moderately soluble in water (about 25 g/100 g of water), and is injected into water for the control of submersed and floating aquatic weeds. See Chapter 29 for details of aquatic-weed control.

Acrolein is a contact herbicide and a general cell toxicant. It acts on enzyme systems through its destructive sulfhydryl reactivity.

Acrolein has an acute oral LD_{50} of 46 mg/kg for rats. It is very toxic to most organisms. It is a flammable liquid subject to explosive reactions under certain conditions. Therefore, it may be applied only by licensed applicators.

SUGGESTED ADDITIONAL READING

Ashton, F. M., and A. S. Crafts, 1981, "Aliphatics," pp. 65–90 and "Glyphosate," pp. 236–253, in *Mode of Action of Herbicides,* Wiley, New York.

Blanchard, F. A., 1954, *Weeds* **3**, 274.

Campbell, F. W., J. O. Evans, and S. C. Reed, 1976, *Weed Sci.* **24**, 22.

Fernandez, C. H., and D. E. Bayer, 1977, *Weed Sci.* **25**, 396.

Foy, C. L., 1975, "The chlorinated aliphatic acids," pp. 399–452, in P. C. Kearney and D. D. Kaufman, Eds., *Herbicides,* Vol. 1, Dekker, New York.

Gottrup, O., P. A. O'Sullivan, R. J. Schraa, and W. H. Vanden Born, 1976, *Weed Res.* **16**, 197.

Ingle, M., and B. J. Rogers, 1961, *Weeds* **9**, 264.

Jaworski, E. G., 1972, *J. Agr. Food Chem.* **20**, 1195.

Marriage, P. B., and S. U. Kahn, 1978, *Weed Sci.* **26**, 374.

Mayer, F., 1957, *Biochem. Z.* **328**, 433.

Meyer, R. E., and K. P. Buchholtz, 1963, *Weeds* **11**, 4.

Prasad, R., and G. E. Blackman, 1964, *J. Exp. Bot.* **15**, 48.

Putnam, A. R., 1976, *Weed Sci.* **24**, 425.

Report of Research Committee, 1954, *Proceedings of the 7th Southern Weed Conference* p. 340.

Weed Science Society of America, 1979, *Herbicide Handbook, 4th ed.,* WSSA, Champaign, IL.

Woolson, E. A., 1976, "Organicarsenical herbicides," pp. 741–776, in P. C. Kearney and D. D. Kaufman, Eds., *Herbicides,* Vol. 2, Dekker, New York.

For chemical use, see the manufacturer's label and follow the directions. Also see the Preface.

9 Amides

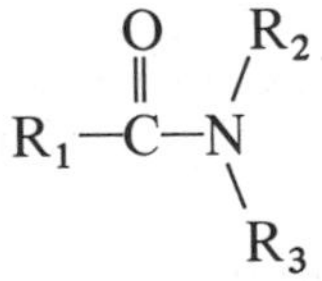

The basic chemical structure of the amide-type herbicides is shown above. However, the substitutions on positions R_1, R_2, and R_3 vary greatly, making this a diverse group of chemicals. Likewise, the weeds that the different amide herbicides control and the crops to which they are selective vary widely. Most amides are used as selective herbicides applied either preemergence or preplant. However, propanil is applied to the foliage of weeds to be controlled.

The amides can be classified into three groups: (1) soil-applied chloroacetamides (alachlor, butachlor, CDAA, metolachlor, and propachlor); (2) other soil-applied amides (diphenamid, naptalam, pronamide, and napropamide); and (3) foliar-applied amides (propanil).

CHLOROACETAMIDES

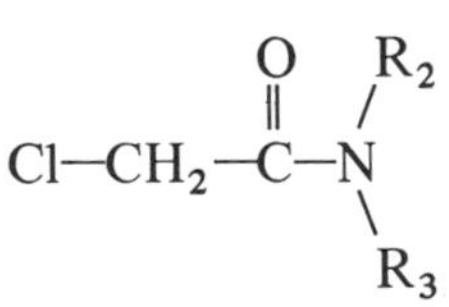

chloroacetamides

The chloroacetamides have a monochlorinated methyl group ($Cl-CH_2-$) in the R_1 position of the amide structure. They are usually applied to the soil as preemergence or preplant soil-incorporation treatments. However, certain ones have postemergence activity on very young weeds.

Alachlor

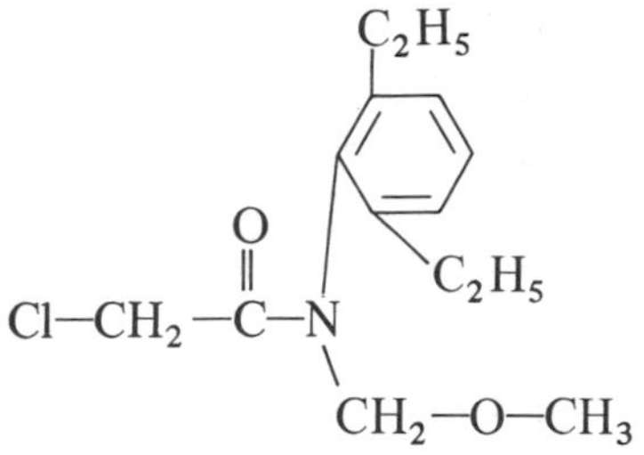

alachlor

Alachlor is the common name for 2-chloro-2′,6′-diethyl-*N*-(methoxymethyl)acetanilide, and the trade name is Lasso®. The chemical is formulated as an emulsifiable concentrate and granules. It is a cream-colored solid, relatively insoluble in water (242 ppm at 25°C). The acute oral LD_{50} of the emulsifiable concentrate is 1800 mg/kg for rats.

Uses

Alachlor is applied as a preemergence, early postemergence, or preplant soil-incorporated treatment to control annual grasses, certain annual broadleaf weeds, and yellow nutsedge in corn, peanuts, soybeans, dry beans, peas, potatoes, cotton, and woody ornamentals. It is often combined with other herbicides to increase the spectrum of weeds controlled (see Figure 9-1). Many uses of alachlor are similar to those of metolachlor. See pages 418 and 420.

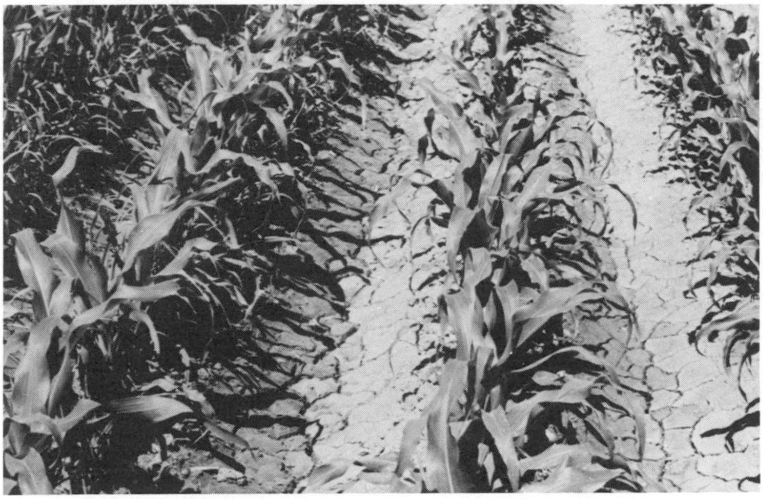

Figure 9-1. Barnyardgrass control in corn with alachlor. *Upper left*: Untreated. (C. L. Elmore, University of California, Davis.)

Soil Influence

Because alachlor is adsorbed by soil clay and organic colloidal particles, it is not subject to excessive leaching in most soils. It is primarily degraded by soil microorganisms. In sandy soils low in organic matter, persistence is shorter than in heavy soils. In medium-textured soils, and with moderate moisture, herbicidal effectiveness usually lasts about 6–10 weeks.

Butachlor

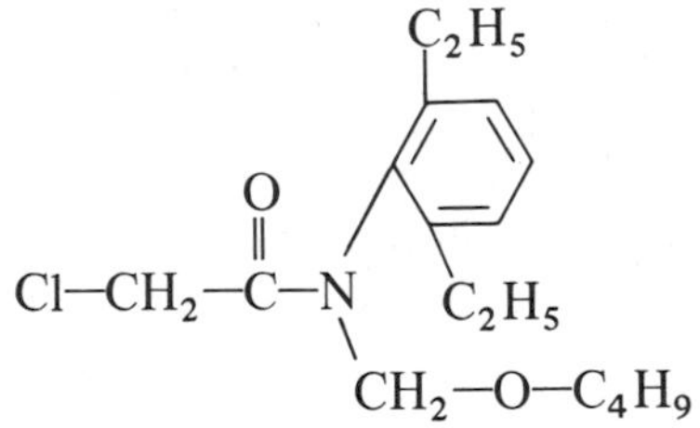

butachlor

Butachlor in the common name for *N*-(butoxymethyl)-2-chloro-2′,6′-diethylacetanilide. The trade name is Machete®. The chemical is formulated as an emulsifiable concentrate, but granule formulations have also been used. It is a slightly sweet, aromatic, amber liquid with low water solubility (23 ppm at 24°C). The acute oral LD_{50} of the emulsifiable concentrate is 3300 mg/kg for rats.

Uses

Butachlor is mainly used as a postemergence treatment to dry-seeded rice in Arkansas. It is commonly used in combination with propanil to control most emerged annual grasses, broadleaf weeds and aquatic weeds. This treatment will also give residual control of grass–aquatic weed complexes for up to six weeks. This tank mix is applied when rice seedlings have emerged to full stand and are at the one- to two-leaf stage of growth. Granule formulations of butachlor have been used in transplanted rice.

CDAA

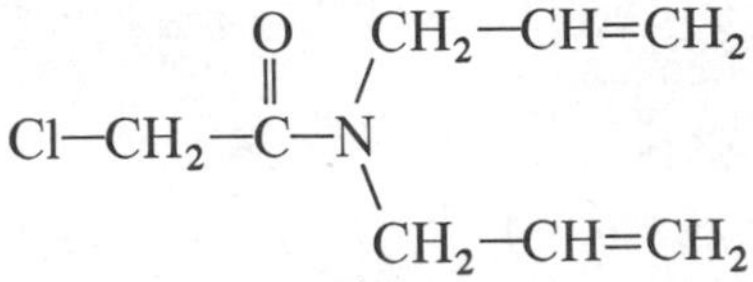

CDAA

CDAA is the common name for *N,N*-diallyl-2-chloroacetamide. The trade name is Randox®. It is formulated as an emulsifiable concentrate and granules, and is an oily, amber-colored liquid. It is quite water-soluble (about 20,000 ppm). The acute oral LD_{50} is 750 mg/kg for rats. If spilled on the skin, it sensitizes the area to temperature changes; this may cause an aching or burning sensation for several hours. The historical development of CDAA, and related products, has been published (P. C. Hamm, 1974).

Uses

CDAA is applied as a preemergence or preplant soil-incorporation treatment for control of many grassy and broadleaf weeds in castor beans, celery, onions, peas, potatoes, sweet potatoes, and tomatoes. It has been used widely in corn, sorghum, and soybeans, but recently other herbicides have largely replaced CDAA. However, it is still registered for use in these three crops. Also see pages 418 and 420.

Soil Influence

As with most organic herbicides, the rate of leaching is correlated with the clay and organic content of the soil. The higher the content of these two factors, the less leaching. CDAA is especially effective on muck and clay loam soils. The relatively short persistence of CDAA, three to six weeks in moist soils, is too short to give seasonlong weed control.

Metolachlor

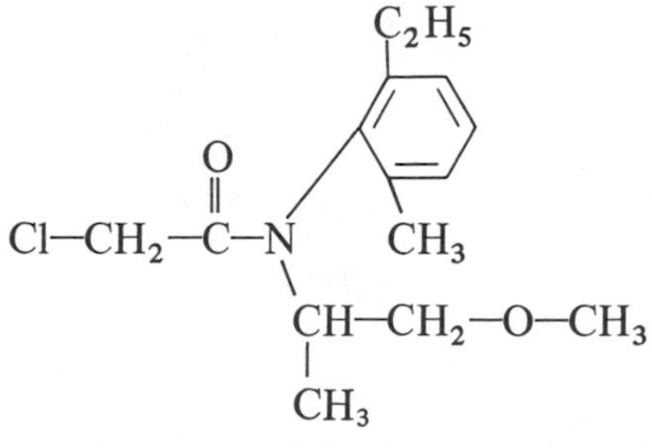

metolachlor

Metolachlor is the common name for 2-chloro-*N*-(2-ethyl-6-methylphenyl)-*N*-(2-methoxy-l-methylethyl)acetamide. The trade name is Dual®. Two emulsifiable concentrate formulations are available. Metolachlor is a white to tan liquid with a water solubility of 530 ppm at 20°C. For rats, the acute oral LD_{50} of technical metolachlor is 2780 mg/kg, and the 6 lb/gal formulation is 2828 to 4286 mg/kg.

Uses

Metolachlor is applied as a preemergence or preplant soil-incorporated treatment for control of annual grasses, yellow nutsedge, and certain broadleaved weeds in corn, soybeans, peanuts, grain sorghum, and woody ornamentals. It is also combined with several other herbicides, as well as applied in liquid fertilizer or as a granule added to certain granular fertilizers. Many uses of metolachlor are similar to those of alachlor.

Soil Influence

Metolachlor appears to behave similar to the other chloroacetamide herbicides in soil. Leaching is restricted by adsorption onto clay and organic colloidal particles, with organic matter being the more important restraint. Dissipation of metolachlor in soils is relatively rapid. The half-life (time required to lose one-half its phytotoxicity) of metolachlor is two to three weeks in the southern United States and four to seven weeks in the north. This is probably related to warmer soil temperatures in the south, which increase microbiological activity. During these weeks, metolachlor and its degradation products combine primarily with soil humic material; a small amount is converted to carbon dioxide.

Propachlor

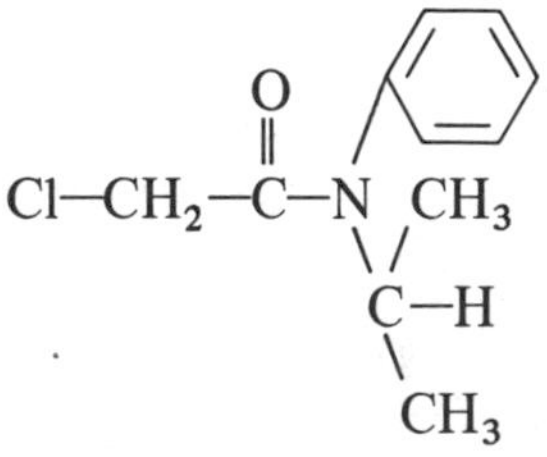

propachlor

Propachlor is the common name for 2-chloro-*N*-isopropylacetanilide. The trade names are Ramrod® and Bexton®. It is a light-tan solid with a water solubility of 580 ppm at 20°C. It is formulated as a wettable powder, flowable liquid, and granules. It is also available as a package mix with atrazine. The acute oral LD_{50} is 710 mg/kg for rats.

Uses

Propachlor is used as a preemergence, preplant soil-incorporated, or early postemergence treatment to control most annual grasses and certain broadleaf weeds in corn, cotton, peas, sorghum, soybeans (seed crop only), sugarcane, and certain vegetable crops. It is active postemergence only on very small weeds before they reach the two-leaf stage. It is commonly used in combination with one of several other herbicides. Also see pages 419 and 421.

Soil Influence

Propachlor is adsorbed by colloidal particles of the soil and is therefore not subject to excessive leaching. There are no soil-persistence problems because the chemical is degraded in four to six weeks in most soils, although it persists somewhat longer in organic soils.

Mode of Action of Chloroacetamides

The mode of action of *all* chloroacetamide herbicides is probably similar; however, not all phenomenona discussed below have been demonstrated for every one.

These compounds inhibit early seedling growth; this effect is most evident on root growth. These responses appear to be associated with an interference with cell division and/or cell enlargement. It is not clear whether germination as such is inhibited, but often the weed seedling does not emerge from the soil.

Results differ on whether these compounds are absorbed predominately by the roots or emerging shoots of the germinating seedling. Some evidence suggests that grasses absorb these herbicides mainly through the shoots, whereas broadleaf plants absorb them through the roots. The compounds themselves may vary in this regard. Translocation appears to be mainly in the apoplast; however, limited symplastic transport may also occur.

Most evidence suggests that these compounds or their degradation products undergo conjugation with glutathione and/or glucose. Hydrolysis of the parent molecule may also occur. These compounds are generally considered to act by inhibiting nucleic acid and protein synthesis, but some evidence suggests that protein synthesis is the primary site of action (Ashton and Crafts, 1981).

OTHER SOIL-APPLIED AMIDES

In contrast to the chloroacetamides, which have a monochloromethyl group ($Cl-CH_2-$) in the R_1 position of the amide structure, other soil-applied amide herbicides have at least one ring structure in the R_1 position of the molecule.

These herbicides are applied to the soil as preemergence or preplant soil-incorporated treatments. They show little if any postemergence activity. They tend to control a broader range of broadleaf weed species than the chloroacetamide herbicides. However, they are somewhat less effective on annual grasses.

Diphenamid

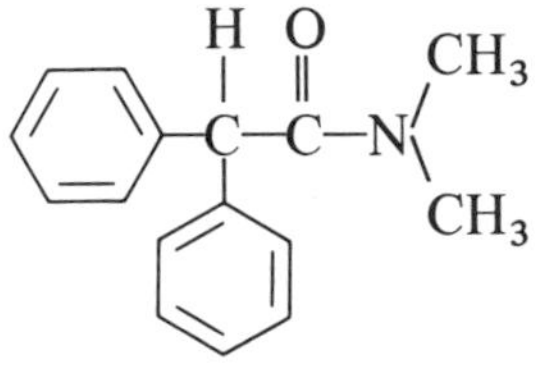

diphenamid

Diphenamid is the common name for *N,N*-dimethyl-2,2-diphenylacetamide. The trade names are Dymid® and Enide®. It is formulated as a wettable powder and granules. It is a white to off-white, crystalline solid with a solubility in water of 261 ppm at 27°C. The acute oral LD_{50} is 1313 mg/kg for rats.

Uses

Diphenamid is used primarily as a selective herbicide to control annual grass and broadleaf weeds in cotton, peanuts, soybeans, tobacco, and several vegetable, horticultural, and orchard crops. These include apples, okra, peaches, peppers, potatoes, strawberries, sweet potatoes, and tomatoes. It may also be used on nonbearing blackberries, citrus, and raspberries. Dichondra and bermudagrass turfs, as well as many ornamentals, are tolerant to diphenamid (see Figure 9-2).

Diphenamid is applied as a preemergence or preplant soil-incorporation treatment. Diphenamid is used as a tank mix with trifluralin, chlorpropham, linuron, or pebulate. These combinations are not necessarily suitable for use on all the crops listed above. Also see page 418.

Soil Influence

Diphenamid is leached fairly rapidly in sandy soils but more slowly in loam or clay soils. It is not tightly adsorbed onto soil colloids. Microorganisms appear to play a significant role in diphenamid degradation in soil. Under warm, moist conditions, diphenamid normally persists three to six months.

Mode of Action

Diphenamid appears to let seeds germinate, but kills the seedling plant before it emerges from the soil. At sublethal concentrations, diphena-

Figure 9-2. *Right*: Tomatoes treated preemergence with diphenamid, followed by a lay-by treatment with trifluralin. *Left*: Not treated. (Elanco Products Company, a division of Eli Lilly and Company.)

mid inhibits root development in many species. In tolerant species such as tomato, marginal-leaf chlorosis may even occur after normal germination at relatively high application rates of diphenamid.

Diphenamid is readily absorbed by roots and rapidly translocated to the tops of plants with accumulation in the leaves (Golab et al., 1966; Lemin, 1966); this suggests apoplastic transport. Foliar absorption appears to be limited. Studies with tomatoes (Lemin, 1966) and strawberries (Golab et al., 1966) show that diphenamid is metabolized by higher plants, yielding *N*-methyl-2,2-diphenylacetamide as the major metabolite. Four additional minor metabolites were tentatively identified. Although diphenamid has been reported to inhibit RNA synthesis (Briquet and Wiaux, 1967), this is not consistent with several studies showing that it does not inhibit protein synthesis (Ashton and Crafts, 1981). It appears that diphenamid inhibits the uptake of inorganic ions by roots and influences the distribution of calcium within the plant (Nashed and Illnicki, 1968).

Naptalam

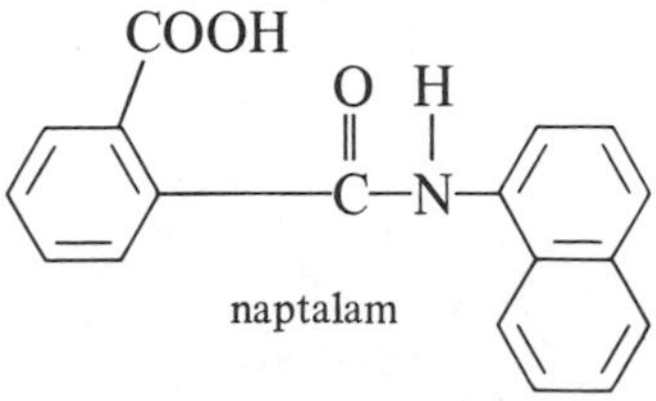

Naptalam is the common name for *N*-1-naphthylphthalamic acid. The trade name is Alanap®. It is formulated as the sodium salt in water-soluble liquid and granule form. It is a purple crystalline solid. The acid form has a water solubility of 200 ppm and the sodium salt 230,800 ppm. Increases of this magnitude in water solubility of herbicides are frequent when acid forms are converted to salts. The acute oral LD_{50} for rats is >8200 mg/kg for the acid and 1770 mg/kg for the sodium salt.

Uses

Naptalam can be used by itself for annual-weed control in cucurbits (cantaloupe, cucumber, muskmelon, watermelon), peanuts, soybeans, and nursery stock. However, it is more frequently used in combination with one or two other herbicides. For cucurbits, it is used with bensulide. For soybeans and peanuts, it can be used with dinoseb. Alachlor, metolachlor, or oryzalin can also be added to the naptalam-dinoseb mixture to make a three-herbicide combination for these two important crops. Giant foxtail, goosegrass, ragweed, and cocklebur are particularly sensitive to naptalam. Also see page 419.

Soil Influence

Naptalam is subject to rapid leaching in porous soils. Thus if heavy rains occur shortly after seeding and treatment, both crop injury and poor weed control may result. Weeds are usually controlled from three to eight weeks after application, and the herbicide presents no soil-residual problem.

Mode of Action

Naptalam has the unique property of acting as an antigeotropic agent; growing shoots and roots have a tendency to lose their ability to grow up or down, respectively. It has not been proven, however, that this is associated with its herbicidal action. Naptalam acts primarily as an inhibitor of seed germination and also inhibits some growth responses induced by the normal plant hormones indole-3-acetic acid and gibberellic acid (Ashton and Crafts, 1981). Naptalam is degraded into α-naphthylamine and phthalic acid.

Pronamide

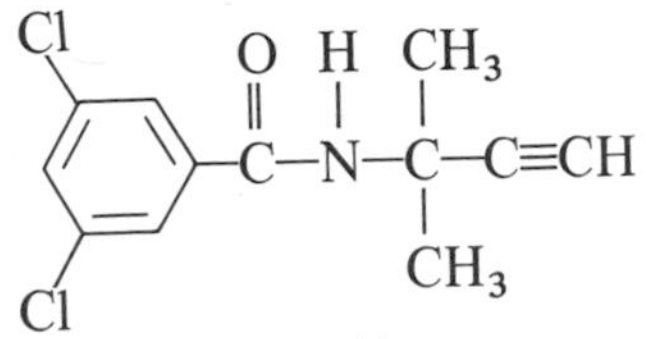

pronamide

Pronamide is the common name for 3,5-dichloro(*N*-1,1-dimethyl-2-propynyl)benzamide. The trade name is Kerb®. It is an off-white solid with a water solubility of 15 ppm, formulated as a wettable powder. The acute oral LD_{50} is 5620–8350 mg/kg for rats.

Uses

As a preemergence or preplant soil-incorporated treatment, pronamide controls many broadleaf and grass weeds in small-seeded legumes, lettuce, established lawns and turf, and certain ornamentals. Control of some perennial grasses, quackgrass and ryegrass, has been observed. Pronamide is most effective with an abundance of water from rainfall or sprinkle irrigation with preemergence treatments. With furrow irrigation, soil incorporation is required. Presumably its relatively low water solubility and considerable affinity for colloidal adsorption sites in soil require considerable water to make it biologically active.

Soil Influence

Pronamide is readily adsorbed on organic matter and other colloidal exchange sites and therefore leaches very little from most soils. It has intermediate persistence in soil, three to eight months. Also see page 419.

Mode of Action

Pronamide inhibits cell division and growth. It is readily absorbed by roots and translocated upward and distributed throughout the plant. Its translocation from leaves is not appreciable (Carlson, 1972). This suggests translocation in the apoplastic system. Pronamide is slowly metabolized by higher plants by means of alterations of the aliphatic side chains (Yih and Swithenbank, 1971).

Napropamide

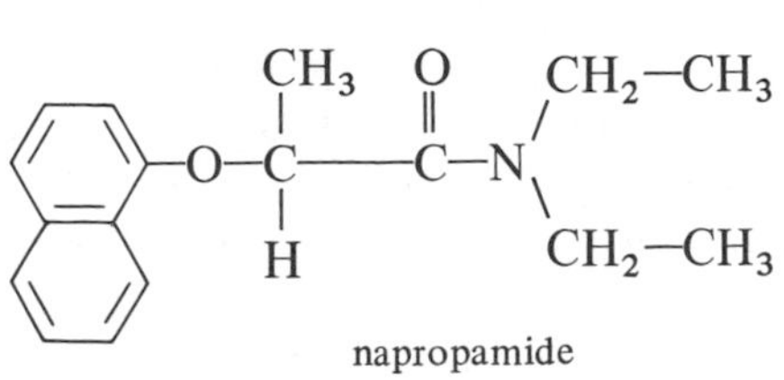

napropamide

Napropamide is the common name for 2-(α-naphthoxy)-*N*,*N*-diethylpropionamide. The trade name is Devrinol®. It is a white crystalline solid with a water solubility of 73 ppm at 20°C. It is formulated as a wettable powder. The acute oral LD_{50} is greater than 5000 mg/kg for rats.

Uses

Napropamide is a preemergence or preplant soil-incorporated herbicide used to control most annual grasses and many broadleaf weeds in several orchard crops including almonds, apricots, cherries, nectarines, peaches, plums, prunes, and oranges. It can be used on newly planted trees as well as established trees, and on grapes, tomatoes, and peppers. (also see page 419).

Soil Influence

Napropamide is quite resistant to leaching in most mineral soils. It is slowly decomposed by soil microorganisms. Its relatively long persistence, more than nine months under some conditions, may cause injury to crops following use. When incorporated into moist, loamy sand and loam soils at 70–90°F, the half-life was 8–12 weeks.

Mode of Action

Limited basic research has been conducted on this herbicide. Napropamide inhibits growth and development of roots. It has been shown to be taken up rapidly by roots of tomatoes and readily translocated upward throughout the stem and leaves. However in corn, upward translocation is much slower. Napropamide is rapidly metabolized to water-soluble metabolites in tomato plants and fruit trees. The major metabolites appear to be hexose conjugates of 4-hydroxynapropamide.

FOLIAR-APPLIED AMIDES

Propanil

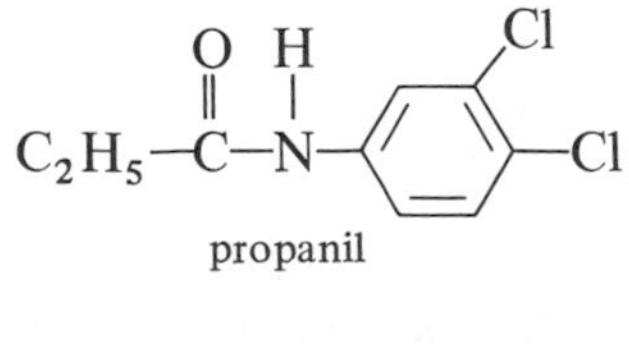

propanil

Propanil is the common name for 3′,4′-dichloropropionanilide. The trade names are Propanil® 3, Propanil® 4, Propanex®-4, Prop-Job®, and Stam®. It is a light-brown to gray-black solid with a water solubility of about 500 ppm. It is formulated as emulsifiable concentrates. The technical material has an acute oral LD_{50} of 1384 mg/kg for rats.

Uses

Propanil is applied postemergence in rice to control several annual grasses, especially barnyardgrass, and a limited number of annual broadleaf weeds and sedges. The time of application is critical, usually depending on the size of barnyardgrass. The ideal time is at the one- to three-leaf stage. It is used in both upland and flooded rice. Both aircraft sprayers and ground sprayers are used on upland rice, but only the former are used on flooded rice. Because propanil is essentially a contact spray, uniform coverage is essential, and the drift of propanil onto other crops is a potential hazard.

Although rice is quite tolerant of propanil, the herbicide may severely injure this crop when certain *insecticides* (carbaryl or any organic phosphate) are applied within 14 days before or after propanil is used. These insecticides inhibit the action of the enzyme in rice that decomposes propanil and thus leads to its selectivity.

Because propanil is applied to the foliage, soil type has no effect on its action; it is rapidly broken down in the soil, therefore presenting no residual problem for later crops. Also see page 419.

Mode of Action

To the leaves of susceptible species, propanil causes chlorosis followed by necrosis. When it remains as discreet, small droplets, a speckled pattern may be observed (Ashton and Crafts, 1981). Propanil is absorbed by leaves, but translocation within the leaf or from the treated leaves to the rest of the plant is very limited (Yih et al., 1968a). The degradation of the chemical in higher plants has been studied rather extensively (Ashton and Crafts, 1981; Casida and Lykken, 1969; Matsunaka, 1969). Propanil inhibits a number of biochemical reactions, especially photosynthesis (Ashton and Crafts, 1981).

SUGGESTED ADDITIONAL READING

Ashton, F. M., and A. S. Crafts, 1981, "Amides," pp. 91–117, in *Mode of Action of Herbicides,* Wiley, New York.

Briquet, M. V., and A. L. Wiaux, 1967, *Meded. Rijksfac. LandbWet. Gent.* **32**, 1040.

Carlson, W. C., 1972, Ph.D. dissertation, University of Illinois, Champaign.

Casida, J. E., and L. Lykken, 1969, *Ann. Rev. Plant Physiol.* **20**, 607.

Golab, T., R. J. Herberg, S. J. Parka, and T. B. Ţepe, 1966, *J. Agr. Food Chem.* **14**, 542.

Hamm, P. C., 1974, *Weed Sci.* **22**, 541.

Jaworski, E. G., 1975, "Chloroacetamides," pp. 349–376, in P. C. Kearney and D. D. Kaufman, Eds., *Herbicides*, Vol. 1, Dekker, New York.

Lemin, A. J., 1966, *J. Agr. Food. Chem.* **14**, 109.

Matsunaka, S., 1969, *Residue Rev.* **25**, 45.

Nashed, R. B., and R. D. Illnicki, 1968, *Proceedings of the 22nd Northeast Weed Control Conference*, p. 500.

Still, G. G., and R. A. Herrett, 1976, "Methylcarbamates, carbanilates, and acylanilides," pp. 609–664, in P. C. Kearney and D. D. Kaufman, Eds., *Herbicides*, Vol. 2, Dekker, New York.

USDA, SEA, 1980, *Suggested Guidelines for Weed Control*, U.S. Government Printing Office, Washington, D.C. (621-220/SEA 3619).

USDA, SEA, 1981, *Compilation of Registered Uses of Herbicides.*

Weed Science Society of America, 1979, *Herbicide Handbook*, 4th Ed., WSSA, Champaign, 14.

Yih, R. Y., D. M. McRae, and H. F. Wilson, 1968a, *Plant Physiol.* **43**, 1291.

Yih, R. Y., D. M. McRae, and H. F. Wilson, 1968b, *Science* **116**, 37.

Yih, R. Y., and C. Swithenbank, 1971, *J. Agr. Food Chem.* **19**, 314.

For chemical use, see the manufacturer's label and follow the directions. Also see the Preface.

10 Benzoics

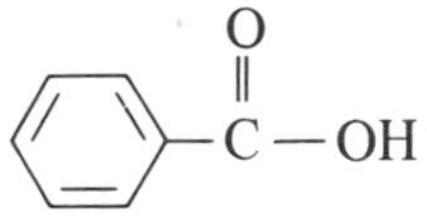

benzoic acid

All benzoic herbicides are derivatives of benzoic acid, shown above. All contain at least two chlorine atoms, whereas one (dicamba) contains a methoxy ($-OCH_3$) group and another (chloramben) contains an amino ($-NH_2$) group. Zimmerman and Hitchcock (1942) first pointed out the growth-regulating properties of the substituted benzoic acids. In 1948 the Jealott's Hill Experiment Station in England evaluated the herbicidal properties of 2,3,6-TBA under field conditions, and similar investigations were simultaneously carried out in the United States. All the benzoic acid herbicides are effective when applied to foliage or soil, except chloramben, which is used as a preemergence herbicide.

2,3,6-TBA

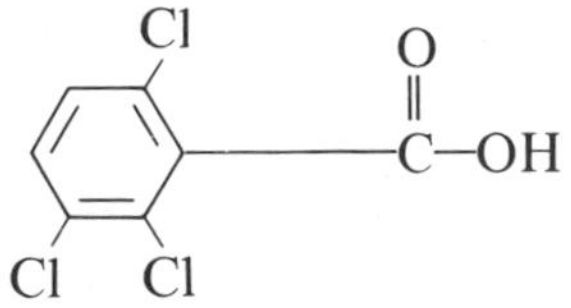

2,3,6-TBA

2,3,6-TBA is the common name for 2,3,6-trichlorobenzoic acid. Trade names are Benzac®, and Trysben®. Most commercial formulations contain a mixture of about 60% 2,3,6-TBA and 40% other chlorinated benzoic acids. Some of these latter derivatives are also phytotoxic, but the 2,3-6-chlorinated derivative is most phytotoxic. They are usually formulated as a liquid dimethylamine salt.

Pure 2,3,6-TBA is a white, crystalline solid, somewhat soluble in water (0.84 g/100 ml). However, the dimethylamine salt is readily soluble in water. The acute oral LD_{50} for rats of 2,3,6-TBA is about 750–1000 mg/kg, and that of its dimethylamine salt 1644 mg/kg.

Uses

2,3,6-TBA is considered a nonselective herbicide and is not used in crops, but it does control several broadleaf perennial weeds such as field bindweed, Canada thistle, and bur ragweed at 10–30 lb/acre, as well as certain woody brush species such as conifers, wild roses, and sassafras at 4–20 lb/acre. It is usually applied to the foliage, but is also taken up by the roots when leached into the rooting zone. Spray drift or possibly volatile fumes may injure susceptible crops such as beans, tomatoes, cotton, and various ornamentals, orchard, and vine crops.

Soil Influence

2,3,6-TBA is not markedly adsorbed by soil colloids, and its high water solubility allows it to be readily leached into the rooting zone. It is not subject to rapid degradation in soil and often persists more than one year.

Mode of Action

2,3,6-TBA produces symptoms similar to those produced by 2,4-D, with epinasty of young shoots and inhibition of growth of apical meristems in dicotyledons. In monocots, it can result in increased lodging caused by its effect on nodal meristems. It promotes cell elongation, proliferation of tissues, and induction of adventitious roots (Zimmerman and Hitchcock, 1951). It also interferes with geotropic and phototropic responses (Vander Beek, 1967).

This herbicide is readily absorbed by leaves and roots and translocated via both the symplastic and apoplastic systems with accumulation in areas of high metabolic activity, such as meristems. It has also been shown to readily "leak" or move out from roots into the surrounding medium when applied to the foliage. It appears to be very stable or resistant to degradation in plants (Balagannis et al., 1965; Mason, 1960).

DICAMBA

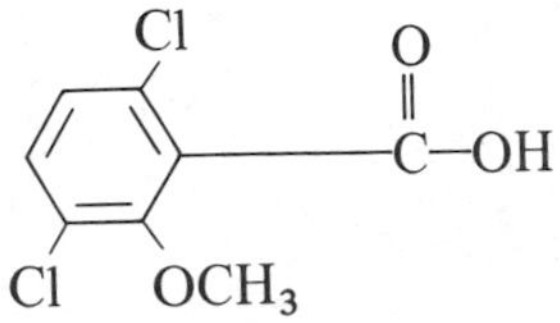

dicamba

Dicamba is the common name for 3,6-dichloro-*o*-anisic acid. The trade name is Banvel®. The chemical structure of dicamba is similar to that of 2,3,6-TBA, except that the chlorine atom at the number 2 position has been replaced by a methoxy($-OCH_3$) group. It is formulated as a liquid in the dimethylamine salt form, and in granular form as the acid or amine salt. It is also formulated in combination with 2,4-D, 2,4,5-T, or MCPA.

Dicamba, as the acid, is a white, crystalline solid somewhat soluble in water (0.45 g/100 ml). The dimethylamine salt is quite soluble in water. The acute oral LD_{50} is about 2900 mg/kg and that of the dimethylamine salt about 1028 mg/kg.

Uses

Dicamba is commonly used to control certain broadleaf weeds and woody plants not readily controlled by phenoxy herbicides. It is somewhat more selective than 2,3,6-TBA and is used on grass crops, including barley, corn, oats, sorghum, wheat, pasture, and range. It can be applied to asparagus beds immediately after cutting. It is also used on fallow land, noncropland, and brush, where it is often combined with 2,4,5-T, 2,4-D, or both, to broaden the spectrum of controlled weed species. It is usually applied to the foliage and stems of these plants, but also is active in the soil. Broadleaf perennial weeds and woody brush or trees are usually the target plant, but most annual broadleaf weeds are also controlled. Also see page 418.

Soil Influence

Dicamba is persistent in soils, and phytotoxicity remains for several months. It is relatively mobile in soils and is subject to leaching.

Mode of Action

Dicamba affects plant growth in much the same way as 2,3,6-TBA, with epinasty of young shoots and proliferative growth (see Figure 10-1). It

Figure 10-1. Modification of leaf structure of melons induced by dicamba. (A. H. Lange, University of California, Parlier.)

is readily absorbed by leaves, stems, and roots and is translocated throughout the plant, accumulating in areas of high metabolic activity (Ashton and Crafts, 1981). This indicates both apoplastic and symplastic transport. It also "leaks" from roots into the surrounding medium. Dicamba is relatively stable in higher plants, but it undergoes hydroxylation or demethylation with hydroxylation. These degradation products may then be conjugated with some endogenous metabolite. The rate of degradation varies greatly with species.

CHLORAMBEN

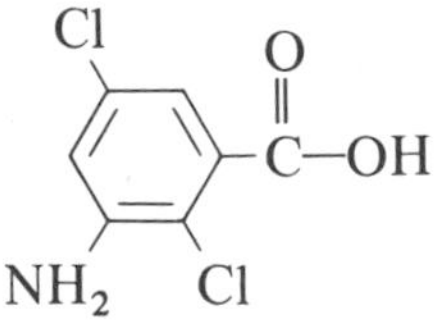

chloramben

Chloramben is the common name for 3-amino-2,5-dichlorobenzoic acid. The trade name is Amiben®. Chloramben is a white amorphous solid, somewhat soluble in water (about 700 ppm), usually formulated in liquid or granular form. As the ammonium salt it is quite soluble in water. The acute oral LD_{50} of chloramben is 3500 mg/kg.

Uses

Chloramben is much more selective than 2,3,6-TBA or dicamba. It has been widely used as a preemergence or preplant incorporated treatment to control many annual broadleaf weeds and grasses in soybeans. It also controls these same weeds in asparagus, beans, corn, peanuts, pumpkins, squash, sunflowers, sweet potatoes, and tomatoes. This herbicide is also formulated or tank-mixed with one of several other herbicides for annual-weed control in several of these crops. Also see page 418.

Soil Interaction

Chloramben is readily leached in sandy soils and is subject to some leaching in other soils, especially after heavy rains. It lasts from six to eight weeks in most soils, and so is much less persistent than 2,3,6-TBA or dicamba.

Mode of Action

Although chloramben is generally less phytotoxic than 2,3,6-TBA or dicamba, it still induces some auxin-like growth abnormalities. It is absorbed by seeds (Haskell and Rogers, 1960) and shows its phytotoxic properties on germinating seeds and young seedlings. Although it is also absorbed by leaves and roots, its translocation is quite limited. Apparently, interaction with some plant constituents prevents its transport. The primary molecular fate of chloramben in higher plants apparently relates to the formation of a glucose conjugate (Colby, 1965; Swanson et al., 1966).

SUGGESTED ADDITIONAL READING

Ashton, F. M., and A. S. Crafts, 1981, "Benzoics," pp. 139–163, in *Mode of Action of Herbicides,* Wiley, New York.

Balagannis, P. G., M. S. Smith, and R. L. Wain, 1965, *Ann. Appl. Biol.* **55**, 149.

Colby, S. R., 1965. *Science* **40**, 619.

Frear, D. S., 1976, "The benzoic acid herbicides," pp. 541–607, in P. C. Kearney and D. D. Kaufman, Eds., *Herbicides,* Vol. 2, Dekker, New York.

Haskell, D. A., and B. J. Rogers, 1960, *Proceedings of the 17th North Central Weed Control Conference,* p. 39.

Mason, G. W., 1960, Ph. D. dissertation, University of California, Davis.

Rieder, G., K. P. Buchholtz, and C. A. Kust, 1970, *Weed Sci.* **18**, 101.

Swan, D. G., and F. W. Slife, 1965, *Weeds* **13**, 133.

Swanson, C. R., R. E. Kadunce, R. A. Hodgson, and D. S. Frear, 1966, *Weeds* **14**, 319.

USDA, SEA., 1980, *Suggested Guidelines for Weed Control,* U.S. Government Printing Office, Washington, D.C. (621-220/SEA 3619).

USDA, SEA., 1981, *Compilation of Registered Uses of Herbicides.*

Vander Beek, L. C., 1967, *Ann. N.Y. Acad. Sci.* **144**, 374.

Weed Science Society of America, 1979, *Herbicide Handbook,* 4th ed., WSSA, Champaign, IL.

Zimmerman, M. H., and A. E. Hitchcock, 1942, *Contr. Boyce Thompson Inst.* **12**, 321.

Zimmerman, M. H., and A. E. Hitchcock, 1951, *Contr. Boyce Thompson Inst.* **16**, 209.

For chemical use, see the manufacturer's label and follow the directions. Also see the Preface.

11 Bipyridyliums

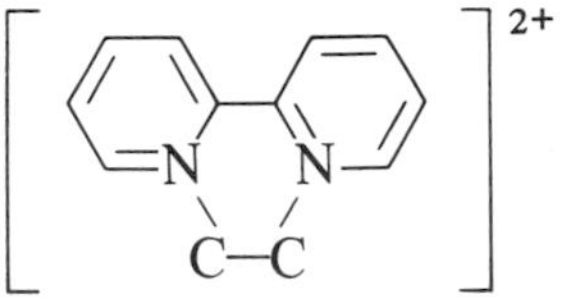

diquat ion

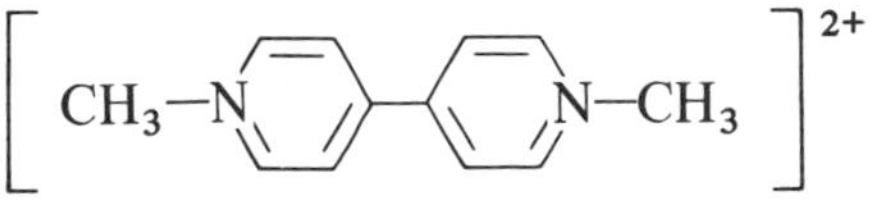

paraquat ion

The above heterocyclic (more than one type of atom in ring) organic compounds belong to the bipyridylium quaternary ammonium class. They were developed by Imperial Chemical Industries in England. Phytotoxic properties of diquat were discovered in 1955 and those of paraquat a few years later. Among several reviews of these compounds, perhaps the most complete are by Calderbank (1968), Akhavein and Linscott (1968), and Calderbank and Slade (1976).

DIQUAT

Diquat is the common name for 6,7-dihydrodipyrido[1,2-α:2′,1′-*c*] pyrazinediium ion. Trade names are Ortho® Diquat and Reglone®. The pure bromine salt is a yellow solid, but in aqueous solution it turns dark reddish-brown. The acute oral LD_{50} of the diquat ion is 230 mg/kg. It should be handled with care, and the user should avoid breathing spray mist or getting the concentrate on skin. When diquat is used as an aquatic herbicide, there is a wide margin of safety between the recommended dosages and the rates necessary to cause toxic symptoms in fish.

PARAQUAT

Paraquat is the common name for 1,1′-dimethyl-4,4′-bipyridinium ion. Trade names are Ortho® Paraquat and Gramoxone®. The pure chloride salt is a white solid, but it forms a dark-red aqueous solution. The acute oral LD_{50} of the paraquat ion is 120 mg/kg. Paraquat may also be absorbed through the skin; the acute dermal LD_{50} of the paraquat ion to rabbits is >480 mg/kg in a single dose. The user should avoid breathing the spray mist, and avoid prolonged skin or eye contact.

Oral ingestion of paraquat represents misuse, yet the serious damage from ingestion deserves special mention. Several poisonings and deaths have been reported as a result of accidentally or intentionally ingesting relatively small amounts of the liquid concentrate (Slade and Bell, 1966). Fatalities resulted mainly from progressive pulmonary (of the lungs) fibrosis with associated liver and kidney damage (Sinow and Wei, 1973). Death may take two to three weeks following ingestion. No cure is known following ingestion of a lethal quantity.

Uses of Each

Diquat is a general contact herbicide used to control aquatic weeds. It is injected into the water for control of submerged weeds or sprayed on foliage to control floating weeds. It is also used as a directed spray for weed control in sugarcane and as a foliar spray to inhibit flowering in this crop. In certain seed crops including alfalfa, castor beans, clover, sorghum, soybeans, and vetch, diquat is applied as a preharvest desiccant.

Paraquat controls emerged annual weeds on noncropland and in many crops, and acts as a plant desiccant. It is a general contact herbicide.

On noncropland, paraquat is widely used to control emerged annual weeds. In many crops it also controls emerged annual weeds using techniques that keep sprays off the leaves and succulent stems of the crop plant. These techniques include (1) preplant or preemergence (to the crop) treatments in many annual crops and certain herbaceous perennial crops; (2) directed sprays in tree crops, berries, grapes, and woody ornamentals; and (3) overall treatments in dormant small-seeded legumes and certain dormant herbaceous perennial seed crops. Users should refer to the manufacturer's label for crops that are registered for use of paraquat as well as the specific method and time of application. These uses are also covered in Chapters 21–31 of this book.

Paraquat is also used as a preharvest desiccant in cotton, potatoes, and soybeans.

Soil Influence on Both

Diquat and paraquat have similar modes of action and are influenced in similar ways by the soil.

An important, unique property of both diquat and paraquat is their speedy inactivation by soils (Brian et al., 1958). This results from reaction between the positively charged herbicide ion and the negatively charged sites on clay minerals. In fact, the herbicide molecule becomes tightly held within the lattice structure of the soil itself, and is adsorbed by soil colloids (Homer et al., 1960). Therefore, these herbicides are essentially nonphytotoxic in most soils. Some phytotoxicity, however, has been demonstrated at high rates in very sandy soils. Although nonphytotoxic, the bound diquat or paraquat ion may persist in soils for long periods of time.

Mode of Action of Both

Bipyridylium-type herbicides cause wilting and rapid desiccation of the foliage to which they are applied, often within a few hours. High light intensities increase the rate of development of phytotoxic symptoms, but are not essential for herbicidal action. Best results in the field have often been obtained by a late-afternoon, rather than a morning or midday application. This appears to allow some internal transport during the night, before development of acute phytotoxicity induced by light, which could limit movement.

Translocation, following a foliar application, appears to be almost solely via the apoplastic system (Baldwin, 1963; Slade and Bell, 1966; Wood and Gosnell, 1965). However, after the loss of membrane integrity, induced by both herbicides, they do move into untreated leaves, presumably along with the flow of other cellular contents. They are poorly translocated from roots (Damonakis et al., 1970) because they are tightly bound to cellular components.

These herbicides are not degraded in higher plants in the usual sense. However, they are reversibly converted from the ion form to the free-

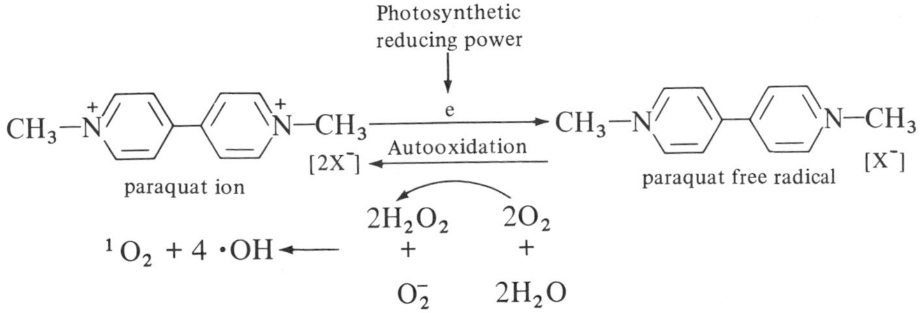

Figure 11-1. Free-radical formation from paraquat ion and autooxidation of free radical yielding H_2O_2 (hydrogen peroxide) and O_2^- (superoxide radical), and subsequently, $\cdot OH$ (hydroxyl radical) and $^1O_2$ (singlet oxygen) (Ashton and Crafts, 1981).

radical form. This interconversion is cyclic and requires light, molecular oxygen, water, and the photosynthetic apparatus. During autooxidation of the paraquat free radical to the ion, four by-products are formed: (1) H_2O_2 (hydrogen peroxide), (2) O_2^- (superoxide radical), (3) •OH (hydroxyl radical), and (4) $^1O_2$ (singlet oxygen) (see Figure 11-1). Each by-product is potentially phytotoxic. However, recent research suggests that the hydroxyl radical and/or the superoxide radical are responsible for the phytotoxic symptoms.

SUGGESTED ADDITIONAL READING

Akhavein, A. A., and D. L. Linscott, 1968, *Residue Rev.* **23**, 97.

Ashton, F. M., and A. S. Crafts, 1981, "Bipyridyliums," pp. 164–79, in *Mode of Action of Herbicides,* Wiley, New York.

Baldwin, B. C., 1963, *Nature* **198**, 872.

Brian, R. C., R. F. Homer, J. Stubbs, and R. L. Jones, 1958, *Nature* **181**, 446.

Calderbank, A., 1968, *Adv. Pest Control Res.* **8**, 129.

Calderbank, A., and P. Slade, 1976, "Diquat and paraquat," pp. 501–540, in P. C. Kearney and D. D. Kaufman, Eds., *Herbicides,* Vol. 2, Marcel Dekker, New York.

Damonakis, M., D. S. H. Drennan, J. D. Fryer, and K. Holly, 1970, *Weed* Res. **10**, 278.

Homer, R. F., G. C. Mees, and T. E. Tomlinson, 1960, *J. Sci. Food Agric.* **11**, 309.

Sinow, J., and E. Wei, 1973, *Bull. Envir. Contamination and Toxicol.* **3**, 163.

Slade, P., and E. G. Bell, 1966, *Weed Res.* **6**, 267.

USDA, SEA, 1980, *Suggested Guidelines for Weed Control,* U.S. Government Printing Office, Washington, D.C. (621-220/SEA 3619).

USDA, SEA, 1981, *Compilation of Registered Uses of Herbicides.*

Weed Science Society of America, 1979, *Herbicide Handbook,* 4th ed, WSSA, Champaign, IL.

Wood, G. H., and J. M. Gosnell, 1965, *Proc. South African Sugar Tech. Assoc.* p. 7.

For chemical use, see the manufacturer's label and follow the directions. Also see the Preface.

12 Carbamates

Carbamate herbicides derive their basic chemical structure from carbamic acid (NH_2COOH). Pure carbamic acid is not stable and quickly decomposes to NH_3 and CO_2. Friesen (1929) first observed the effect of esters of carbamic acid on plants. Herbicidal properties of the carbamate herbicide propham were described by Templeman and Sexton (1945), following the discovery of 2,4-D. The discovery of the herbicidal effect of carbamates thus followed closely the discovery of the prime forerunner of modern chemical weed control. Numerous carbamate compounds have been evaluated since 1945; many have been used as herbicides. However, only six are currently available in the United States. These can be classified into two groups: *soil-applied*—propham and chlorpropham—or *foliar-applied*—asulam, barban, phenmedipham, and desmedipham.

As more carbamates were developed, the selectivity of specific carbamate compounds on particular crops has led to their increased use on a wide variety of crops. Selectivity has also resulted in more-effective weed control. Whereas the first carbamate herbicide, propham, is effective primarily as a preemergence herbicide, some of the newer compounds are applied almost exclusively to emerged weeds. The chemical structure of these compounds has become more and more complicated (see Table 12-1).

PROPHAM

Propham is the common name for isopropyl carbanilate. Although it has several trade names worldwide, the most common name in the United States is Chem-Hoe®. In pure form it is a white solid and has a water solubility of about 250 ppm. It is formulated as an emulsifiable concentrate, flowable suspension, or granules. The acute oral LD_{50} of technical propham is about 9000 mg/kg for rats.

Table 12-1. Common Name, Chemical Name, and Chemical Structure of the Carbamate Herbicides

$$R_1-\overset{H}{\overset{|}{N}}-\overset{O}{\overset{\|}{C}}-O-R_2$$

Common Name	Chemical Name	R_1	R_2
Asulam	Methyl sulfanilylcarbamate	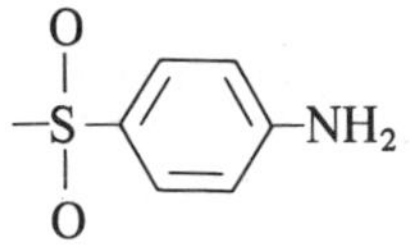	$-CH_3$
Barban	4-Chloro-2-butynyl *m*-chlorocarbanilate	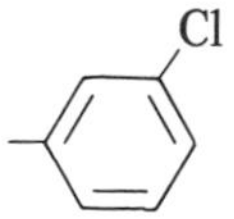	$-CH_2C{\equiv}CCH_2Cl$
Chlorpropham	Isopropyl *m*-chlorocarbanilate	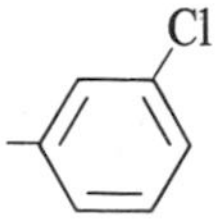	$-\overset{H}{\overset{\vert}{\underset{\underset{CH_3}{\vert}}{C}}}-CH_3$
Desmedipham	Ethyl *m*-hydroxycarbanilate carbanilate	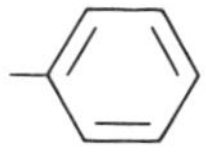	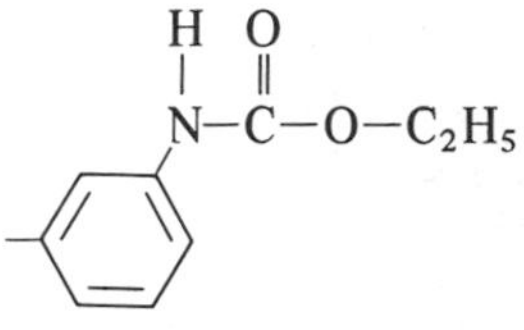
Phenmedipham	Methyl *m*-hydroxycarbanilate *m*-methylcarbanilate	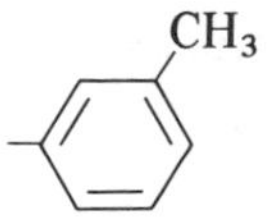	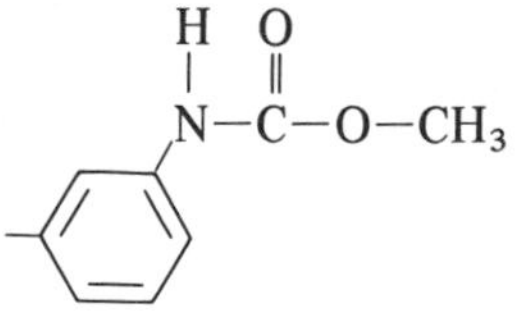
Propham	Isopropyl carbanilate	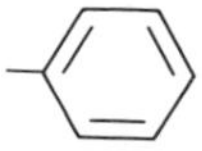	$-\overset{H}{\overset{\vert}{\underset{\underset{CH_3}{\vert}}{C}}}-CH_3$

Uses

Propham is used primarily as a preemergence herbicide to control annual grasses including volunteer small grains. It also controls a few broadleaf weeds such as chickweed and burning nettle. Preplant and postemergence applications can be used in certain crops. However, when propham is applied on emerged sensitive weeds, they should be very small, in the one- to two-leaf stage. Propham is used on alfalfa, clovers, flax, lentils, lettuce, peas, safflower, spinach, sugar beets and established grasses grown for seed. Propham also is combined with 2,4-D for established lawn grasses grown for seed, and combined with benefin for lettuce. Also see page 419.

Soil Influence

Propham is moderately leached in most soils and moves slightly more than chlorpropham. When applied to moist soil followed by rainfall, it may penetrate to a depth of 5–8 in. It degrades rapidly in most soils, usually remaining phytotoxic for less than one month.

Mode of Action

When annual grasses are treated before emergence, they develop much thicker and shorter coleoptiles. The herbicide blocks cell division and induces polyploid nuclei (Ennis, 1948). Propham is absorbed mostly by roots, but very little by leaves (Baldwin et al., 1954). It is translocated primarily in the apoplast (Ashton and Crafts, 1981) and appears to be rapidly degraded in higher plants.

CHLORPROPHAM

Chlorpropham is the common name for isopropyl *m*-chlorocarbanilate. It was formerly called Chloro-IPC and now is trade-named Furloe® in the United States. Pure chlorpropham is a white solid with a water solubility of 88 ppm; it is formulated as an emulsifiable concentrate and in granular form. The acute oral LD_{50} of technical chlorpropham is 3800 mg/kg for albino rats.

Chlorpropham was developed soon after propham was introduced. It has many of the same herbicidal characteristics, but it differs from propham in two ways: It persists longer in the soil, thus giving longer weed control; and it is less selective than propham to certain crop species, such as lettuce.

Figure 12-1. Shepherdspurse control in garlic with chlorpropham. *Left*: Treated. *Right*: Untreated.

Uses

Like propham, chlorpropham is applied primarily as a preemergence herbicide to control annual grasses and a few annual broadleaf weeds. It is also used as a postemergence treatment because it has higher herbicidal efficacy than propham.

Chlorpropham is used in alfalfa, beans, blackberries, blueberries, carrots, clovers, cranberries, garlic (see Figure 12-1), grass seed crops, onion, peas, raspberries, safflower, southern peas, soybeans, spinach, sugar beets, tomatoes, and woody ornamentals. For weed control in soybeans, it is frequently used in combination with one of several other herbicides. Chlorpropham also has growth-regulating properties and is used as a postharvest sprout inhibitor on potato tubers. Also see p. 418.

Soil Influence

Chlorpropham is tightly bound to soil colloids and in general does not leach below the upper inch in nonsandy soil types. In part, this property is the basis of selectivity in certain large, deep-seeded crops such as beans. Chlorpropham persists in most soils at phytotoxic levels for one to two months.

Mode of Action

Chlorpropham inhibits cell division, thus causing multinucleate root cells and inhibited root elongation (Ennis, 1949). It is readily translocated throughout the plant when absorbed by roots (Prendeville et al., 1968) but not when applied to leaves, suggesting almost exclusive apoplastic transport. It also appears to be absorbed by emerging shoots as they pass through the treated soil (Knake and Wax, 1968). The vapors of chlorpropham have been shown to be absorbed by seeds (Ashton and Helfgott, 1966) and emerged dodder plants (Slater et al., 1969). In the latter case, chlorpropham prevents the parasitic dodder from attaching to the host plant.

Chlorpropham is rapidly degraded in higher plants. It has been reported to inhibit photosynthesis, respiration, and synthesis of ATP, RNA, proteins, and lipids. All of these may contribute to its interference with cell division; however, a direct cause-and-effect relationship has not been demonstrated with any of these processes.

ASULAM

Asulam is the common name for methyl sulfanilycarbamate. The trade name is Asulox®. It is a white crystalline solid with a water solubility of 5000 ppm, and is formulated as an aqueous solution of the sodium salt. The acute oral LD_{50} is greater than 8000 mg/kg for rats.

Uses

Asulam is used for postemergence weed control in sugarcane and noncrop areas. In sugarcane the primary target weeds are johnsongrass, paragrass (*Brachiaria mutica*), alexandergrass (*Brachiaria plantaginea*), crabgrass (*Digitara* spp.), foxtail (*Setaria* spp.), goosegrass, broadleaf panicum (*Panicum adspersum*), and barnyardgrass. The height of the sugarcane at the time of application depends on geographical location; it should be at least 12 in. tall in Hawaii, at least 14 in. tall in Florida, at least 18 in. tall in Louisiana, and about 3 ft. tall in Texas. It is also used in Christmas tree plantings (douglas fir, grand fir, noble fir, scotch pine) as an aerial or ground application. The target weed is usually western bracken. On noncropland, asulam controls these same weeds. Also see page 418.

Soil Influence

Since asulam is applied to the foliage of the target species, soil type does not affect its performance. It is not persistent in soils, it has a half-life of only 6 to 14 days.

Mode of Action

Asulam is readily absorbed by susceptible species, but absorption is increased by the addition of certain surfactants in some, but not all, species (Catchpole and Hibbitt, 1972; Babiker and Duncan, 1975). It appears to be translocated primarily in the symplast.

Asulam inhibits bud growth but does not appear to significantly reduce photosynthesis. Its action appears to involve an inhibition of RNA and protein synthesis (Ashton and Crafts, 1981).

BARBAN

Barban is the common name for 4-chloro-2-butynyl *m*-chlorocarbanilate. The trade name is Carbyne®. It is a colorless solid with a water solubility of 11 ppm, and is formulated as an emulsifiable concentrate. The acute oral LD_{50} is 1350 mg/kg.

Uses

Barban is a postemergence herbicide used almost exclusively to control wild oat. However, several members of the buckwheat family are also controlled, and reed canarygrass is highly sensitive. Time of application is critical to obtain control of wild oat. Barban must be applied at the two-leaf stage of wild oat–from the appearance of the second leaf until the appearance of the third leaf; earlier or later applications will give poor control. However, if wild oat is growing slowly because of cold weather, low soil moisture, or low fertility, and if it does not reach the two-leaf stage within nine days after emergence, it should be sprayed before the fourteenth day after emergence.

Barban is used in barley, flax, lentils, peas, safflower, soybeans, sugar beets, sunflower, and wheat–as well as mustard as an oil crop.

Soil Influence

Since barban is applied to the foliage, its interaction with the soil is of limited significance. However, it is adsorbed by soil colloids and is rapidly degraded in less than one month.

Mode of Action

Substantial amounts of barban are absorbed by leaves within a few hours, but it continues to be absorbed for at least one week (Foy, 1961). Translocation is limited and appears to be mostly in the apoplast. It is degraded rapidly in most plant species (Riden and Hopkins, 1962). Barban has been reported to inhibit several biochemical processes; however, its interference with nucleic acid metabolism would appear to be most closely associated with its effect on cell division and growth.

PHENMEDIPHAM

Phenmedipham, trade named Betanal®, is the common name for methyl *m*-hydroxycarbanilate *m*-methylcarbanilate. It has two carbamate groups in its chemical structure (see Table 12-1). It is a colorless solid with a water solubility of less than 10 ppm. Phenmedipham is formulated as an emulsifiable concentrate, and its acute oral LD_{50} is greater than 8000 mg/kg.

Uses

Phenmedipham is a postemergence herbicide that controls annual broadleaf weeds in sugar beets. Weeds should be small, with the rate of application depending on the size of the weeds: 1–1$\frac{1}{4}$ lb/acre at the cotyledon to two-true-leaf stage and 1$\frac{1}{4}$–1$\frac{1}{2}$ lb/acre at the two-to four-true-leaf stage; sugar beets should be past the two-leaf stage. Sensitivity of sugar beets to phenmedipham increases at temperatures over 85°F under moisture stress, or if preplant or preemergence herbicides are used before the phenmedipham application.

Soil Influence

Phenmedipham is not subject to excessive leaching. It has a half-life of about 25 days in most soils.

Mode of Action

Phenmedipham is readily absorbed by leaves and appears to be translocated via the apoplastic system. Studies conducted on sugar beets have shown that the chemical is degraded within several days after application, and traces (0.1 ppm) of the breakdown product, 3-methylaniline, were detected at harvest upon hydrolysis of the tissue (Kassenbeer, 1969). Phenmedipham inhibits the Hill reaction of photosynthesis.

DESMEDIPHAM

Desmedipham is the common name for ethyl *m*-hydroxycarbanilate carbanilate with a trade name of Betanex®. It is a colorless or light-yellow crystalline solid with a water solubility of about 7 ppm. It is formulated as an emulsifiable concentrate. The pure compound has an LD_{50} of 10,250 mg/kg; the formulated product (16% desmedipham) shows an LD_{50} of 3720 mg/kg for rats.

Uses

Like phenmedipham, desmedipham is used in sugar beets as a postemergence treatment to control several annual weeds. However, it is unique because it controls a serious weed in sugar beets, redroot pigweed, which is not controlled by phenmedipham. Its behavior in soils and its mode of action are similar to those of phenmedipham as discussed above.

SUGGESTED ADDITIONAL READING

Ashton, F. M., and A. S. Crafts, 1981, "Carbamates," pp. 180–200, in *Mode of Action of Herbicides,* Wiley, New York.

Ashton, F. M., and S. Helfgott, 1966, *Proceedings of the 18th California Weed Control Conference,* p. 8.

Babiker, A. G. T., and H. J. Duncan, 1975, *Pestic. Sci.* **6**, 655.

Baldwin, R. E., V. H. Freed, and S. C. Fang, 1954, *J. Agr. Food Chem.* **2**, 428.

Catchpole, A. H., and C. J. Hibbitt, 1972, *Proceedings of the 11th British Crop Protection Council,* p. 77.

Ennis, Jr., W. B., 1948, *Amer. J. Bot.* **35**, 15.

Ennis, Jr., W. B., 1949, *Amer. J. Bot.* **36**, 823.

Foy, C. L., 1961, *Res. Progress Report, Western Weed Control Conference,* p. 96.

Friesen, G., 1929, *Planta* **8**, 666.

Kassenbeer, H., 1969, *Schering AG Conference, Berlin,* p. 55.

Knake, E. L., and L. M. Wax, 1968, *Weed Sci.* **16**, 393.

Prendeville, G. N., Y. Eshel, C. S. James, G. F. Warren, and M. M. Schreiber, 1968, *Weed Sci.* **16**, 432.

Riden, J. R., and T. R. Hopkins, 1962, *J. Agr. Food Chem.* **10**, 455.

Slater, C. H., J. H. Dawson, W. R. Furtick, and A. P. Appleby, 1969, *Weed Sci.* **17**, 238.

Still, G. G., and R. A. Herrett, 1976, "Methylcarbamates, carbanilates, and acylanilides," pp. 609–664, in P. C. Kearney and D. D. Kaufman, Eds., *Herbicides,* Vol 2, Marcel Dekker, New York.

Templeman, W. G., and W. A. Sexton, 1945, *Nature* **156**, 630.

USDA, SEA, 1980, *Suggested Guidelines for Weed Control,* U.S. Government Printing Office, Washington, D.C. (621-220/SEA 3619).

USDA, SEA, 1981, *Compilation of Registered Uses of Herbicides.*

Weed Science Society of America, 1979, *Herbicide Handbook,* 4th Ed., WSSA, Champaign, IL.

For chemical use, see the manufacturer's label and follow the directions. Also see the Preface.

13 Dinitroanilines

The herbicidal properties of 2,6-dinitroanilines were first reported by Eli Lilly and Company scientists in 1960 (Alder et al., 1960), and other companies have also developed herbicides from this class of compounds. Most pure dinitroanilines are yellow-orange crystalline solids with a low water solubility. They are generally volatile and more or less susceptible to photodegradation. Most are selective herbicides used as a preplant soil-incorporation treatment.

Chemical structures shown below are those basic to anilines and 2,6-dinitroanilines.

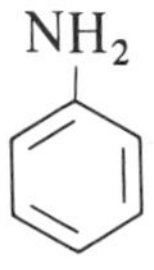

aniline

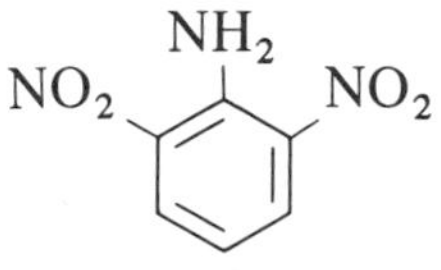

dinitroaniline

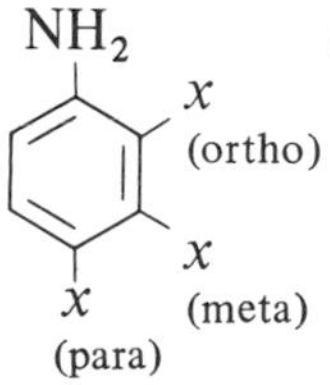

(toluidine)

$x = CH_3$ in one position

Table 13-1 gives specific structures of 13 dinitroaniline herbicides. All of these except the following five are commercially available in the United States. Dinitramine (Cobex®) and nitralin (Planavin®), which were commercial products, are no longer manufactured or sold. Ethalfluralin (Sonalan®) and prodiamine (Rydex®) are still considered to be experimental compounds. Therefore, these four compounds are not covered in detail. Butralin (Amex®, Tamex®) is not marketed in the United States, but there is significant use worldwide.

TRIFLURALIN

Trifluralin is the common name for α,α,α-trifluoro-2,6-dinitro-*N,N*-dipropyl-*p*-toluidine. Trade names include Treflan®, Trefanocide®, and Elancolan®. Trifluralin is a yellow-orange crystalline solid with a water

Table 13-1. Common Name, Chemical Name, and Chemical Structure of the Dinitroaniline Herbicides

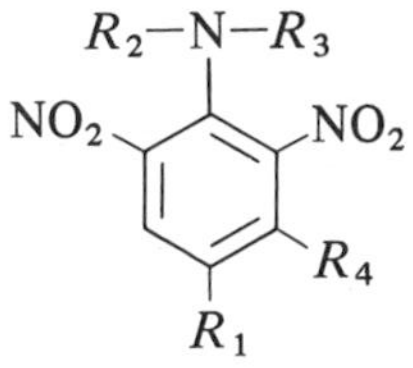

Common Name	Chemical Name	R_1	R_2	R_3	R_4
Benefin	*N*-Butyl-*N*-ethyl-α,α,α-trifluoro-2,6-dinitro-*p*-toluidine	$-CF_3$	$-C_2H_5$	$-C_4H_9$	—H
Butralin	4-(1,1-Dimethylethyl)-*N*-(1-methylpropyl)-2,6-dinitrobenzenamine	$-C-(CH_3)_3$	—H	$-CH(CH_3)C_2H_5$	—H
Dinitramine	N^4,N^4-Diethyl-α,α,α-trifluoro-3,5-dinitrotoluene-2,4-diamine	$-CF_3$	$-C_2H_5$	$-C_2H_5$	$-NH_2$
Ethalfluralin	*N*-Ethyl-*N*-(2-methyl-2-propenyl)-2,6-dinitro-4-(trifluoromethyl)benzenamine	$-CF_3$	$-C_2H_5$	$-CH_2-C(CH_3)=CH_2$	—H
Fluchloralin	*N*-(2-Chloroethyl)-2,6-dinitro-*N*-propyl-4-(trifluoromethyl)aniline	$-CF_3$	$-C_3H_7$	$-CH_2CH_2Cl$	—H

Isopropalin	2,6-Dinitro-*N*,*N*-dipropylcumidine	iso-C_3H_7	$-C_3H_7$	$-C_3H_7$	$-H$
Nitralin	4-(Methylsulfonyl)-2,6-dinitro-*N*,*N*-dipropylaniline	$-S(\rightarrow O)(\rightarrow O)-CH_3$	$-C_3H_7$	$-C_3H_7$	$-H$
Oryzalin	3,5-Dinitro-N^4,N^4-dipropylsulfanilamide	$-S(\rightarrow O)(\rightarrow O)-NH_3$	$-C_3H_7$	$-C_3H_7$	$-H$
Pendimethalin	*N*-(1-Ethylpropyl)-3,4-dimethyl-2,6-dinitrobenzenamine	$-CH_3$	$-CH(C_2H_5)_2$	$-H$	$-CH_3$
Prodiamine	2,4-Dinitro-N^3,N^3-dipropyl-6-(trifluoromethyl)-1,3-benzenediamine	$-CF_3$	$-C_3H_7$	$-C_3H_7$	$-NH_2$
Profluralin	*N*-(Cyclopropylmethyl)-α,α,α-trifluoro-2,6-dinitro-*N*-propyl-*p*-toluidine	$-CF_3$	$-CH_2-\triangleleft$	$-C_3H_7$	$-H$
Prosulfalin	*N*-[[4-(Dipropylamino)-3,5-dinitrophenyl] sulfonyl]-*S*,*S*-dimethylsulfilimine	$-S(\rightarrow O)(\rightarrow O)-N=S(CH_3)_2$	$-C_3H_7$	$-C_3H_7$	$-H$
Trifluralin	α,α,α-Trifluoro-2,6-dinitro-*N*,*N*-dipropyl-*p*-toluidine	$-CF_3$	$-C_3H_7$	$-C_3H_7$	$-H$

solubility of less than 1 ppm. The compound is formulated as an emulsifiable concentrate and as granules. The acute oral LD_{50} in mice is about 5000 mg/kg.

Uses

Trifluralin, the first 2,6-dinitroaniline herbicide developed and marketed, is the most widely used herbicide of this class and one of the most important herbicides used selectively in crops. Although the major use has been in cotton and soybeans, trifluralin is used on more than 40 other crops including established alfalfa, several types of beans, barley, cole crops, carrots, celery, collards, cucurbits, grapes, guar, hops, kale, mint, mustard greens, okra, peanuts, peas, peppers, potatoes, safflower, southern peas, sugar beets, sugarcane, sunflower, tomatoes, turnip greens, wheat, and many fruit and nut trees.

In most of these crops, trifluralin is applied as a preplant or preemergence soil-incorporation treatment. In some crops, such as tomatoes, potatoes, sugar beets, cantaloupe, cucumbers, and watermelon, the herbicide is too phytotoxic to be applied preplant on seeded crops, but can be used on transplants or on established plants.

Trifluralin controls most weed seeds as they germinate. This includes nearly all grass weed seeds and also many broadleaf weeds including pigweeds, purslane, lambsquarters, kochia, Russian thistle, henbit, knotweed, and chickweed. Also see page 419.

Trifluralin also controls certain perennial weeds (e.g., field bindweed and johnsongrass from rhizomes) when used in accordance with special rate recommendations and application techniques.

Soil Influence

Four factors are responsible for the disappearance or degradation of trifluralin in soil. These include volatilization, photodecomposition, microbial decomposition, and chemical decomposition. The relative importance of each factor is influenced by soil type, moisture content, temperature, microflora, rate of application, and method of incorporation. In warm moist soils, trifluralin at recommended rates of application is decomposed in less than 12 months (see Figure 13-1).

Studies on loss of trifluralin from soil surfaces have shown that volatilization and photodecomposition are important sources of herbicide loss. The rate of loss was highest in the few hours immediately after application. Wet soil surfaces and high soil temperature increased the rate of loss. The more adsorptive soils showed the longest retention time for the herbicide. Losses through volatilization and photodecom-

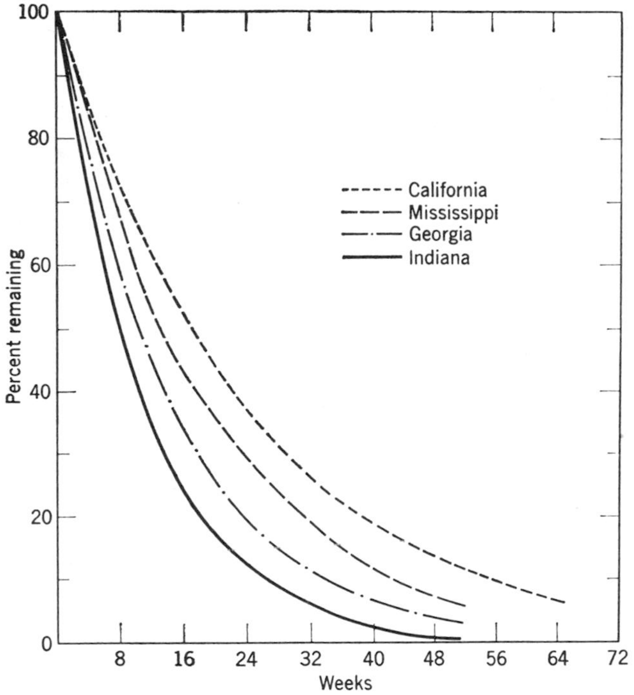

Figure 13-1. Field persistence of trifluralin under different climatic conditions. Note that after 48 weeks, the trifluralin was nearly gone, with the longest persistence in dryland conditions in California. (Lilly Research Laboratories.)

position are minimized when trifluralin is mixed into the soil after application.

In greenhouse studies, the rate of trifluralin degradation in soil was compared under anaerobic and aerobic conditions. After 40 days, with moisture at 200% of the field's water-holding capacity (anaerobic), 98% of the trifluralin had degraded. With aerobic moisture conditions (0.0%, 50%, and 100% of field capacity) less than 25% of the trifluralin had decomposed in the same time period.

In soil saturated with water, oxygen will most likely become a limiting factor. Under such conditions, anaerobic degradation of organic compounds can be expected. It has not been established whether this is chemical or microbial, but both are probably involved. In nonautoclaved, underwater soil, trifluralin degradation was complete in seven days at 76°F, whereas only 10% had degraded at 38°F (Probst et al., 1967) (see Figure 13-2). Under warm anaerobic conditions, the trifluralin was rapidly degraded, whereas under aerobic conditions degradation was much slower. Under field conditions, trifluralin degradation was nearly complete after 48 weeks under four different climatic conditions (see Figure 13-1). More detailed discussions of degradation of trifluralin

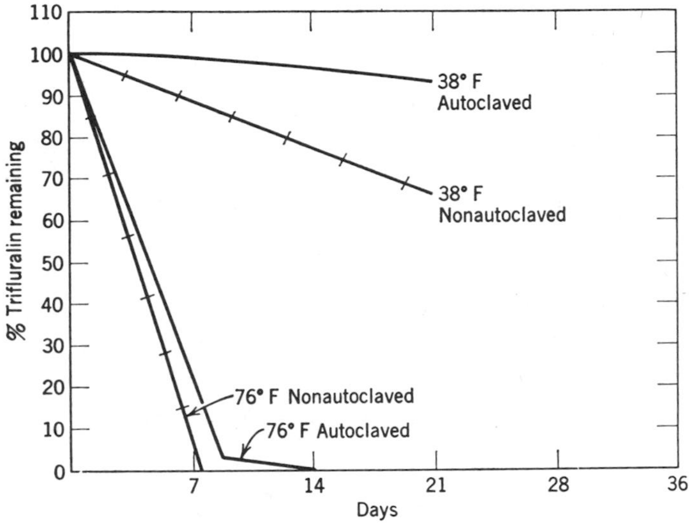

Figure 13-2. Anaerobic degradation of trifluralin in autoclaved and nonautoclaved soil as a function of temperature in soil saturated with water (200% field capacity). (Probst et al., 1967.)

in soil have been published (Golab et al., 1967; Probst et al., 1967; Probst et al., 1975; Golab et al., 1979).

Trifluralin is strongly adsorbed onto the soil and is extremely resistant to movement by water. By mixing the herbicide with soil, an effective concentration of herbicide is formed in the zone of weed-seed germination. Despite heavy rainfall, trifluralin is not leached from this weed-seed-germination zone. With furrow irrigation, there is little, if any, lateral movement.

The availability of trifluralin to germinating weed seeds is influenced by its adsorption onto soil. The content of sand, silt, and clay influences the amount required for weed control, with the most trifluralin required on the clay (heavy) soils. Adsorption on organic matter appears to strongly restrict availability of trifluralin, thus limiting its efficacy as a herbicide. Trifluralin is not usually recommended for use on peat and muck soils of greater than 10% organic matter.

Mode of Action

Most research on the mode of action of dinitroaniline herbicides has been conducted with trifluralin. Although these herbicides are basically similar in activity, varying degrees of tolerance by several crop species can be demonstrated.

Many studies have shown that trifluralin inhibits growth of roots.

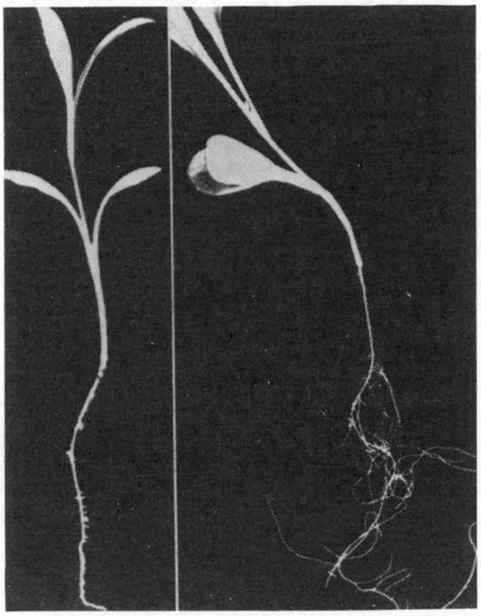

Figure 13-3. Inhibition of lateral root formation is typical of the dinitroaniline herbicides. *Left*: Treated. *Right*: Untreated. (D. E. Bayer, University of California, Davis.)

Characteristically, the root increases in diameter or swells in the active meristematic region near the root tip. This is caused by the disruption of cell division. Lateral or secondary root development is also inhibited (see Figure 13-3).

Trifluralin appears to be absorbed primarily by emerging grass shoots as they pass through the treated soil; however, some root absorption may also occur. There is no significant translocation of trifluralin into the stem or foliage of higher plants, nor is it found in the harvested seed crop.

Trifluralin is not degraded to a significant degree by higher plants.

The dinitroaniline herbicides alter the chemical composition and several biochemical processes in higher plants. Changes in sugar, amino acid, and nucleic acid content have been reported. Inhibitions of photosynthesis, RNA synthesis, protein synthesis, lipid synthesis, and oxidative phosphorylation, as well as shifts in the activity of various enzymes have also been observed.

Detailed information on the mode of action of trifluralin and other dinitroaniline herbicides is given by Probst et al. (1975), Parka and Soper (1977), Golab et al. (1979) and Ashton and Crafts (1981).

BENEFIN

Benefin is the common name for *N*-butyl-*N*-ethyl-α,α,α-trifluoro-2,6-dinitro-*p*-toluidine. Trade names include Balan®, Balfin®, Banafine®, and Quilan®. Pure benefin is a yellow-orange crystalline solid, and its solubility in water is less than 1 ppm at 25°C. Benefin is susceptible to decomposition by ultraviolet light. In the adult rat, the LD_{50} is greater than 10,000 mg/kg. The herbicide is formulated as a $1\frac{1}{2}$ lb/gal concentrate and as granules.

Uses

Benefin controls most annual grasses and many broadleaf weeds including carpetweed, chickweed, knotweed, lambsquarters, pigweeds, purslane, redmaids, and Florida pusley. Benefin is recommended for alfalfa, birdsfoot trefoil, clovers, lettuce, peanuts, and tobacco. It is applied as a preplant soil-incorporated application in all of these crops except tobacco, in which it is soil-incorporated prior to transplanting. It is used as a tank mix with vernolate in peanuts. Granular benefin is also surface-applied to turf grasses for annual-grass-weed control. Also see page 418.

Soil Influence

Benefin is strongly adsorbed on soil colloids and is not subject to leaching. Microorganisms appear to be responsible for the major loss of benefin from soils. Usually benefin persists in soils for four to five months at recommended rates.

Mode of Action

Although little information is available on the mode of action of benefin, presumably its action is similar to that of trifluralin. It is absorbed by plant organs but translocation is limited. Benefin does not appear to be degraded to a significant degree by higher plants.

BUTRALIN

Butralin is the common name for 4-(1,1-dimethylethyl)-*N*-(1-methylpropyl)-2,6-dinitrobenzenamine. The trade names are Amex®, and Tamex®. It is a yellow-orange solid, with a water solubility of about 1 ppm. It is formulated as an emulsifiable concentrate. It has an acute oral LD_{50} of 12,600 mg/kg for rats.

Uses

Butralin is approved for use in cotton, soybeans, watermelons, and cucumbers in the United States, but is not marketed. In other countries it is sold for use in these crops. Butralin controls most annual grasses and several broadleaf weeds. It is usually applied as a preplant soil-incorporation treatment.

Soil Influence

Soil colloids are strong adsorptive sites for butralin, and it is not subject to leaching. It undergoes slow degradation by soil microorganisms but persists several months.

Mode of Action

Limited research suggests that butralin has some foliar activity that is enhanced by wetting agents. Translocation is limited. Apparently it is slowly degraded in plants.

FLUCHLORALIN

Fluchloralin is the common name for *N*-(2-chloroethyl)-2,6-dinitro-*N*-propyl-4-(trifluoromethyl)aniline. The trade name is Basalin®. It is a yellow-orange crystalline solid. Water solubility is probably less than 1 ppm. The compound is formulated as an emulsifiable concentrate and as granules. It has an acute oral LD_{50} of 1550 mg/kg for rats.

Uses

Fluchloralin controls most annual grasses and several annual broadleaf weeds in soybeans and cotton. Rhizome johnsongrass can also be suppressed by chisel plowing and later discing before applying the herbicide. These cultural practices have two goals: (1) to place the rhizome in the soil layer to be treated with the herbicide and (2) to cut the rhizomes into short pieces. Fluchloralin is applied as a preplant soil-incorporation treatment. It can be applied with most fluid fertilizers or impregnated on dry bulk fertilizers. In addition to being used alone in soybeans, it may be combined with metribuzin as a tank mix.

Soil Influence

Fluchloralin is tightly adsorbed to soil colloids and is not subject to leaching. It is degraded by soil microorganisms but may persist more than one season after application.

Mode of Action

Fluchloralin is absorbed by roots and translocated throughout the plant. However, there is little foliar absorption. Fluchloralin appears to be extensively metabolized by higher plants. These two qualities—considerable translocation and large-scale metabolism—appear to be unique to fluchloralin; with most other dinitroaniline herbicides these processes are minimal.

Little has been reported on its mechanism of action; apparently it affects seed germination.

ISOPROPALIN

Isopropalin is the common name for 2,6-dinitro-*N,N*-dipropylcumidine. The trade name is Paarlan®. This red-orange oil-like chemical has a water solubility of less than 1 ppm. Isopropalin is formulated as an emulsifiable concentrate. The acute oral LD_{50} is about 5000 mg/kg for rats.

Uses

Isopropalin controls most annual grasses and several annual broadleaf weeds starting from seed. These include pigweed, lambsquarters, purslane, carpetweed, Florida purslane, and poorjoe. It is used on transplant tobacco. It is applied as a preplant soil-incorporation treatment. Also see page 419.

Soil Influence

Isopropalin is strongly adsorbed by soil and shows negligible leaching. It appears to be degraded in soils mainly by microorganisms and usually drops to nonphytotoxic levels within a year under warm, moist soil conditions.

Mode of Action

Research on its mode of action is very limited. Apparently it affects seed germination and is poorly absorbed and translocated; no significant terminal residues or specific metabolites have been detected in the plants studied.

ORYZALIN

Oryzalin is the common name of 3,5-dinitro-N^4,N^4-dipropylsulfanilamide. The trade names are Surflan® and Ryzelan®. Oryzalin is a yellow-orange crystalline solid with a water solubility of 2.5 ppm. In contrast to the other dinitroaniline herbicides, oryzalin has a low vapor pressure and is relatively stable in sunlight; thus oryzalin can be applied to the soil surface for later movement into the upper soil zone by rain. It is formulated as a wettable powder. The acute oral LD_{50} is greater than 10,000 mg/kg for rats.

Uses

Oryzalin is usually used as a preemergence herbicide. Shallow incorporation usually increases its herbicidal activity. It controls most annual grasses and several broadleaf weeds. Also see page 419.

Figure 13-4. Oryzalin was applied to this row of grapes for three years, consecutively. Shallow cultivation or the use of a contact herbicide further increases the effectiveness of the treatment. (Elanco Products Company, a division of Eli Lilly and Co.)

In soybeans oryzalin can be used alone but is often used as a tank mix with metribuzin, metribuzin plus paraquat, linuron plus paraquat, or naptalam plus dinoseb. In Texas, oryzalin can be used in cotton. In California, it can be used in established, nonbearing horticultural crops including nut trees (almond, pistachio, walnut), deciduous fruit trees (apple, sweet cherry, peach, pear, plum, prune), oranges, and grapes (see Figure 13-4). It is also used in ornamentals.

Soil Influence

Laboratory studies indicate that oryzalin is biodegradable in soil. In laboratory leaching studies, oryzalin applied to a soil surface remained primarily in the top 2 in. when 6 in. of water was passed through an 8-in. column. In field studies, at least 0.5 in. of rainfall or overhead irrigation water was necessary to position oryzalin in the weed-seed-germinating layer of the soil. Excessive rainfall or irrigation did not leach oryzalin out of the weed-seed-germination zone.

In areas with more than 25 in. of rainfall, oryzalin applied at the recommended rate has not been persistent from one year to the next. Cultivation does not reduce, and in fact may increase, the herbicidal efficacy of the compound.

Mode of Action

Limited research suggests that oryzalin is not readily absorbed and translocated in plants. No significant terminal residues nor specific metabolites have been detected. It appears to affect seed germination.

PENDIMETHALIN

Pendimethalin is the common name for *N*-(1-ethylpropyl)-3,4-dimethyl-2,6-dinitrobenzenamine. The trade name is Prowl®; Stomp and Herbadox are also used as product names. It is a crystalline, orange-yellow solid. The water solubility is about 0.5 ppm. The compound is formulated as an emulsifiable concentrate. The acute oral LD_{50} for technical pendimethalin in rats is 1050 (female) and 1250 (male) mg/kg.

Uses

Pendimethalin controls most annual grasses and certain broadleaf weeds. It is usually applied as a preplant soil-incorporation treatment.

But it can be used as a preemergence application when leached into the soil within seven days after application by adequate rainfall or overhead irrigation. It can also be applied in certain liquid fertilizers.

Pendimethalin is used in corn, cotton, soybeans, and transplanted tobacco. It can be used in corn as a tank mix with atrazine, cyanazine, dicamba, paraquat, atrazine plus paraquat, and cyanazine plus paraquat. A tank mix with chloramben or metribuzin can also be used in soybeans. Special local-need permits for certain states have been granted for postemergence applications in corn and preplant soil-incorporation treatments in transplanted tobacco. Also see page 419.

Soil Influence

Pendimethalin is strongly adsorbed on soil clay and organic matter and is not subject to leaching. In contrast to the case with most other dinitroaniline herbicides, soil microorganisms do not appear to play a significant role in degradation of pendimethalin. Under normal environmental conditions, it persists less than one year at phytotoxic levels in soil.

Mode of Action

Limited research suggests that pendimethalin is absorbed more readily by very small broadleaf weeds than by grasses. Very small amounts are absorbed by roots, with negligible transport to other organs. Herbicidal effect appears to be related to inhibition of cell division. Residue studies indicate that pendimethalin is degraded in plants, animals, and soil.

PROFLURALIN

Profluralin is the common name for *N*-(cycloprophylmethyl)-α,α,α-trifluoro-2,6-dinitro-*N*-propyl-*p*-toluidine. The trade name is Tolban®. It is a yellow-orange crystalline solid, or deep-orange liquid with a water solubility of 0.1 ppm. The compound is formulated as an emulsifiable concentrate, but granular formulations are also available. The acute oral LD_{50} is 2200 mg/kg in rats.

Uses

Profluralin controls annual grasses from seed and several broadleaf weeds and has shown tolerance for many crops. It is registered for use in alfalfa, beans (several types), chickpeas, cotton, okra, peas, safflower, southern peas, soybeans, and sunflowers. Certain states are excluded from some of these uses, so the user should check the product label.

Profluralin is applied as a preplant soil-incorporation treatment. It may be mixed with most liquid or dry bulk fertilizers for application. It may be applied as a tank mix with EPTC for certain types of beans. Following the preplant soil-incorporation of profluralin, preemergence applications of other herbicides are sometimes made: prometryn or fluometuron in cotton, or linuron, metribuzin, or naptalam plus dinoseb in soybeans.

Soil Influence

Profluralin is strongly adsorbed to organic matter and clay, and is therefore not readily leachable. It is degraded by soil microorganisms and has a soil half-life of 80–160 days; 70–80% is degraded in six months.

Mode of Action

Absorption by leaves appears to be negligible, and that amount absorbed by roots and emerging shoots is translocated to other organs in insignificant amounts. It is degraded by higher plants, but the quantity of metabolites varies considerably with the species used as well as the organ sampled and the temperature.

PROSULFALIN

Prosulfalin is the common name for *N*-[[4-(dipropylamino)-3,5,-dinitrophenyl] sulfonyl]-*S,S*-dimethylsulfilimine. The trade name is Sward®. It is a crystalline, yellow-orange solid with a water solubility of 5.6 ppm. The compound is formulated as a wettable powder. The 50% wettable powder formulation has an acute oral LD_{50} of 2000 mg/kg in mice.

Uses

Prosulfalin is a selective preemergence herbicide used in established grass turf to control annual grasses such as crabgrass and annual bluegrass. Some annual broadleaf weeds are also controlled. It inhibits seed germination. It is strongly adsorbed on soils and shows negligible leaching. It has a half-life of about one month and dissipates to no detectable residue in six months.

SUGGESTED ADDITIONAL READING

Alder, E. F., W. L. Wright, and Q. F. Soper, 1960, *Proceedings of the 17th North Central Weed Control Conference,* p. 23.

Ashton, F. M., and A. S. Crafts, 1981, "Dinitroanilines," pp. 201–223, in *Mode of Action of Herbicides,* Wiley, New York.

Golab, T., W. A. Althaus, and H. L. Wooten, 1979, *J. Agr. Food Chem.* **27**, 163.

Golab, T., R. J. Herberg, S. J. Parka, and J. B. Tepe, 1967, *J. Agr. Food Chem.* **15**, 638.

Parka, S. J., and Q. F. Soper, 1977, *Weed Sci.* **25**, 79.

Probst, G. W., T. Golab, R. J. Herberg, F. J. Holzer, S. J. Parka, C. Van der Schans, and J. B. Tepe, 1967, *J. Agr. Food Chem.* **15**, 592.

Probst, G. W., T. Golab, and W. L. Wright, 1975, "Dinitroanilines," pp. 453–500, in P. C. Kearney and D. D. Kaufman, Eds., *Herbicides,* Vol. 1, Dekker, New York.

USDA, SEA, 1980, *Suggested Guidelines for Weed Control,* U. S. Government Printing Office, Washington, D. C. (621-220/SEA 3619).

USDA, SEA, 1981, *Compilation of Registered Uses of Herbicides.*

Weed Science Society of America, 1979, *Herbicide Handbook,* 4th Ed., WSSA, Champaign, IL.

For chemical use, see the manufacturer's label and follow the directions. Also see the preface.

14 Diphenyl Ethers

Diphenyl ether herbicides control many annual weeds in certain vegetable crops, agronomic crops, and orchards. They are effective both as foliar and soil treatments. The first compound of this class was nitrofen, developed by Rohm and Hass Company in the mid-1960s. Nitrofen and CNP (*p*-nitrophenyl 2,4,6-trichlorophenyl ether) were applied to about 96% of the cultivated paddy rice in Japan in 1974 (Matsunaka, 1976). Other compounds developed later have shown promise on widely differing crops.

Table 14-1 gives the common name, chemical name, and chemical structure of the five diphenyl ethers discussed in this chapter.

ACIFLUORFEN

Acifluorfen is the common name for 5-[2-chloro-4-(trifluoromethyl) phenoxy]-2-nitrobenzoic acid. The trade name is Blazer®. The technical product, a viscous sodium-salt solution, contains 45% of the active ingredient. It has a faint brown color and is infinitely soluble in water. It is formulated as a 2 lb/gal liquid concentrate. The acute oral LD_{50} for rats is 1300 mg/kg for the technical product and 3330 mg/kg for the formulation.

Uses

Acifluorfen is used largely as a postemergence treatment to control many broadleaf weeds in soybeans. It also controls certain annual grasses. Best control is obtained when weeds are less than 4 in. high and actively growing. Acifluorfen can be sprayed over the top of soybeans, usually at the one- to two-trifoliate leaf stage. Soybean leaves actually sprayed may show some cupping, crinkling, and speckling; but later growth is normal. It may also be suitable for use in peanuts and other large-seeded legume crops. Also see page 418.

Soil Influence

Acifluorfen is mainly applied to foliage of weeds; therefore, soil type has little influence on its herbicidal effectiveness. It is strongly adsorbed by soil and not subject to leaching. The half-life is 30 to 60 days.

Mode of Action

Acifluorfen is considered a contact herbicide since little translocation occurs, and leaf injury occurs rather quickly, often within 24–48 hours. This injury includes chlorosis and necrosis. Light is required for these injury symptoms to develop, and membrane damage is suggested. It has been reported to interfere with photosynthesis.

BIFENOX

Bifenox is the common name for methyl 5-(2,4-dichlorophenoxy)-2-nitrobenzoate. The trade name is Modown®. It is a yellow-tan solid with a water solubility of 0.35 ppm at 25°C. It is formulated as a wettable powder, flowable liquid, and emulsifiable concentrate. The acute oral LD_{50} for rats is greater than 6400 mg/kg.

Uses

Bifenox is mainly used as a preemergence application to control many broadleaf weeds and certain grasses in sorghum and soybeans. It also controls certain sedges and aquatic weeds. It is used in drilled or dry-seeded rice and forest tree nursery seedbeds. Combinations of bifenox with propachlor in sorghum, alachlor or trifluralin in soybeans, and propanil in rice are also used. The method of application with these combinations vary; the user should read the label. Also see page 418.

Soil Influence

Soil colloids hold bifenox strongly, and little leaching occurs. It is degraded primarily by microbial activity with a half-life of 7 to 14 days. Bifenox retains its herbicidal activity in soil for 6 to 8 weeks.

Table 14.1. Common Name, Chemical Name, and Chemical Structure of Several Diphenyl Ether Herbicides

Common Name	Chemical Name	Chemical Structure
Acifluorfen[1]	5-[2-Chloro-4-(trifluoromethyl) phenoxyl-2-nitrobenzoic acid	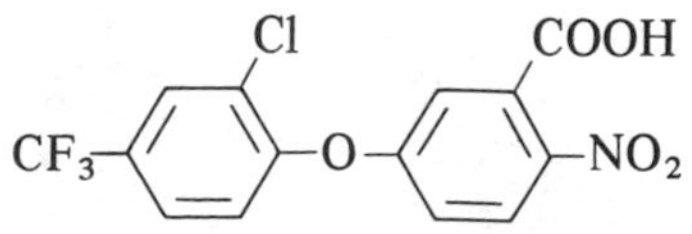
Bifenox	Methyl 5-(2, 4-dichlorophenoxy)-2-nitrobenzoate	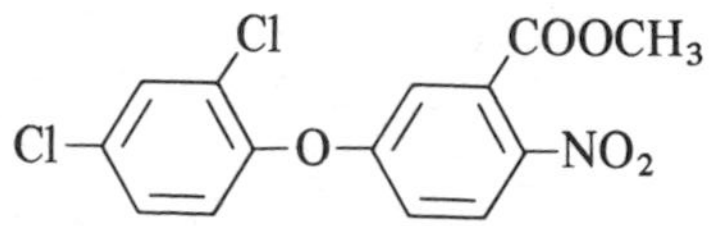
Diclofop-methyl	Methyl 2-[4-(2, 4-dichlorophenoxy) phenoxy] propanoate	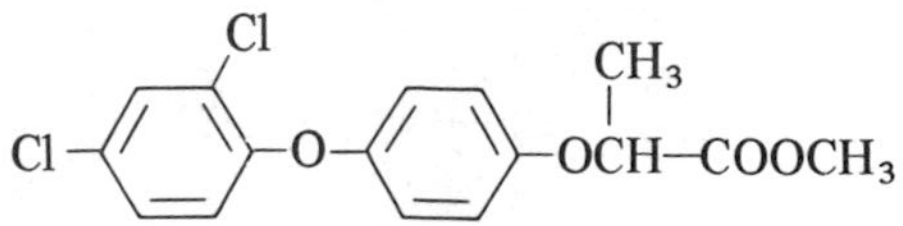

Fluorodifen	*p*-Nitrophenyl α,α,α-trifluoro-2-nitro-*p*-tolyl ether	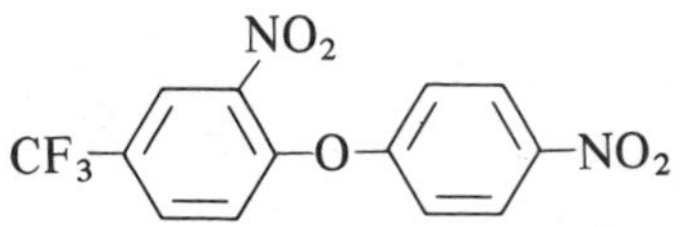
Nitrofen	2,4-Dichlorophenyl *p*-nitrophenyl ether	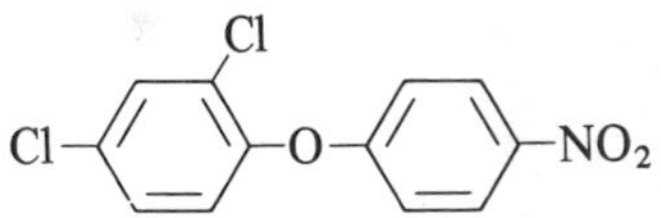
Nitrofluorfen	2-Chloro-1-(4-nitrophenoxy)-4-(trifluoromethyl) benzene	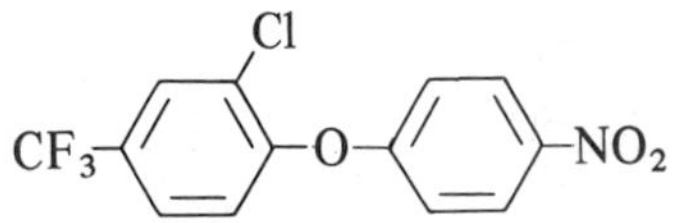
Oxyfluorfen	2-Chloro-1-(3-ethoxy-4-nitrophenoxy)-4-trifluoromethyl) benzene	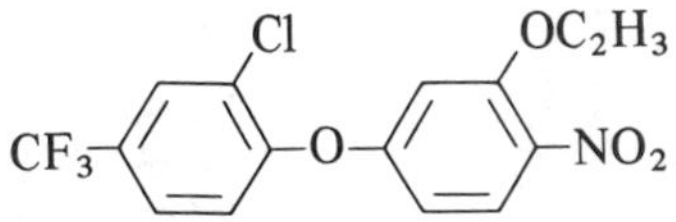

[1] Formulated as the sodium salt.

Mode of Action

Bifenox is rapidly absorbed, but little translocation occurs after leaf or root uptake. Limited research suggests that bifenox inhibits photosynthesis. Higher plants appear to degrade it by ring hydroxylation and later conjugation or binding to normal plant constituents. The rate of degradation appears to vary considerably in different species.

DICLOFOP-METHYL

Diclofop-methyl is the common name for methyl 2-[4-(2,4-dichlorophenoxy)phenoxy]propanoate. Its trade names are Hoelon®, Hoe-Grass®, Illoxan®, and Iloxan®. It is a colorless solid with a water solubility of 3000 ppm at 22°C. It is formulated as an emulsifiable concentrate. The acute oral LD_{50} of the technical product is 557–580 mg/kg for rats.

Uses

Prime use of diclofop-methyl is to control wild oats and many other annual grasses in wheat and barley by postemergence treatment. It also has controlled these weeds plus volunteer corn in soybeans. Best results are obtained when most wild oat and annual-grass plants are in the one- to three-leaf stage. For volunteer corn, diclofop-methyl should be applied after essentially all corn has emerged, but before it grows large enough to prevent thorough coverage, including the whorl. It does not control broadleaf weeds or perennial grasses.

Soil Influence

Studies have shown that diclofop-methyl does not leach downward nor move laterally in soils; therefore, it appears similar to other diphenyl ether herbicides, which are readily adsorbed by soil colloids. It dissipates relatively rapidly in soil; its half-life is 10 days in sandy soils and 30 days in sandy clay soils under aerobic conditions. Under anaerobic conditions it disappears even faster; up to 86% of the parent compound is metabolized within two days.

Diclofop-methyl appears to have little if any effect on soil microorganisms or their activity. It caused no change of nitrogen fixation or nitrification. Use of cellulose, starch, and protein by soil microorganisms is not altered.

Mode of Action

Leaves readily absorb diclofop-methyl from postemergence applications. Soil zonal placement studies show that it is more toxic when placed in the root zone than in the shoot zone of emerging seedlings. These results suggest differential absorption by these two organs. Translocation appears to be limited.

With a preemergence application, symptoms develop slowly; in wild oat and other susceptible grasses, both shoot and root growth are suppressed. This is followed by wilting and scattered chlorotic mottling; these spots unite some time later. Symptoms may differ with species and environmental conditions. In the Pacific Northwest, wild oat may show leaf-tip burn with necrosis developing toward the leaf base.

Ryegrass develops a rusty color before death. Corn undergoes little color change, but develops a weak stem near the soil surface, which causes later lodging.

According to other reports, diclofop-methyl inhibits meristematic activity. Wild oat develops excessive adventitious roots after a soil-applied treatment. But this reaction has not been seen after shoot treatment. This difference may be related to its limited translocation.

The difference in diclofop-methyl selectivity between wheat and wild oat may be based on differential metabolism between these two species. Diclofop-methyl undergoes demethylation to form diclofop, which undergoes ring hydroxylation in wheat; this hydroxylated form is nonphytotoxic. But in wild oat, diclofop-methyl forms a diclofop conjugate through an ester linkage. This reaction is reversible, and the phytotoxic diclofop is later released in the wild oat plant.

NITROFEN

Nitrofen is the common name for 2,4-dichlorophenyl-*p*-nitrophenyl ether. The trade name is TOK®. It is a dark-brown solid, essentially insoluble in water. It is formulated as an emulsifiable concentrate or wettable powder. The acute oral LD_{50} of the technical material is about 2360 mg/kg in rats.

Uses

As a preemergence herbicide, nitrofen controls annual weeds in cole crops (broccoli, brussels sprouts, cabbage, cauliflower), carrots, celery, horseradish, onions (dry bulb), parsley, and sugar beets. In all of these crops except sugar beets, it may also be applied postemergence, post-transplant, or both, to the crop to control emerged annual weeds and those that may emerge later.

Since crop tolerance varies with formulation, emulsifiable concentrate or wettable powder, the labels should be checked for the formulation to be used on each crop at various stages of growth.

Nitrofen is also used on certain ornamentals, fallow land, and noncropland. Also see page 419.

Soil Influence

Nitrofen is readily adsorbed by most soils and is not subject to leaching. It is not recommended for preemergence use on muck or high-organic-matter soils since weed control cannot be obtained with recommended dosages. Incorporation into the soil also reduces its effectiveness as a preemergence treatment. Slow but complete biodegradation to CO_2 occurs in typical soils. However, herbicidal effectiveness in soils is relatively short, perhaps in part due to binding.

Mode of Action

Foliar-applied nitrofen acts as a contact herbicide, causing chlorosis and necrosis; light is required. Soil-applied nitrofen inhibits seed germination and growth; light is not required. These facts suggest that nitrofen has two mechanisms of action.

Translocation of nitrofen is quite limited; somewhat more moves from roots to shoots than vice versa. It apparently moves largely through the apoplast.

Degradation of nitrofen in higher plants is relatively rapid. The ester linkage is cleaved and ring substitutions may be modified. Conjugates of the parent molecule, degradation products, or both, have been detected.

OXYFLUORFEN

Oxyfluorfen is the common name for 2-chloro-1-(3-ethoxy-4-nitrophenoxy)-4-(trifluoromethyl)benzene. The trade name is Goal®. It is a red-brown solid at room temperature, with a melting point 65–80°C. It has a water solubility of 0.1 ppm. It is formulated as an emulsifiable concentrate. The acute oral LD_{50} of the technical product is greater than 5000 mg/kg for rats.

Uses

Oxyfluorfen effectively controls many annual weeds by preemergence or postemergence application. It is registered for use *in California only* for almonds, apricots, nectarines, peaches, plums, prunes, and grapes. It can also be used as a tank mix with paraquat or napropamide in these crops. A three-way mix, oxyfluorfen with paraquat plus oryzalin has been used on nonbearing almonds, peaches, plums, and prunes. Oxyfluorfen also looks promising on soybeans, alone or in combination with alachlor. Oxyfluorfen alone is also used in conifer seedbeds. In North Carolina and South Carolina, this herbicide has been used as a directed spray in soybeans to control witchweed, a serious parasitic weed. Also see page 418.

Soil Influence

Like the other diphenyl ether herbicides, oxyfluorfen is strongly adsorbed on soil colloids and is not subject to leaching. It undergoes detoxification in soils during its half-life of 30 to 40 days, but microbial degradation apparently is not a major factor.

Mode of Action

Oxyfluorfen is readily absorbed by both leaves and roots but translocation from these sites is very limited. Reports to date suggest that it degrades quite slowly in higher plants. It is considered to be a contact herbicide and exposure to light is required before herbicidal symptoms develop following a foliar application.

SUGGESTED ADDITIONAL READING

Ashton, F. M., and A. S. Crafts, 1981, "Diphenyl ethers," pp. 224–235, in *Mode of Action of Herbicides,* Wiley, New York.

Matsunaka, S., 1976, "Diphenyl ethers," pp. 709–739, in P. C. Kearney and D. D. Kaufman, Eds., *Herbicides,* Vol. 2, Marcel Dekker, New York.

USDA, SEA, 1980, *Suggested Guidelines for Weed Control,* U. S. Government Printing Office, Washington, D. C. (621-220/SEA 3619).

USDA, SEA, 1981, *Compilation of Registered Uses of Herbicides.*

Weed Science Society of America, 1979, *Herbicide Handbook,* 4th Ed., WSSA, Champaign, IL.

For chemical use, see the manufacturer's label and follow the directions. Also see the Preface.

15 Nitriles

Nitriles are organic compounds containing a —C≡N group. Those used as herbicides are benzonitriles with OH^-, Cl^-, and/or Br^- substitutions on the benzene ring. Herbicidal properties of these compounds were discovered in the late 1950s and early 1960s. Dichlobenil is applied to the soil, whereas bromoxynil and ioxynil are applied to the foliage of weeds to be controlled.

DICHLOBENIL

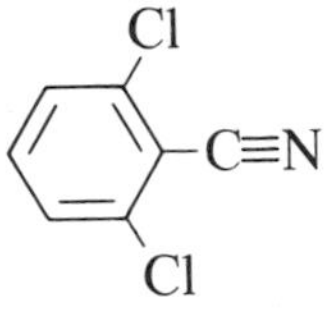

dichlobenil

Dichlobenil is the common name for 2,6-dichlorobenzonitrile. The trade name is Casoron®. It is a white crystalline solid with a water solubility of 18 ppm at 20°C. It is formulated as a wettable powder, dispersible powder, and in granular forms. Dichlobenil is also formulated with dalapon as granules and with monolinuron as a dispersible powder. Its acute oral LD_{50} is greater than 3160 mg/kg in rats.

Uses

Dichlobenil inhibits germination of seeds of both grass and broadleaf plants, but it does not control emerged weeds. It controls annual weeds in tree fruit and nut crops including almonds, apples, avocados, cherries, citrus, figs, filberts, mangoes, nectarines, peaches, pears, plums, and English walnuts. It is also used on small fruits (blackberries, blueberries, cranberries, grapes, and raspberries) as well as on many woody ornamentals. And it can be used to control aquatic weeds in ponds, reservoirs, and lakes.

In none of these crops can dichlobenil be applied immediately after transplanting. The waiting period after transplanting ranges from four weeks for most species to six months for pecans and one year for citrus. In certain crops, dichlobenil must not be applied within a specified number of days before harvest. Treated orchards should not be grazed.

Dichlobenil is relatively volatile and therefore needs to be leached into the soil by rainfall or sprinkler irrigation soon after application to prevent volatility losses. It can be mixed in by shallow incorporation. Volatility losses from granular formulations are usually less than those from the wettable powder formulation.

Dichlobenil has shown promise for control of certain perennial weeds (e.g., field bindweed) in orchards and vineyards when applied to the soil as a subsurface layer. Emerging weed shoots are not able to penetrate the concentrated layer of dichlobenil.

Soil Influence

Dichlobenil appears to be adsorbed tightly by soil colloids. This, together with its relatively low water solubility, greatly restricts its leaching. Dichlobenil is relatively persistent in soils, usually two to six months, but has been detected for as long as one year after application. As a concentrated subsurface layer, it probably persists even longer.

Mode of Action

Dichlobenil acts primarily on apical growing points and root tips. This inhibition of growth is followed by a gross disruption of tissues, mostly in the meristems and phloem. This may cause swelling or collapse of stem, root, and leaf petioles. Apical meristems and leaves may show dark discoloration (Milborrow, 1964).

Because dichlobenil is applied to soil, its rapid absorption by roots and seeds is of particular significance. However, it is also absorbed by leaves in vapor form (Massini, 1961). Translocation upward from roots to shoot in the apoplast can be rapid, but transport downward from leaves to roots is limited. Dichlobenil has been reported to greatly restrict the translocation of photosynthate from leaves to other plant parts. It is metabolized by higher plants by hydroxylation of the ring. These hydroxylation products are phytotoxic. Although dichlobenil appears to have little effect on electron transport and phosphorylation in chloroplasts and mitochondria, the hydroxy degradation products are strong inhibitors of these reactions.

BROMOXYNIL

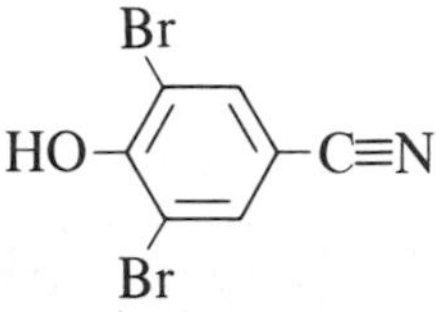

bromoxynil

Bromoxynil is the common name for 3,5-dibromo-4-hydroxybenzonitrile. The trade names are Brominal® and Buctril®. It is a light-buff solid with a water solubility of less than 0.02%, formulated in liquid form as the octanoic acid ester. It is also formulated in combination with MCPA. The acute oral LD_{50} of bromoxynil for adult rats is 440 mg/kg.

Uses

Bromoxynil is used on barley, oats, wheat, flax, and established grasses for seed. It is applied as a postemergence spray to control annual broadleaf weeds not readily controlled by 2,4-D or MCPA (e.g., fiddleneck, resistant mustards, and wild buckwheat). It is applied when the grain is in the two-leaf to boot stage. It is also used on noncropland. It is more effective on small weeds than on large weeds.

Bromoxynil is often mixed with MCPA to control a greater variety of broadleaf weeds, but this combination must be applied at the two- to four-leaf stage of the grain. It is also combined with dicamba or difenzoquat for use in wheat. Also see page 418.

Soil Influence

Since bromoxynil is applied to leaves, little research has been conducted on the influence of soil on its performance. However, recent observations suggest that it may also be active through the soil.

Mode of Action

The contact action of bromoxynil appears as blistered or necrotic spots on leaves within 24 hr. Later, extensive leaf tissue is destroyed, and the plant dies. Bromoxynil is both a photosynthetic and respiratory inhibitor, readily absorbed by leaves, but with limited translocation. Therefore, complete coverage of the foliage is essential to good weed control.

It is metabolized in higher plants by undergoing ring hydroxylation and probably later conjugation.

IOXYNIL

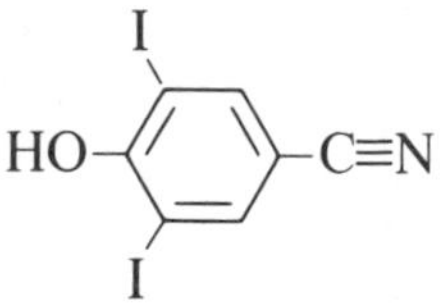

ioxynil

Ioxynil is the common name for 4-hydroxy-3,5-diiodobenzonitrile. The trade names are Actrilawn® and Totril®. It is a light-buff solid with a water solubility of 130 ppm at 25°C. The acute oral LD_{50} for rats is 110 mg/kg. Ioxynil is similar to bromoxynil and is often used in combination with other herbicides. Since it is not marketed in the United States, additional details are not given.

SUGGESTED ADDITIONAL READING

Ashton, F. M., and A. S. Crafts, 1981, "Nitriles," pp. 254–271, in *Mode of Action of Herbicides,* Wiley, New York.

Frear D. S., 1976, "The benzoic acid herbicides," pp. 541–607, in P. C. Kearney and D. D. Kaufman, Eds., *Herbicides,* Vol. 2, Marcel Dekker, New York.

Massini, P., 1961, *Weed Res.* **1**, 142.

Milborrow, B. V., 1964, *J. Exp. Bot.* **15**, 515.

USDA, SEA, 1980, *Suggested Guidelines for Weed Control,* U.S. Government Printing Office, Washington, D.C. (621-220/SEA 3619).

USDA, SEA, 1981, *Compilation of Registered Uses of Herbicides.*

Weed Science Society of America, 1979, *Herbicide Handbook,* 4th ed., WSSA, Champaign, IL.

For chemical use, see the manufacturer's label and follow the directions. Also see the Preface.

16 Phenoxys

Phenoxy compounds have the phenyl ring attached to an oxygen, which in turn is attached to an aliphatic acid. Using the phenyl ring, oxygen, and butyric acid, this would be represented by the structural formula of:

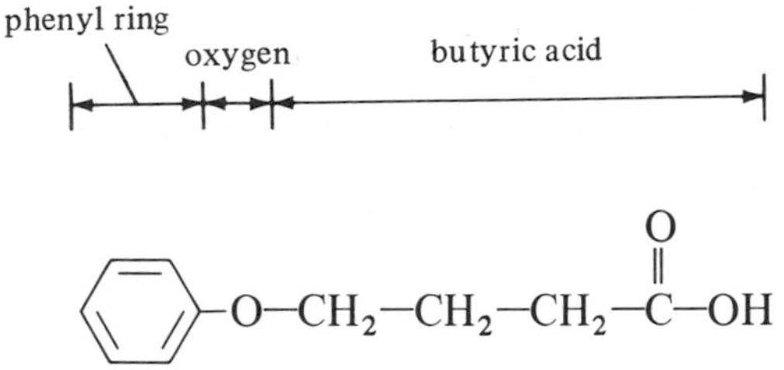

The butyric acid, in this case, represents the aliphatic acid portion of the molecule. The aliphatic acid with one carbon atom is named formic acid; two carbons, acetic acid; three carbons, propionic acid; four carbons, butyric acid; and so on. The longest continuous chain of carbon atoms is usually selected to determine the name.

Some examples of phenoxy compounds are 2,4-D, MCPA, 2,4-DB, dichlorprop, 2,4,5-T, and silvex. Each will be discussed, with the major emphasis on 2,4-D. Scientists believe that most phenoxy herbicides have effects similar to those of 2,4-D.

Because of World War II security regulations, research on selective herbicides was not reported as it progressed. Technical papers were often published several years after the work was actually done. See *Botanical Gazette,* volume **107** (1946), for several of these papers. In England, Slade et al. (1945) of Imperial Chemical Industries, Ltd., reported that they used α-naphthaleneacetic acid to control yellow charlock (wild mustard) with slight injury to oats. They later used MCPA without injury.

2,4-D

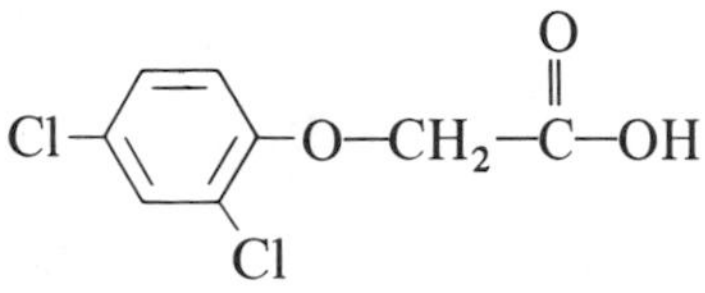

2,4-D or (2,4-dichlorophenoxy)acetic acid

2,4-D is the common name for (2,4-dichlorophenoxy)acetic acid. There are numerous trade names. It is a white crystalline solid with a water solubility of about 600 ppm. Its salts (sodium, lithium, amine), however, are quite soluble in water, and the acute oral LD_{50} of its various formulations range from 300 to 1000 mg/kg.

The first reference to 2,4-D in the literature is an article by Pokorny in 1941. He outlined a very simple method of preparing 2,4-D but made no reference to its use. In 1942 Zimmerman and Hitchcock of Boyce Thompson Institute first described the use of 2,4-D as a plant-growth regulator. In 1944 Marth and Mitchell of the U.S. Department of Agriculture reported that 2,4-D killed dandelion, plantain, and other weeds from a bluegrass lawn, and Hamner and Tukey (1944) described successful field trials using 2,4-D as a herbicide. An excellent paper on the discovery and development of 2,4-D has been written by Peterson (1967).

Common Forms

2,4-D is formulated as an emulsifiable acid, amine salts, mineral salts, and esters. The esters are oil-soluble and emulsifiable with water. The concentration of essentially all phenoxy herbicide formulations is expressed as *acid equivalents* in pounds per gallon. Recommendations are also made on this basis. *Acid equivalent* refers to that part of a formulation that theoretically can be converted to the acid.

Acid

The acid form of 2,4-D is only slightly soluble in water, and the emulsifiable concentrate is relatively expensive. Thus the acid form is not commonly used in commercial formulations. However, at least one emulsifiable concentrate formulation of 2,4-D acid is available. It is used mostly to control hard-to-kill perennial broadleaved weeds such as field bindweed, Canada thistle, and Russian knapweed.

Amines

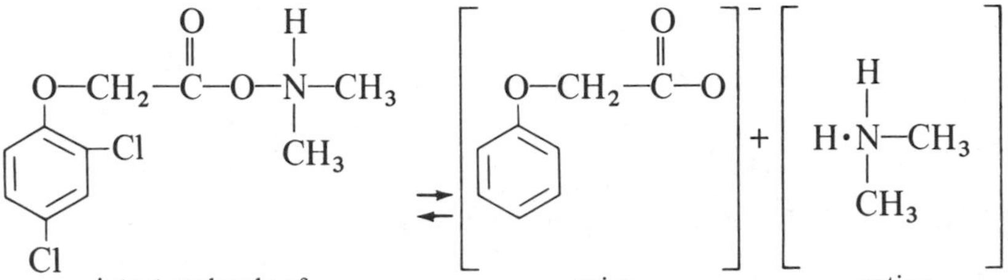

intact molecule of dimethylamine salt of 2,4-D

anion cation

anion and cation of ionized molecule of dimethylamine salt of 2,4-D

Amine salts of 2,4-D are the most commonly used forms of 2,4-D. The dimethylamine salt and its ionization in water is shown above. A mixture of alkanolamine salts (of the ethanol and isopropanol series) is the most widely used, but other amine salts are also available. The dimethylamine salt of 2,4-D is a white crystalline solid. The amine-salt formulations of 2,4-D are usually liquids and represent no volatility hazard to sensitive plants.

Amines are soluble in all proportions in water and therefore are well adapted to low-gallonage spray equipment, even at high rates of application. Most amine salts of 2,4-D dissolved in water form a clear solution, though it may be colored. Most amine salts are not soluble in petroleum oils.

An exception, however, is the *N*-oleyl-1,2-propylenediamine salt of 2,4-D; it is oil-soluble and is essentially insoluble in water.

Esters

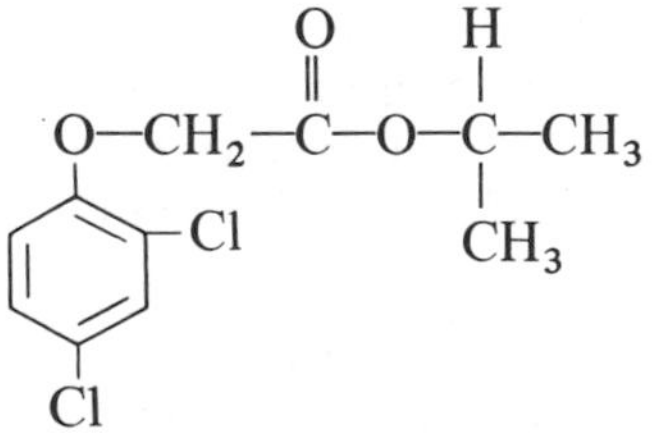

isopropyl ester of 2,4-D (a volatile form)

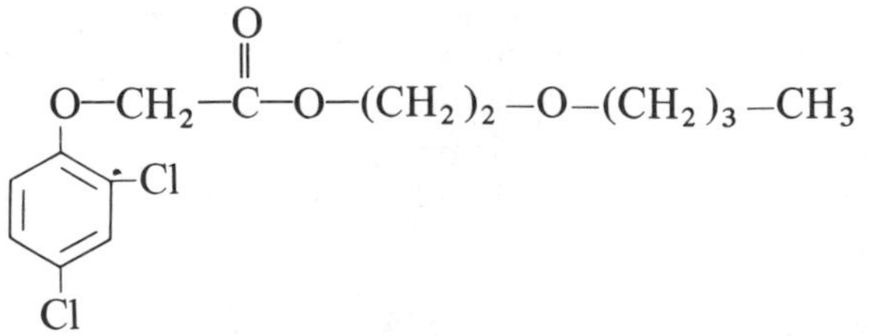

butoxyethyl ester of 2,4-D (a low volatile form)

Esters of 2,4-D are colorless liquids. In contrast to amines, esters are essentially insoluble in water. Esters of 2,4-D are synthesized by a reaction between 2,4-D acid and an alcohol. One molecule of water is eliminated. The alkyl group of the alcohol replaces the hydrogen of the carboxyl group of the 2,4-D acid.

The 2,4-D ester is identified by the name of the alcohol used. The cheaper and more abundant alcohols like methyl (one carbon) ethyl (two carbons), isopropyl (three carbons), and butyl (four carbons) are no longer used because of their high volatility. However, the long-chain alcohols with an ether linkage (—O—) have a lower volatility hazard than the short-chain alcohols when used to formulate 2,4-D. Therefore, long-chain alcohols, even though more expensive, have assumed increasing importance (see the section on volatility in this chapter).

2,4-D ester compounds are only slightly soluble in water, but they are soluble in some petroleum oils. The ester is usually slightly diluted in oil, after which an emulsifying agent is added. Esters are usually sold as a liquid. When mixed with water, the emulsifying agent keeps the tiny, oil-like droplets suspended for a time, much the way butterfat is suspended in milk. Water is the continuous phase and oil-like droplets are dispersed; thus it is an oil-in-water (O/W) type of emulsion.

When mixed with water, the 2,4-D emulsion appears milky. If properly emulsified, the ester can be mixed in all proportions with water, but if allowed to stand, the oil-like droplets may separate. Mixing will reform the emulsion. Once applied, the oil-like droplet is not easily washed from plants by rain. The oil-like characteristics make it difficult to remove ester forms from spray equipment.

So far we have considered an O/W emulsion. But if the oil is the continuous phase and the water the dispersed phase, the emulsion is a water-in-oil (W/O) type. This is often referred to as an invert emulsion. (For further details, see Chapter 6.)

Ester formulations of 2,4-D are generally considered the most toxic to plants. There are at least three possible explanations: (1) Volatility permits absorption of the gases through the stomates, (2) wetting action of the oil-like ester and the oil carrier may actually aid penetration of the stomates, and (3) ester forms, with their low polarity, are compatible with the cuticle and aid penetration directly through the cuticle.

Because of its greater toxicity, the ester form is often more effective on resistant species, especially on woody plants. It is also more likely to injure crop plants when excessive amounts are applied. Lower rates of 2,4-D esters—about two-thirds as much—than of salts are usually suggested for postemergence application to crop plants.

Mineral Salts

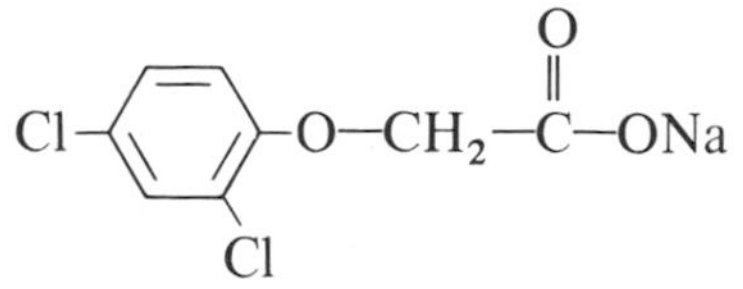

sodium salt of 2,4-D

If sodium replaced the hydrogen atom of the carboxyl group of 2,4-D acid, the sodium salt of 2,4-D is produced. It is a white crystalline solid. Lithium, potassium, and ammonium salts of 2,4-D have been manufactured but are no longer used, and the sodium salt has only a limited use.

The sodium salt is sold as a water-soluble powder. Some stirring or mixing is usually required to get it into solution. It is not soluble enough in water to be applied in low-gallonage equipment at high rates of application. In most other respects it resembles the amine salts. The amine salts have gradually replaced other salts because the amines are more easily dissolved in water.

Precipitate Formation

When a salt of 2,4-D is dissolved in water, some of the molecules are dissociated into ions, which are then free to combine with other ions in the solution. This creates a problem when the water is "hard," having a high calcium or magnesium content. The calcium and magnesium salts of 2,4-D are only slightly soluble in water–the calcium salt at 2.5 g/liter and the magnesium salt at 17.4 g, both at a temperature of 20°C. Thus if calcium or magnesium ions are present in the solution, a precipitate may be formed that will clog the filters and nozzles (Zussman, 1949).

Esters cause little or no clogging in hard water. The 2,4-D ester is not in solution nor is it ionized. It is emulsified in the water with the oil-like droplets dispersed or suspended in the water. 2,4-D ester droplets are "insulated" from the water by an emulsifying agent; therefore, there is little opportunity for a reaction to take place and a precipitate to form. (For further details, see Chapter 6.)

Mixtures of chemicals may form complex molecules resulting in a precipitate. Before a mixture is placed in the sprayer, it should be made in a small container in the exact proportions to be used. Then it should be watched over a period of time. If no precipate forms then, it is likely that none will form in the sprayer. This precaution may save the user from the job of removing a congealed mass from the sprayer or the difficult task of cleaning clogged filters and nozzles.

Volatility

Vapors of 2,4-D esters may kill or injure susceptible plants, and where the vapors are confined, they may inhibit the germination of seed (see Figure 6-9). The amount of fumes or vapors given off is related to the vapor pressure of the chemical (Mullison and Humner, 1949); 2,4-D acid, amine salts, and sodium salts have very low volatility characteristics and cause little or no volatility hazard (see Figure 6-8 and Table 16-1).

Table 16-1. General Characteristics of Different Forms of 2,4-D

Form of 2,4-D	Soluble in Water	Soluble in Oil	Appearance When Mixed With Water	Precipitates Formed in Hard Water	Volatility[1]	General Remarks
Acid	No	No	Milky	Yes	No hazard	Only for specialized use
Amine salt	In all proportions	Not soluble, usually	Clear	Yes	No hazard	Good for general farm use, lawns, turf, some woody plants
Sodium salt	Medium solubility	Not soluble	Clear	Yes	No hazard	Only for specialized use
Esters						
Volatile forms	No, but can be emulsified	Yes	Milky	No	Volatile	Dangerous to use near susceptible crops due to volatility
Low-volatile forms	No, but can be emulsified	Yes	Milky	No	Medium to low volatility	Some danger from volatility if used near susceptible plants, especially with high temperature

[1] The tendency to form volatile fumes or gases that can injure plants.

As previously discussed under the section on 2,4-D esters, high-molecular-weight alcohols are used to make 2,4-D esters with reduced volatility. The butoxyethyl ester and the isooctyl ester (eight carbons) are both low-volatile esters. The high-molecular-weight alcohols are relatively expensive; therefore, their 2,4-D esters are more expensive than the volatile formulations.

It should be emphasized that the above esters are *low-volatile* and not *nonvolatile.* Under hot, humid conditions they can volatilize enough to injure susceptible plants.

Uses

2,4-D is used to control annual and perennial broadleaf weeds on noncroplands as well as in tolerant crops. It is applied to the foliage or stem (or both) of the plant to be controlled. At rates above 1 lb/acre it may serve as a soil-applied preemergence herbicide, with the effects lasting for 30 days or less. The usual rate is 0.5–2 lb/acre; however, for woody-plant control, rates as high as 6 lb/acre are used. At high rates, it is likely that plants will be controlled through both root and foliar absorption.

The highly versatile 2,4-D is used on barley, corn, oats, pastures, rangeland, rice, rye, sorghum, sugarcane, and wheat, as well as for aquatic-weed control. In low concentrations it is used as a growth regulator to reduce fruit drop, increase fruit size, and increase storage life in certain citrus crops. For woody-plant control it is often combined with 2,4,5-T.

The numerous details of timing, method of application, and rates are discussed in later chapters on specific crops. Also see page 418.

Soil Influence

Soil type and formulation of 2,4-D influences its leaching in soil. Water-soluble forms leach more readily than those that are slightly soluble. It is adsorbed by soil colloids, and less leaching occurs in clay and organic soils than in sandy soils.

Chapter 5 covers factors affecting the length of time that 2,4-D may remain toxic in the soil. Of major importance is the decomposition by microorganisms, the growth and activity of which is affected principally by food supply, soil temperature, soil moisture, and soil aeration. Low rates of 2,4-D will normally be decomposed in one to four weeks in a warm, moist loam soil. There is no risk of the chemical accumulating in the soil from one year to the next under such conditions. In very dry soils or in frozen soils, the rate of decomposition may be inhibited considerably.

At normal dosages, 2,4-D does not generally reduce the total number of microorganisms in the soil. When heavy rates of treatment are used, however, some may be inhibited and others stimulated. Aerobic microorganisms are more sensitive, and 2,4-D may seriously inhibit their growth. Anaerobic organisms may not be significantly affected, and some facultative anaerobic organisms may even be stimulated. Therefore, those organisms requiring free oxygen for respiration may be hindered by 2,4-D, whereas those not requiring free oxygen may actually be stimulated.

The effect of 2,4-D on legume nodulation is of vital importance to the vigor of the legume plant. Minute rates of 2,4-D slow down the formation of nodules on the bean plant (Payne and Fults, 1947), but these plants are far more sensitive to 2,4-D than are the rhizobia bacteria (Carrol, 1952). In the soil solution, 0.21 lb of 2,4-D per acre seriously restricted the germination and growth of beans, peas, red clover, and alfalfa. In the latter study as well, rhizobia were grown independently on culture media to learn the effects of 2,4-D on the organisms. The bacteria grew almost normally until equivalent test rates of 2,4-D reached 200 lb/acre. Hence the lessened nodulation associated with 2,4-D is primarily a plant response.

Mode of Action

Plant Structure

Twisting and curvature (epinasty) are among the most obvious effects of 2,4-D treatment on broadleaf plants. These plants generally develop grotesque and malformed leaves, stems, and roots when treated with 2,4-D (see Figures 16-1, 16-2, 16-3, and 16-4). The chemical appears to concentrate in young embryonic or meristematic tissues that are growing rapidly. It affects these tissues more than more-mature or relatively inactive young tissues.

Histological studies with the kidney bean showed that the cambium, endodermis, embryonic pericycle, phloem parenchyma, and phloem rays all showed active cell division; the cortex and xylem parenchyma exhibited little response to 2,4-D treatment; and the epidermis, pith, mature xylem, sieve tubes, and the differentiated pericycle gave no response (Swanson, 1946). The types of tissue affected in field bindweed and sow thistle (Tukey et al., 1946) were much the same as those affected in the bean.

In further studies on the bean leaf, it was shown that the malformed leaves resulted from continued growth of the apical meristem and failure of the lateral meristem derivatives to develop normally. Such growth reduced expansion of the areas between the veins so the veins became elongated and grew abnormally close together (see Figure 16-3). Buds

Figure 16-1. Scientists studying the selective killing of giant ragweed in corn in Henderson County, Kentucky, in 1947. This was one of the first large-scale tests made with 2,4-D in the United States. (Sherwin-Williams Co.)

Figure 16-2. A common burdock plant twisted and curled following treatment with 2,4-D.

Figure 16-3. Bean leaves showing the effect of 2,4-D. The leaf that is second from the right is normal.

Figure 16-4. Abnormal corn brace roots as a result of an excess of 2,4-D during a susceptible stage of growth. Corn yields were not reduced.

that were undergoing rapid differentiation, and thus were in a state of great physiological activity when treated, showed the greatest effects (Zussman, 1949). Similar effects have been noted in grapes and cotton (Dunlap, 1948). (See Figure 4-7.)

Most adventitious roots develop from the pericycle, and thus if the pericycle of the root is developing rapidly when treated with 2,4-D, the number and structure of the new roots may be changed. Proliferation of the brace roots of corn occurred when 1–$1\frac{1}{2}$ lb of 2,4-D per acre were applied in the eight-leaf stage (1 ft high). When the corn root had outgrown the meristematic stage, similar applications did not produce malformed brace roots (Rodgers, 1952) (see Figure 16-4).

Absorption

Plant roots readily absorb 2,4-D, probably absorbing polar forms (salts) most readily. The leaves most readily absorb nonpolar forms of 2,4-D (the acid and ester forms), whereas the salt formulations are absorbed more slowly. The use of surfactants increases the rate of foliar absorption. Chapters 4 and 6 give more details.

The length of time that 2,4-D must be on a plant prior to a rain varies with the formulation of 2,4-D, the rate of 2,4-D application, and the temperature, humidity, and the susceptibility of the plants. In most cases, however, a rain-free period of 6–12 hr is adequate for effective weed control. The esters, being oil-like, have a tendency to resist washing from the plant, even that not absorbed.

Translocation

Translocation of a herbicide applied to the foliage is essential if plant roots are to be killed. This is especially important for control of perennial weeds. (The general principles of translocation are the same as those discussed in Chapter 4.) After 2,4-D migrates through the leaf cuticle, it moves to the phloem; it is then moved up and down through the phloem with the photosynthate. Little translocation from the leaves takes place if the readily available food supply has been diminished by continued darkness or reduced light.

Because 2,4-D moves through the phloem with the photosynthate, the treatment of rapidly developing leaves of a perennial weed in the early spring causes little or no translocation of 2,4-D to the roots. Also, excessive rates of application would kill the living phloem cells, stopping translocation to the roots.

Perennial weeds are treated most effectively when large amounts of foods are being translocated to the roots, such as late spring or early fall. Furthermore, low rates of chemical application when applied repeatedly may give better perennial-weed control than a single heavy application. The above assumptions have proven to be generally valid, as tested through research and practical applications. The rates of chemical application and timing of treatments are different for different species.

Translocation of 2,4-D from the soil upward follows the transpiration stream, with the movement of water and soil nutrients through the xylem. Translocation up or down is favored by sufficient soil moisture to favor rapid plant growth. Dry soils may slow translocation.

Many techniques are used to trace translocation through the plant. Radiolabeled herbicide molecules have proven especially valuable in translocation studies (Yamaguchi and Crafts, 1958).

Maturity of the Plant

Maturity of the plant directly influences its susceptibility to 2,4-D. In general, all plants are most susceptible during the time of seed germination. Even the grasses are affected by low concentrations of 2,4-D at this time. The plant gains tolerance with age; some are tolerant while still small, and others never gain more than a slight tolerance for 2,4-D.

Some plants may develop a second period of susceptibility. This usually coincides with a period of rapid growth. At this time the meristems are metabolically active and very susceptible to 2,4-D. Small grains, therefore, may be very susceptible to 2,4-D in the germinating and small-seedling stages. The plant becomes tolerant in the fully tillered stage, becomes susceptible again in the jointing and heading stage, and is very tolerant as the grain reaches the "soft-dough" stage (see Figure 22-3).

Molecular Fate

The degradation of the phenoxy herbicides by higher plants has been reviewed (Loos, 1975). 2,4-D is degraded to nonphytotoxic forms in higher plants, undergoing decarboxylation and demethylation of the side chain as well as dechlorination and hydroxylation of the ring. 2,4-D also forms conjugates with glucose and certain amino acids; 2,4-D metabolites also conjugate with glucose. The resistance of certain species to 2,4-D has been in part attributed to their ability to rapidly degrade 2,4-D to nontoxic molecules.

Mechanism of Action

2,4-D does not appear to act as a simple inhibitor. Although certain enzymes can be inhibited *in vitro* by 2,4-D, there is no firm evidence that it acts *in vivo* by directly interfering with intermediary metabolism, respiration, or photosynthesis. It appears to be acting as an auxin, but accumulates to higher concentrations than the native auxin, indoleacetic acid, because it is degraded more slowly. Both inhibition and promotion of growth are involved when susceptible plants respond to 2,4-D, depending on the organ and tissue examined, and the amount of 2,4-D in them.

Two plant processes are most relevant to the mechanism of action of the phenoxy herbicides–judging by growth responses of plants. These processes are nucleic acid metabolism and cell-wall plasticity. Low levels of 2,4-D stimulate RNA polymerase, which results in an increase in RNA and protein synthesis. Low levels of 2,4-D also induce cell enlargement; the herbicide increases the activity of certain enzymes responsible for loosening cell walls and the formation of new cell-wall material. Turgor pressure per se causes actual cell enlargement. Abnormal stimulation of these processes by low levels of 2,4-D leads to uncontrolled growth.

High levels of 2,4-D act just the opposite–they inhibit these processes and growth.

The level of 2,4-D in the intact plant is low at first and increases with time to a high level. Thus at first a stimulation of the above processes cause uncontrolled growth, and later these are inhibited.

Hanson and Slife (1969) proposed that the immediate cause of death is physiological disfunction of the plant brought about by abnormal growth. In turn, the abnormal growth is believed to be based on an abnormal nucleic acid metabolism.

MCPA

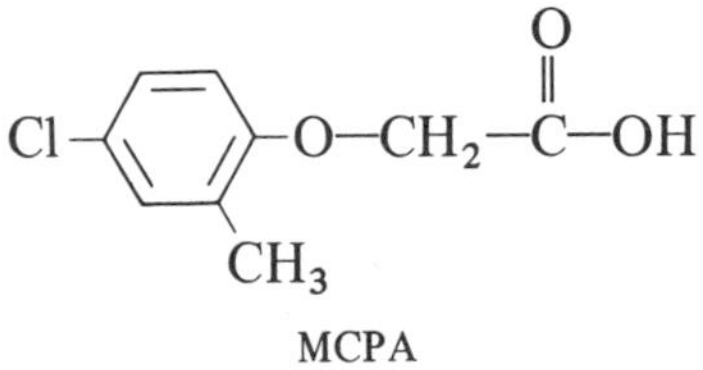

MCPA

MCPA is the common name for [(4-chloro-*o*-tolyl)oxy] acetic acid. Another name that has been used is (2-methyl-4-chlorophenoxy)acetic acid. It has several trade names. MCPA is identical to 2,4-D except it has a methyl ($—CH_3$) group at the number 2 position of the ring instead of a chlorine atom. It is a light-brown solid, essentially insoluble in water. The acute oral LD_{50} is about 700 mg/kg.

Its herbicidal properties, length of residual toxicity in the soil, and toxicity to man and animals are similar to 2,4-D. It can be formulated as the acid, salt, and ester forms much the same as 2,4-D.

MCPA was one of the first hormonelike herbicides tested in England. It still is an important herbicide in England, Sweden, and other northern European countries. It remains a specialized chemical in the United States, used on a relatively small scale.

MCPA is less injurious to crops such as oats and rice, and it is more effective on some weed species. The reverse is also true. In the northern European countries, MCPA is used for weed control in small grains. It has also been used for weed control in small-seeded legumes, small grains, pastures, flax, and peas. In the United States, MCPA has remained somewhat more expensive to use than 2,4-D, with a higher cost per pound of active ingredient, and larger quantities are usually needed for effective weed control. Also see page 419.

2,4-DB

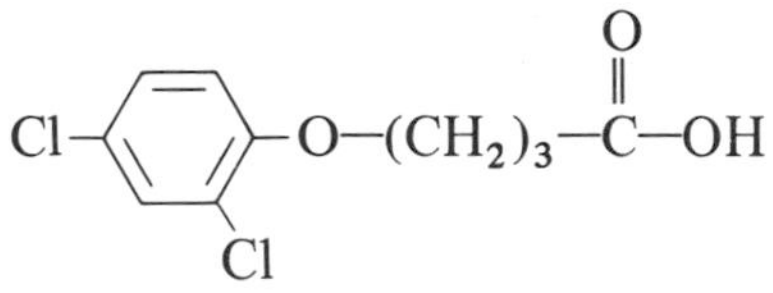

2,4-DB

2,4-DB is the common name for 4-(2,4-dichlorophenoxy)butyric acid. The trade names are Butoxone® and Butyrac®. It is a white crystalline solid, practically insoluble in water. It is formulated as amine salts and low-volatile esters. The amine salts are water-soluble and nonvolatile; the esters are oil-soluble and usually emulsifiable in water.

2,4-DB has been particularly effective (1) as a postemergence treatment to seedlings of small-seeded legumes to control small broadleaf weeds or (2) in established legumes before flowering. It has also been used to control cocklebur (less than 3 in. high) in soybeans when the beans are 8–12 in. high. Also see page 418.

2,4-DB is not highly phytotoxic per se; however, it undergoes beta-oxidation in plants and soils to form 2,4-D. Some plants are able to make this conversion rapidly whereas in others (e.g., small-seeded legumes) this reaction takes place slowly. Thus most broadleaf weeds are controlled by 2,4-DB, whereas the small-seeded legumes are not injured.

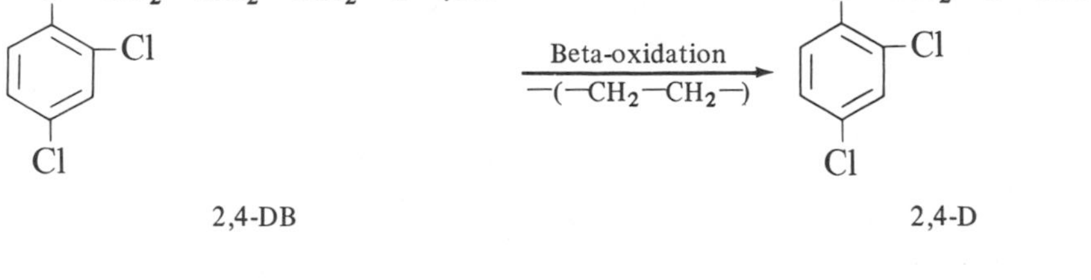

2,4-DB 2,4-D

DICHLORPROP

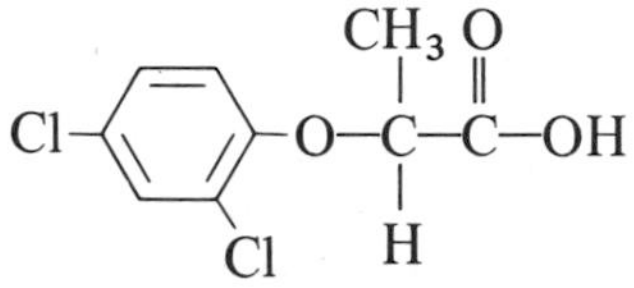

dichlorprop

Dichlorprop is the common name for 2-(2,4-dichlorophenoxy)propionic acid. It has also been referred to as 2,4-DP. It has several trade names. Dichlorprop varies from a white to tan color, and is a crystalline solid with a water solubility of 710 ppm. Although it could be formulated in various salts and esters common to 2,4-D, the available commercial formulations are usually low-volatile esters. The actue oral LD_{50} is 800 mg/kg.

In the United States, the use of dichlorprop is limited to noncropland. It has been primarily used for woody-plant control. It may be used in combination with 2,4-D to control 2,4-D resistant species. See page 418.

2,4,5-T

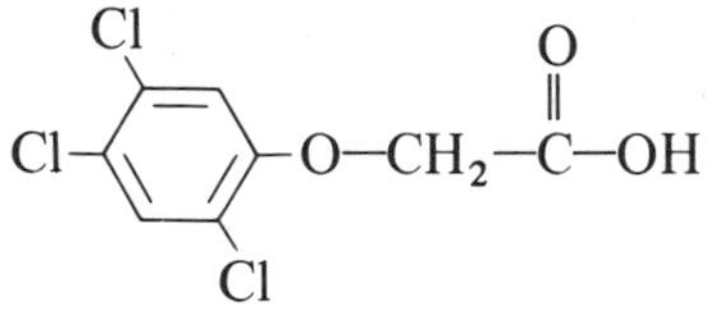

(2,4,5-trichlorophenoxy)acetic acid

2,4,5-T is the common name for (2,4,5-trichlorophenoxy)acetic acid. It has several trade names. It is identical with 2,4-D except for an additional chlorine atom at the number 5 position of the ring. A white solid with a water solubility of 238 ppm, it is usually formulated as the amine salts and esters, much the same as 2,4-D. Amine salts of 2,4,5-T are water-soluble and nonvolatile. Esters are oil-soluble and usually emulsifiable in water; esters are classified as either volatile or low-volatile.

2,4,5-T is effective on many woody species that are resistant to 2,4-D, but is less effective on many other plants. Therefore, it is prudent to consider relative susceptibility and cost in selecting between the two herbicides.

Mixtures of 2,4-D and 2,4,5-T are frequently sold as "brush killers." These are especially effective on a mixture of brush species. 2,4-D may be more effective on some species and 2,4,5-T on others. 2,4,5-T has also been used as a plant-growth regulator for preharvest fruit-drop control in apples.

If temperatures rise too high during synthesis of 2,4,5-T, a dioxin material is formed. This dioxin is 2,3,7,8-tetrachlorodibenzoparadioxin. It is regarded as a potential carcinogenic and teratogenic toxicant. The use of 2,4,5-T (containing as much as 28 ppm of dioxin) as a defoliant during the war in Vietnam aroused differences of opinion over the continued use of 2,4,5-T.

With effective manufacturing control procedures, 2,4,5-T can be manufactured essentially free of dioxin (less than 0.1 ppm). Free of the dioxin, and used according to labeled instructions approved by the Environmental Protection Agency, 2,4,5-T is not considered a human health hazard.

SILVEX

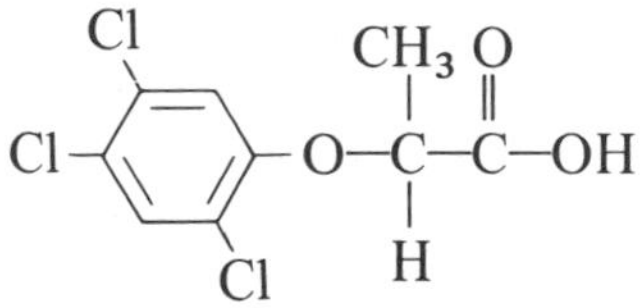

silvex

Silvex is the common name for 2-(2,4,5-trichlorophenoxy)propionic acid. It has also been referred to as 2,4,5-TP. There are several trade names. It is a white solid with a water solubility of about 180 ppm. It is usually formulated into various amine salts and esters. These forms have physical and chemical characteristics similar to their 2,4-D equivalents; thus salts of silvex are water-soluble and have very low volatility characteristics. Esters are oil-soluble and are usually emulsifiable in water; esters also are classed as either volatile or low-volatile.

The chemical controls some plant species that are resistant to both 2,4-D and 2,4,5-T. Some species especially susceptible to silvex are chickweed, henbit, wild strawberry, and a number of the oaks and maples. 2,4,5-TP has shown considerable promise against some aquatic weeds, and it has also been used as a plant-growth regulator for preharvest-drop control in apples and pears.

SUGGESTED ADDITIONAL READING

Ashton, F. M., and A. S. Crafts, 1981, "Phenoxys," pp. 272–302, in *Mode of Action of Herbicides*, Wiley, New York.

Carlyle, R. E., and J. D. Thorpe, 1947, *J. Amer. Soc. Agron.* **39**, 929.

Carrol, R. B., 1952, *Contr. Boyce Thompson Inst.* **16**, 409.

Dunlap, A. A., 1948, *Phytopath.* **38**, 638.

Hamner, C. L., and H. B. Tukey, 1944, *Bot. Gaz.* **106**, 232.

Hanson, J. B., and F. W. Slife, 1969, *Residue Rev.* **25**, 59.

Loos, M. A., 1975, "Phenoxyalkanoic acids," pp. 1-128, in P. C. Kearney and D. D. Kaufman, Eds., *Herbicides,* Vol. 1, Marcel Dekker, New York.

Marth, P. C., and J. W. Mitchell, 1944, *Bot. Gaz.* **106**, 224.

Mullison, W. R., and R. W. Humner, 1949, *Bot. Gaz.* **111**, 77.

Payne, M. G., and J. L. Fults, 1947, *J. Amer. Soc. Agron.* **39**, 52.

Peterson, G. E., 1967, *Agric. History* **41**, 243.

Penner, D., and F. M. Ashton, 1966, *Residue Rev.* **14**, 39.

Pokorny, R., 1941, *J. Amer. Chem. Soc.* **63**, 1768.

Rodgers, E. G., 1952, *Plant Physiol.* **27**, 153.

Slade, R. E., W. G. Templeman, and W. A. Sexton, 1945, *Nature* **155**, 497.

Swanson, C. P., 1946, *Bot. Gaz.* **107**, 522.

Tukey, H. B., C. L. Hamner, and B. Imkoffe, 1946, *Bot. Gaz.* **107**, 62.

USDA, SEA, 1980, *Suggested Guidelines for Weed Control,* U.S. Government Printing Office, Washington, D.C. (621-220/SEA 3619).

USDA, SEA, 1981, *Compilation of Registered Uses of Herbicides.*

Weed Science Society of America, 1979, *Herbicide Handbook,* 4th Ed., WSSA, Champaign, IL.

Yamaguchi, S., and Crafts, 1958, *Hilgardia* **28**, 161.

Zimmerman, P. W., and A. E. Hitchcock, 1942, *Contr. Boyce Thompson Inst.* **12**, 321.

Zussman, H. W., 1949, *Agric. Chem.* **4**, 27–29, 73.

For chemical use, see the manufacturer's label and follow the directions. Also see the Preface.

17 Thiocarbamates

Thiocarbamate herbicides include butylate, cycloate, diallate, EPTC, molinate, pebulate, thiobencarb, triallate, and vernolate. Their common names, chemical names, and chemical structures are given in Table 17-1. They are derivatives of carbamic acid with one of the oxygen atoms replaced by a sulfur atom as well as other substitutions. CDEC and metham are *di*thiocarbamates with both oxygen atoms of carbamic acid substituted by sulfur atoms. All thiocarbamate herbicides were developed by Stauffer Chemical Company, except diallate and triallate from Monsanto.

Most thiocarbamate herbicides are relatively volatile. If not immediately mixed into the soil by tillage equipment or applied in irrigation water, much of the applied herbicide will be lost. They are primarily used as selective herbicides in a wide variety of crops.

BUTYLATE

Butylate is the common name for *S*-ethyl diisobutylthiocarbamate, with the trade name Sutan®. This amber liquid has a water solubility of 44 ppm at 22°C and is relatively volatile. It is formulated as an emulsifiable concentrate as well as granular forms. All formulations also contain a chemical compound (R-25788) that reduces the toxicity of butylate to corn, but not to weeds. These fortified forms are trade named Sutan®+. R-25788, *N,N*-diallyl-2,2-dichloroacetamide, is discussed later. The acute oral LD_{50} of technical butylate is 4659 and 5431 mg/kg for male and female rats, respectively.

Uses

Butylate is used as a preplant, soil-incorporation treatment to control annual weeds, especially grasses, in corn (field, sweet, pop, and silage).

Table 17-1. Common Name, Chemical Name, and Chemical Structure of the Thiocarbamate Herbicides

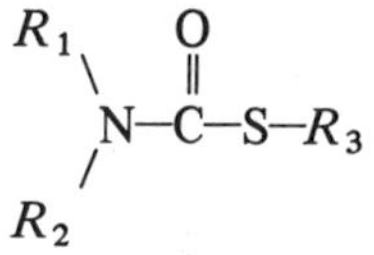

Common Name	Chemical Name	Chemical Structure		
		R_1	R_2	R_3
Butylate	*S*-Ethyl diisobutylthiocarbamate	$CH_3-CH(CH_3)-CH_2-$	$CH_3-CH(CH_3)-CH_2-$	C_2H_5-
Cycloate	*S*-Ethyl *N*-ethylthiocyclohexanecarbamate	C_2H_5-	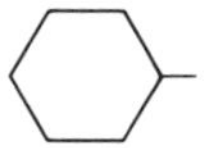	C_2H_5-
Diallate	*S*-(2,3-Dichloroallyl)diisopropylthiocarbamate	$CH_3-CH(CH_3)-$	$CH_3-CH(CH_3)-$	$Cl-C(H)=C(Cl)-CH_2-$
EPTC	*S*-Ethyl dipropylthiocarbamate	C_3H_7-	C_3H_7-	C_2H_5-

Molinate	*S*-Ethyl hexahydro-1*H*-azepine-1-carbothioate	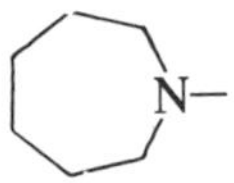 [1]		C_2H_5-
Pebulate	*S*-Propyl butylethylthiocarbamate	C_2H_5-	C_4H_9-	C_3H_7-
Thiobencarb	*S*[(4-Chlorophenyl) methyl] diethylcarbamothioate	C_2H_5-	C_2H_5-	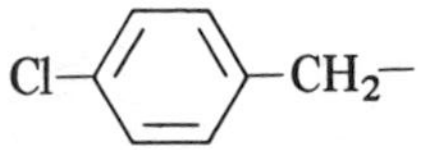
Triallate	*S*-(2,3,3-Trichloroallyl)diisopropylthiocarbamate	$CH_3-CH(CH_3)-$	$CH_3-CH(CH_3)-$	$Cl-C(Cl)=C(Cl)-CH_2-$
Vernolate	*S*-Propyl dipropylthiocarbamate	C_3H_7-	C_3H_7-	C_3H_7-

[1] The nitrogen atom in the molinate ring structure is the nitrogen atom of the parent thiocarbamic acid molecule; there is only one nitrogen atom in molinate.

Certain annual broadleaf weeds and purple and yellow nutsedge are also controlled. It also partially controls (suppresses) rhizome johnsongrass. Butylate can be applied when combined with certain fluid fertilizers or impregnated on dry bulk fertilizers. It has also been applied with center-pivot-sprinkler irrigation equipment. Butylate may also be used with atrazine or cyanazine to increase the range of weeds controlled. Also see page 418.

Soil Influence

Butylate, like the other thiocarbamate herbicides, is adsorbed by the clay and organic matter in soils and is not readily leached. The relative leachability of some of these compounds is molinate > EPTC > pebulate > cycloate > butylate. Vernolate is less subject to leaching than EPTC, but data on its relationship to the other compounds are not available.

Microbial breakdown of butylate and most other thiocarbamate herbicides plays an important role in their disappearance from soils. Thiocarbamates are not persistent in soils; their half-life is from 1 to 3.5 weeks in moist loam soil at 70–80°F. The half-life of butylate is three weeks.

Table 17-2. Half-Life of Six Thiocarbamate Herbicides in Moist Loam Soil at 70 to 80° F

Herbicide	Weeks
EPTC	1.0
Vernolate	1.5
Pebulate	2.0
Butylate	3.0
Molinate	3.0
Cycloate	3.5

Mode of Action

Butylate is absorbed by both leaves and roots and is translocated upward throughout the plant following root uptake. It appears to move via the apoplast system. It inhibits growth in the meristematic tissues, apparently by interfering with both cell division and enlargement. It has been reported that butylate is rapidly metabolized in corn to CO_2, diisobutylamine, fatty acids, conjugates of amines and fatty acids, and certain natural plant constituents. It has also been suggested that butylate forms a phytotoxic sulfoxide that is later combined with glutathione to form a nontoxic product.

Figure 17-1. Protection of corn from EPTC injury by the antidote R-25788. *Left to right*: Control, EPTC, EPTC plus R-25788. (G. R. Stephenson.)

R–25788 is an antidote or protectant for corn when applied with butylate, EPTC, or vernolate. It is commonly present in the formulation of these herbicides when they are to be used in corn. It is also effective as a seed treatment for corn prior to planting. The control of weeds is not affected by combining R–25788 with the thiocarbamate herbicides. Crops other than corn or wheat are not protected from thiocarbamate herbicide injury by R–25788 (see Figure 17–1).

Two broad concepts have developed to explain the biochemical mechanism of the action of this antidote: (1) increased formation of a nontoxic glutathione-thiocarbamate (sulfoxide) conjugate (Lay and Casida, 1976; Carringer et al., 1978) and (2) competitive inhibition (Stephenson et al., 1979). Also see Pallos and Casida (1978) for a detailed coverage of herbicide antidotes.

CYCLOATE

Cycloate is the common name for *S*-ethyl *N*-ethylthiocyclohexanecarbamate. The trade name is Ro-Neet®. This colorless liquid has a water solubility of 85 ppm at 22°C and is relatively volatile. It is formulated as an emulsifiable concentrate and in granular form. The acute oral LD_{50} of technical cycloate ranges from 2000 to 3190 mg/kg (male) and 3160 to 4100 mg/kg (female) for rats.

Uses

Cycloate controls most annual grasses including volunteer barley, several broadleaf weeds, and yellow and purple nutsedges in sugar beets, table beets, and spinach. It is usually applied by preplant soil incorporation; in California it has also been applied in sprinkler irrigation water. It may be combined with compatible fluid fertilizers for application or injected into the soil for sugar beets with special equipment at planting time. Also see page 418.

Soil Influence

Cycloate is resistant to leaching in heavy clay and high-organic-matter soils. It is less subject to leaching than molinate, EPTC, or pebulate. In loamy sand, cycloate will leach 3–6 in. with 8 in. of water. Its half-life in moist loam soil at 70–80°F is 3.5 weeks.

Mode of Action

Like other thiocarbamates cycloate inhibits growth of meristematic tissue. It is absorbed by leaves and roots and is readily translocated to leaves and stems of plants after root uptake. Cycloate has been reported to be rapidly metabolized by sugar beets to ethylcyclohexylamine, CO_2, amino acids, sugars, and other plant constituents. Another report suggests that cycloate forms a sulfoxide that is later combined with glutathione.

DIALLATE

Diallate is the common name for *S*-(2,3-dichloroallyl)diiopropylthiocarbamate. The trade name is Avadex®. This oily liquid with a water solubility of 14 ppm at 25°C is formulated as an emulsifiable concentrate as well as in granular form. The acute oral LD_{50} of diallate is 395 mg/kg for rats.

Uses

Diallate is primarily used to control wild oat. It can be applied in the spring in barley, corn, flax, forage legumes (alfalfa and sweet, red, and alsike clovers), lentils, peas, potatoes, soybeans, and sugar beets. In the fall it is also applied for spring-planted corn, flax, and forage legumes. But it can be applied on corn, potatoes, and soybeans only in certain states; the label should be checked.

Diallate is applied as a preplant soil-incorporation treatment. The granular form is only used in sugar beets and such use is limited to certain states.

Soil Influence

Diallate, as well as other thiocarbamate herbicides, competes with moisture for the adsorption sites on soil particles. It is adsorbed on clay and organic colloids, and therefore, less leaching occurs in clay and organic soils than in sandy soils. Microbial breakdown is of major importance. Under most conditions diallate has a half-life of about 30 days and persists one to three months in soil.

Mode of Action

Diallate appears to have more effect on cell division than on cell enlargement. At acute dosages lethal to wild oat, the first leaf does not emerge from the coleoptile, but at sublethal dosages, the first leaf emerges from the coleoptile and is somewhat distorted, dark green, and glossy in appearance (Parker, 1963) as well as brittle (Banting, 1967). In wild oats, absorption is primarily through the emerging coleoptile (Appleby et al., 1965; Parker, 1963).

EPTC

EPTC is the common name for *S*-ethyl dipropylthiocarbamate; its trade name is Eptam®. It is a light, yellow-colored liquid with a water solubility of 370 ppm; it is formulated both as an emulsifiable concentrate and in granular form. Eradicane® is the trade name for the formulation containing EPTC and R–25788, the antidote for corn. The acute oral LD_{50} of EPTC is 1652 mg/kg (male) and 3160 mg/kg (female) for rats.

Uses

EPTC was the first thiocarbamate herbicide developed. Its volatile nature resulted in highly variable weed control when it was first used as a surface-applied preemergence herbicide. Soil incorporation corrected this defect and provided the first general use of that technique. Pre-1960 research had definitely established that soil incorporation increased weed control of several herbicides. The power-driven rotary hoe was most effective. Soil incorporation by the disc, sweep-type cultivator, or drag harrow was also used. This method has been used widely with many herbicides. Soil incorporation places the herbicide in the seed-germinating area of the soil. Thus the method is not dependent on rainfall or irrigation for leaching the herbicide into the seed-germinating area of the weed.

EPTC is also applied through the sprinkler system and by metering into furrow-irrigation water. It may be combined with compatible fluid fertilizers or impregnated on dry bulk fertilizers for application.

EPTC is used against a wide array of weeds including annual grasses and a variety of annual broadleaf weeds. It also controls yellow and purple nutsedges. Also see page 419.

EPTC can be used in alfalfa, almonds, beans (dry, green), birdsfoot trefoil, castor beans, clovers, cotton, flax, grapefruit, lespedeza, oranges, peas, potatoes, safflower, sugar beets, sunflower, sweet potatoes, table beets, tangerines, tomatoes, walnuts, and certain ornamentals and nursery stock. Some of these uses are restricted to specific states and special application techniques. EPTC is also used as a tank mix with trifluralin in beans.

Soil Influence

EPTC is adsorbed by dry soil, but it is subject to some leaching. The amount of leaching decreases as the clay and organic content of the soil increase. It is used only on mineral soils. EPTC, like most other thiocarbamate herbicides, is readily volatilized from wet soil if not mixed in immediately after application. EPTC does not persist long in soils; the half-life in moist loam soil is one week at 70–80°F.

Mode of Action

EPTC inhibits growth of the meristematic region of grass leaves. The first leaf of grasses is often trapped in the coleoptile and emerges from the side of the coleoptile grossly distorted (Dawson, 1963). Broadleaf weeds may develop cupped leaves with necrotic tissue around the edges of the leaf (see Figure 17–2).

EPTC is absorbed by seeds, roots, and emerging shoots in contact with treated soil (Gray and Weierich, 1969). The relative importance of these three sites of entry varies with the plant species. EPTC is rapidly degraded in higher plants. Initially a sulfoxide appears to be formed that is combined with glutathione (Lay and Casida, 1976). The sulfoxide may be more phytotoxic than EPTC. The glutathione conjugate is not phytotoxic. EPTC and butylate react alike with the antidote R–25788 (see previous section "Butylate—Mode of Action").

EPTC has been reported to inhibit photosynthesis and respiration as well as lipid, protein, and RNA synthesis (Ashton et al., 1977). Certain symptoms of EPTC injury are characteristic of gibberellic acid (GA) deficiency and are decreased by a GA application (Donald et al., 1979; Wilkinson and Ashley, 1979).

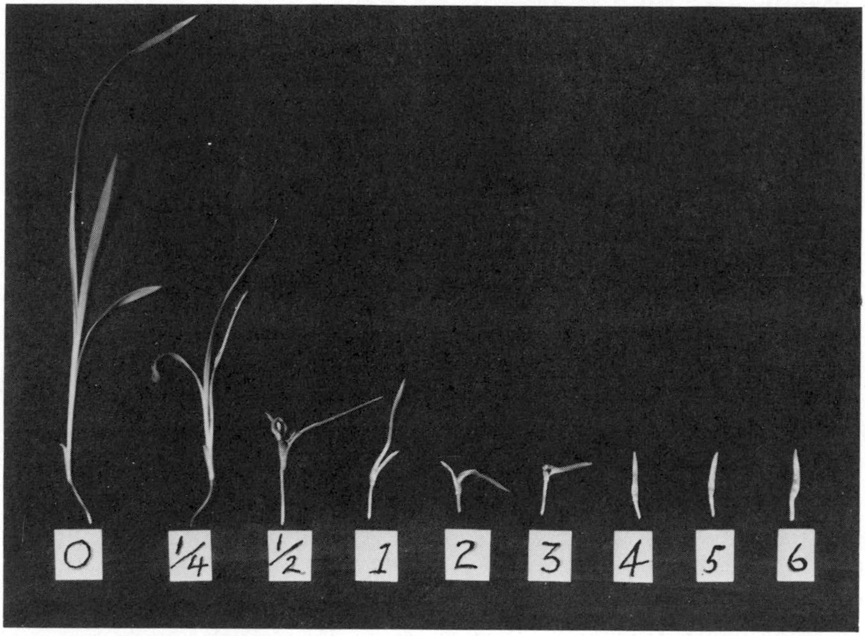

Figure 17-2. Barnyardgrass seedlings grown in sandy loam containing the indicated parts per million of EPTC. (J. H. Dawson, USDA, Prosser, WA.)

MOLINATE

Molinate is the common name for *S*-ethyl hexahydro-1*H*-azepine-1-carbothioate, with the trade name of Ordram®. It is a liquid with a water solubility of about 800 ppm at 20°C; it is formulated as an emulsifiable concentrate and in granular form. The acute oral LD_{50} is 720 mg/kg for rats.

Uses

Molinate controls barnyardgrass and dayflower in rice and suppresses growth of annual sedges and spikerush. It is applied in several ways—preplant soil incorporation, postemergence, at flooding, and at post-flooding—depending on the stage of rice growth and local cultural practices. In combination with propanil, it is applied postemergence-preflood to control dayflower and sprangletop. Molinate also has a special local-need registration in Arkansas and Louisiana for reduction of competition of red rice in cultivated rice. Also see page 419.

Soil Influence

Molinate is adsorbed onto dry soil but is subject to leaching, particularly in mineral soils. It is somewhat more leachable than EPTC. It is subject to volatility losses when applied to the surface of wet soil. Microbial degradation is the major means of molinate loss from soil. It has a half-life of three weeks in loam soil at 70–80°F.

Mode of Action

Molinate is readily absorbed by roots and translocated upward to the leaves presumably in the apoplastic system. It is rapidly metabolized by plants. Molinate suppresses plant growth and development generally.

PEBULATE

Pebulate is the common name for *S*-propyl butylethylthiocarbamate; the trade name is Tillam®. It is a yellow liquid with a water solubility of 60 ppm, and is formulated as an emulsifiable concentrate or in granular form. The acute oral LD_{50} is 921–1120 mg/kg for male rats.

Uses

Pebulate is used as a preplant, soil-incorporation treatment to control most annual grasses, yellow and purple nutsedge, and several broadleaf weeds. It is used in sugar beets, tomatoes, and tobacco. It is more effective for control of hairy nightshade than other herbicides applied to tomatoes. However, it does not give adequate control of black nightshade at registered application rates. It also controls dodder. Also see page 419.

Soil Influence

Pebulate is adsorbed by dry soil and under these conditions volatility losses are minimal. However, when applied to the surface of wet soil volatilization losses can be substantial. It leaches less than EPTC but more than cycloate under similar conditions of soil type and rainfall. Microbial degradation is the major means of loss from soil. The half-life of pebulate is two weeks in moist loam soil at 70–80°F.

Mode of Action

Pebulate acts by inhibiting plant growth and development. It is readily absorbed by roots and translocated upward, presumably in the apoplastic system, and then distributed throughout the stems and leaves. It is rapidly metabolized by plants.

THIOBENCARB

Thiobencarb is the common name for *S*[(4-chlorophenyl)methyl] diethylcarbamothioate. The trade names are Bolero® and Saturn®. It is a light-yellow or brownish-yellow liquid with a water solubility of 30 ppm at 20°C. It is formulated as an emulsifiable concentrate. Technical thiobencarb has an acute oral LD_{50} of 920–1903 mg/kg for rats.

Uses

Thiobencarb is applied as a preemergence to early postemergence spray on rice to control barnyardgrass and sprangletop.

Soil Influence

Thiobencarb is rapidly adsorbed by soil and not readily leached with water. Adsorption onto soil minimizes its loss from volatilization and photodegradation. It has a half-life of two to three weeks under aerobic conditions and six to eight months under anaerobic conditions. Microbial degradation accounts for most of this loss.

TRIALLATE

Triallate is the common name for *S*-(2,3,3-trichloroallyl)diisopropylthiocarbamate. Far-Go® is its trade name. Triallate is an amber oily liquid with a water solubility of only 4 ppm at 25°C; it is formulated as an emulsifiable concentrate and granules. It has an acute oral LD_{50} of 1675–2165 mg/kg for rats.

Uses

Triallate controls wild oats in barley, lentils, peas, and wheat. It is used as a preplant or postplant, soil-incorporation treatment. In the postplant treatment, the seed has already been planted below the depth of herbicide incorporation that follows. It is also applied as a mixture with compatible sprayable fluid fertilizer.

A combination of triallate and trifluralin controls foxtails, pigeongrass, and wild oats in spring wheat. This use is restricted to Minnesota, Montana, and North Dakota. Also see page 418.

Soil Influence

The adsorption of triallate onto soil colloids limits its leaching. Improper incorporation may result in volatility losses at high temperatures. It is mainly degraded by soil microorganisms. It persists in soil up to six weeks under Northern Great Plains conditions.

Mode of Action

Triallate is mainly absorbed by coleoptiles of grasses emerging from the treated layer of soil; there is only minor translocation at early stages of seedling development. It acts mostly by inhibiting cell division but also affects cell elongation. Wild oats are usually controlled before they emerge through the soil.

VERNOLATE

Vernolate is the common name for *S*-propyl dipropylthiocarbamate; the trade name is Vernam®. It is a liquid with a solubility in water of 90 ppm at 20°C. It is formulated as an emulsifiable concentrate and in granular form. Technical vernolate has an oral LD_{50} of 1780 mg/kg for rats.

Uses

Vernolate controls annual grasses, many annual broadleaf weeds, and yellow and purple nutsedge in peanuts, potatoes, soybeans, and sweet potatoes. It is usually applied as a preplant, soil-incorporation treatment, but can be injected below the soil surface for peanuts and soybeans. Also see page 419.

In soybeans, tank mixes of vernolate with chloramben, fluchloralin, profluralin, or trifluralin can be used. Also in soybeans, overlay sprays of bentazon, chloramben, linuron, or naptalam plus dinoseb can be applied after a vernolate preplant treatment.

With peanuts, similar overlay treatments are used (after vernolate preplant treatment) with diphenamid, naptalam, or dinoseb plus alachlor. Tank mixes of vernolate with benefin or trifluralin are also used in peanuts.

For selected uses, vernolate can be applied with compatible fluid fertilizers or impregnated on certain dry bulk fertilizers.

Soil Influences

Vernolate is adsorbed by dry soil but is leachable. It is more resistant to leaching than EPTC. When the soil is wet at the time of application, vernolate is readily lost by volatilization; but it is held by dry soil. Vernolate is lost from soil largely by microbial degradation. Its half-life in moist loam soil is 1.5 weeks at 70–80°F.

Mode of Action

Vernolate is absorbed by plant roots and translocated throughout the stems and leaves. It is rapidly metabolized in plants. Vernolate inhibits growth in meristematic tissues, suggesting it interferes with cell division and/or cell elongation.

CDEC

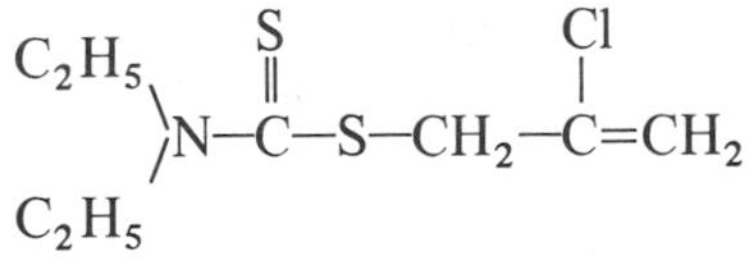

CDEC

CDEC is a *di*thiocarbamate. Both oxygen atoms of the parent carbamate molecule are replaced by sulfur atoms. This is in contrast to the *thio*carbamates previously discussed in this chapter; they contain only one sulfur atom.

CDEC is the common name for 2-chloroallyl diethyldithiocarbamate. As the trade name, Vegadex®, implies, it is primarily used in vegetable crops. CDEC is an amber, oily liquid with a water solubility of 92 ppm; it is formulated both as an emulsifiable concentrate and in granular form. Its acute oral LD_{50} is 850 mg/kg for rats.

Uses

CDEC controls most annual grasses and several of the most common annual broadleaf weeds found in vegetable crops. These weeds include pigweed, henbit, chickweed, and purselane. It also controls dodder.

This herbicide can be used on 25 vegetables: beans (lima and snap), broccoli, brussel sprouts, cabbage, cantaloupe, cauliflower, celery, chicory, collards, corn, cucumbers, endive, escarole, hanover salad, kale, lettuce, mustard greens, okra, potatoes, soybeans, spinach, tomatoes, turnips, and watermelon. It is also used on some nursery stock, including hydrangea, euonymus, potentilla spirea, azaleas, junipers, yews, and privet. CDEC may also be combined with CDAA to broaden the spectrum of weed species controlled in celery and cabbage.

CDEC should be applied at seeding time or before crop and weed emergence. It does not control emerged annual weeds or perennial weeds. CDEC may also be applied to transplants or on certain of these crops following a clean cultivation. For best results, the user should sprinkle-irrigate immediately after application. When CDEC is used with furrow irrigation, it must be incorporated into the soil. See page 418.

Soil Influence

CDEC is most effective on "light" soils that are low in clay and organic content, although with the use of sprinkle irrigation it can be used on muck soils. CDEC is reversibly adsorbed on soil colloids. Moderate rainfall or sprinkle irrigation after application enhances its performance, but heavy precipitation after application will cause excessive leaching in sandy soils and may produce substantial crop injury. CDEC is not persistent in soils and lasts from four to six weeks under most conditions.

Mode of Action

CDEC is readily absorbed by roots, but little, if at all, by foliage. Following root absorption it appears to be translocated throughout the plant in the apoplastic system. The chemical is rapidly metabolized by higher plants into CO_2 and other metabolites, including diethylamine as an intermediate.

METHAM

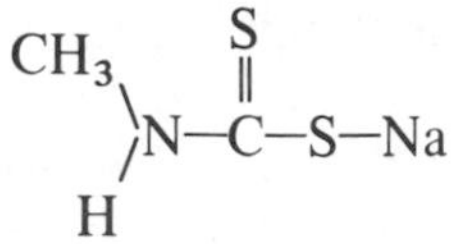

metham

Metham, like CDEC, is a dithiocarbamate.

Metham is the common name for sodium methyldithiocarbamate. The trade name is Vapam®. Pure metham is a white crystalline solid, but since it is unstable in this form, it is formulated as a water-soluble solution. Its acute oral LD_{50} is 820 mg/kg for male rats.

Uses

Metham is a temporary soil fumigant used to control nematodes, garden centipedes, soil-borne disease organisms, and most germinating weed seeds and seedlings. It may also be used to control certain shallow perennial weeds (e.g., nutsedge) limited to small patches.

This herbicide is used in the field as well as in potting soil. When used in the field, the soil should be cultivated before application to allow diffusion of the gaseous toxicant.

Metham may be applied in various ways depending on the size of the area to be treated and the equipment available. For small areas, a sprinkling can or hose proportional diluter may be used. For large areas, soil injection, spray application with immediate rotary-tiller incorporation, or application through a sprinkler-irrigation system may be used.

Metham is most effective when it is possible to confine the vapors with a plastic tarp; however, the water-seal method (saturating the top 2 in. of soil with water) may also be used. When using a tarp, the treated area should be kept covered for 48 hr or longer. Seven days after treatment, the area should be cultivated to a depth of 2 in., and at least 14–21 days should pass after application before the treated area is seeded.

Metham has also been used to kill roots in sewers.

SUGGESTED ADDITIONAL READING

Appleby, A. P., W. R. Furtick, and S. C. Fang, 1965, *Weed Res.* **5**, 115.

Ashton, F. M., and A. S. Crafts, 1981, "Thiocarbamates," pp. 303–327, in *Mode of Action of Herbicides,* Wiley, New York.

Ashton, F. M., O. T. de Villiers, R. K. Glenn, and W. B. Duke, 1977, *Pestic. Biochem. Physiol.* **7**, 122.

Banting, J. D., 1967, *Weed Res.* **7**, 302.

Carringer, R. D., C. E. Rieck, and L. P. Bush, 1978, *Weed Sci.* **26**, 167.

Dawson, J. H., 1963, *Weeds* **11**, 60.

Donald, W. W., R. S. Fawcett, and R. G. Harver, 1979, *Weed Sci.* **27**, 122.

Fang, S. C., 1975, "Thiocarbamates," pp. 323–348, in P. C. Kearney and D. D. Kaufman, Eds., *Herbicides,* Vol. 1, Marcel Dekker, New York.

Gray, R. A., and A. J. Weierich, 1969, *Weed Sci.* **17**, 223.

Lay, M. M., and J. E. Casida, 1976, *Pestic. Biochem. Physiol.* **6**, 442.

Pallos, F. M., and J. E. Casida, 1978, *Chemistry and Action of Herbicide Antidotes,* Academic, New York.

Parker, C., 1963, *Weed Res.* **3**, 259.

Stephenson, G. R., N. J. Bunce, R. I. Makowski, M. D. Bergsma, and J. C. Curry, 1979, *J. Agr. Food Chem.* **27**, 543.

Weed Science Society of America, 1979, *Herbicide Handbook,* 4th Ed., WSSA, Champaign, IL.

Wilkinson, R. E., and D. Ashley, 1979, *Weed Sci.* **27**, 270.

For chemical use, see the manufacturer's label and follow the directions. Also see the Preface.

18 Triazines

In 1952, J. R. Geigy, S.A., of Basel, Switzerland, started investigations with triazine derivatives as potential herbicides (Gysin, 1974). Herbicidal properties of chlorazine were reported in 1955 by Gast et al. and by Antognini and Day. Since then, numerous triazine derivatives have been synthesized and screened for their herbicidal properties. Eleven triazine herbicides are now being used in the United States, and more are probably employed worldwide.

Chemically, the triazines are heterocyclic nitrogen derivatives. The word *heterocyclic* is used to designate a ring structure composed of atoms of different kinds. In this case the ring is composed of nitrogen and carbon atoms. Most triazine herbicides are symmetrical, that is, they have alternating carbon and nitrogen atoms in the ring. However, one exception is metribuzin, which is asymmetrical.

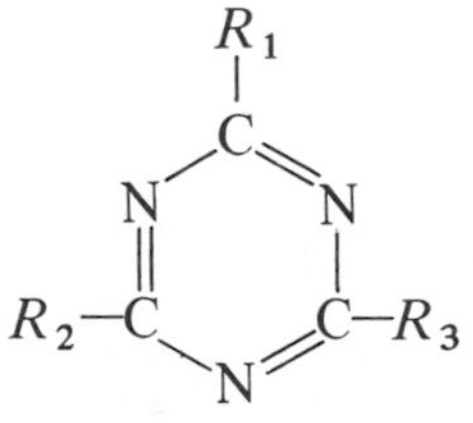

symmetrical triazine

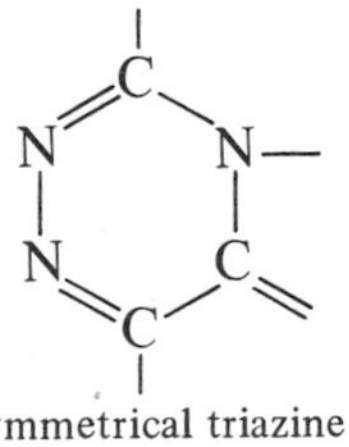

asymmetrical triazine

Structures of symmetrical triazine herbicides are given in Table 18-1. The substitution on the R_1 position of the ring determines the ending of the common name: *-azine* = chlorine atom; *-tryn* = methylthio group ($-SCH_3$); and *-ton* = methoxy group ($-OCH_3$). Water solubility of a series of triazines is markedly influenced by the substitution of R_1: prometon ($-OCH_3$), 750 ppm; prometryn ($-SCH_3$), 48 ppm; and propazine ($-Cl$), 8.6 ppm.

The greatest use of the triazine herbicides has been as a selective herbicide on cropland. Several excellent reviews have been written on vari-

Table 18-1. Common Name, Chemical Name, and Chemical Structure of the Symmetrical Triazine Herbicides[1]

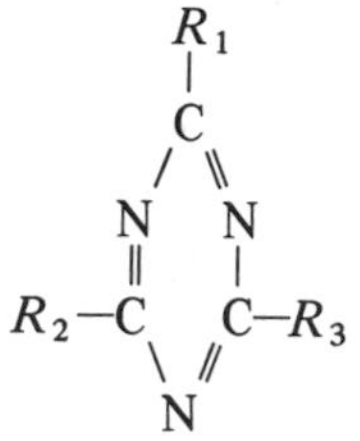

Common Name	R_1	R_2	R_3
Ametryn	$-SCH_3$	—NH·*iso*—C_3H_7	$-NHC_2H_5$
Atrazine	—Cl	—NH·*iso*—C_3H_7	$-NHC_2H_5$
Cyanazine	—Cl	$-NHC_2H_5$	H CH_3 | | —N—C—C≡N | CH_3
Cyprazine	—Cl	—NH·*iso*—C_3H_7	—NH— ◁
Dipropetryn	$-SC_2H_5$	—NH·*iso*—C_3H_7	—NH·*iso*—C_3H_7
Prometon	$-OCH_3$	—NH·*iso*—C_3H_7	—NH·*iso* C_3H_7
Prometryn	$-SCH_3$	—NH·*iso*—C_3H_7	—NH·*iso*—C_3H_7
Propazine	—Cl	—NH·*iso*—C_3H_7	—NH·*iso*—C_3H_7
Simazine	—Cl	$-NHC_2H_5$	$-NHC_2H_5$
Terbutryn	$-SCH_3$	—NH·*tert*—C_4H_9	$-NHC_2H_5$

[1] With a symmetrical molecule, the numbering can start from any one of the nitrogen atoms in the ring.

ous aspects of triazine herbicides, including their history (Gysin, 1974), uses (Gast, 1970), degradation (Esser et al., 1975), molecular structure and function (Gysin, 1960, 1974), mode of action (Ashton, 1965; Ashton and Crafts, 1981; Esser et al., 1975; Ebert and Dumford, 1976), and comprehensive (Gysin and Knüsli, 1960). Volume 32 of *Residue Reviews* contains 15 chapters dealing with triazine–soil interactions.

AMETRYN

Ametryn is the common name for 2-(ethylamino)-4-(isopropylamino)-6-methylthio-*s*-triazine. The trade name is EVIK®. It is a white crystalline solid with a water solubility of 185 ppm at 20°C. It is formulated as a wettable powder. The acute oral LD_{50} is 1110 mg/kg for rats.

Uses

Ametryn is a selective herbicide for control of annual grass and broadleaf weeds in bananas, corn, grapefruit, oranges, pineapple, sugarcane, and noncrop areas. It is effective when applied preemergence to annual weeds. It also has foliar activity, thus can also be applied postemergence to weeds. In pineapple it can be applied immediately after planting or after harvest is completed and before weeds emerge. Further applications may be made at one to two month intervals prior to differentiation. In sugarcane, ametryn can be applied at planting or after ratooning. Directed postemergence application may also be used in Florida. In bananas it is used both preemergence and postemergence to annual weeds. Also see page 418.

Ametryn is used as a directed postemergence application in corn when the crop is at least 12 in. high. It is also used as a directed spray in grapefruit and oranges. And it can be used as a potato-vine desiccant.

ATRAZINE

Atrazine is the common name for 2-chloro-4-(ethylamino)-6-(isopropylamino)-*s*-triazine. The trade name is AAtrex®. It is a white crystalline solid with a water solubility of 33 ppm at 27°C. It is formulated as a wettable powder, a flowable liquid suspension, and in a granular form. The LD_{50} of atrazine is 3080 mg/kg for rats.

Uses

Atrazine is widely used to control annual grass and broadleaf weeds in corn, macadamia orchards, pineapples, perennial grasses, sorghum, and sugarcane. It is effective when applied preemergence to annual weeds. When applied with an emulsifiable oil it controls many emerged annual weeds, but in this case crop selectivity may be reduced.

In corn, atrazine may be combined with alachlor, metolachlor, paraquat, propachlor, or simazine, or may be combined with simazine-plus-paraquat.

Atrazine combined with paraquat or combined with paraquat-plus-simazine controls existing vegetation and gives residual control where corn will be planted directly into cover crop, established sod, or in previous crop residues. In sorghum, atrazine may be combined with metolachlor, paraquat, propachlor, or terbutryn.

Atrazine is used in some areas for selective weed control on conifer reforestation, Christmas tree plantations, perennial range grasses, and grass-seed fields, as well as chemical fallow.

Atrazine is also widely used as a nonselective herbicide on noncropland. It is also formulated with prometon as a wettable powder for this same purpose.

CYANAZINE

Cyanazine is the common name for 2-[[4-chloro-6-(ethylamino)-*s*-triazin-2-yl]amino]-2-methylpropionitrile. The trade name is Bladex®. It is a white crystalline solid with a water solubility of 171 ppm at 25°C. It is formulated as a flowable liquid, wettable powder, and in granular form. The acute oral LD_{50} is 334 mg/kg for rats.

Uses

Cyanazine controls many annual grasses and broadleaf weeds in corn, sorghum, and cotton. It is usually applied as a preemergence treatment. However, when it is combined with certain other herbicides in corn or cotton, other application methods are required. Also see page 418.

In corn, cyanazine may be used alone but is often applied in combination with alachlor, atrazine, butylate (plus protectant), EPTC (plus protectant), or paraquat. Three-way combinations are also used, that is, cyanazine with atrazine plus any of these: alachlor, butylate (plus protectant), EPTC (plus protectant), or metolachlor.

When cyanazine is used in combination with the thiocarbamate herbicides butylate or EPTC, soil incorporation is required to prevent volatility losses. When used as a tank mix, both two-way combinations are incorporated before planting the crop. Cyanazine can also be applied as a preemergence treatment after a preplant incorporation of the thiocarbamate herbicide.

In sorghum, cyanazine may be used alone or as a tank mix with propachlor or propazine. These are usually preemergence applications. In cotton, cyanazine is applied alone as a preemergence treatment or as a directed postemergence application in combination with MSMA.

CYPRAZINE

Cyprazine is the common name for 2-chloro-4-(cyclopropylamino)-6-(isopropylamino)-*s*-triazine. The trade name is Outfox®. It is a white crystalline solid with a water solubility of 6.9 ppm at 25°C and 195 ppm at 40°C. It is formulated as an emulsifiable liquid. The acute oral LD_{50} of cyprazine is about 1200 mg/kg for rats.

Cyprazine is used as a postemergence treatment to control annual grass and broadleaf weeds in corn. It should be applied on rapidly growing, emerged weeds before they are 2 in. high.

DIPROPETRYN

Dipropetryn is the common name for 2-(ethylthio)-4,6-bis(isopropylamino)-*s*-triazine. The trade name is Sancap®. It is a white crystalline solid with a water solubility of 16 ppm at 20–25°C. It is formulated as a wettable powder. The acute oral LD_{50} is 5000 mg/kg for rats.

Uses

Dipropetryn controls annual broadleaf weeds and grasses in cotton. It is applied as a preemergence treatment and requires rainfall or overhead irrigation to make it effective. It resists leaching and is therefore suitable for use on the sandy soils of Arizona, New Mexico, Oklahoma, and Texas where other herbicides may cause crop injury or result in poor weed control.

METRIBUZIN

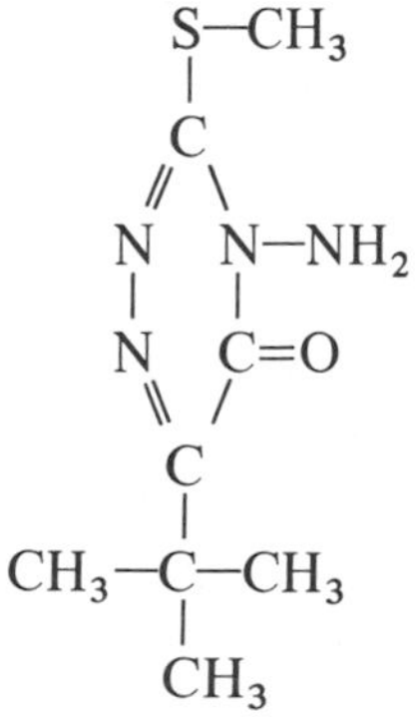

metribuzin

Because metribuzin is an asymmetrical triazine, its structural formula is given here rather than in Table 18-1. Metribuzin is the common name for 4-amino-6-*tert*-butyl-3-(methylthio)-*as*-triazin-5(4*H*)-one. It has two trade names—Sencor® and Lexone®. It is a white crystalline solid with a water solubility of 1220 ppm at 20°C. It is formulated into wettable-powder, liquid-flowable, and dispersible-granule forms. The acute oral LD_{50} of technical metribuzin is 1090 and 1206 mg/kg for male and female rats, respectively.

Uses

Metribuzin is registered for use on dormant established alfalfa, established asparagus, established legumes for hay or seed, potatoes, soybeans, sugarcane, and transplanted tomatoes. It is usually applied preemergence to both crop and annual weeds, but preplant soil incorporation is sometimes used. Rainfall or overhead irrigation is required for effective preemergence treatment. Also see page 419.

Numerous tank-mix combinations with metribuzin, including alachlor, chloramben, oryzalin, pendimethalin, profluralin, or trifluralin are used in soybeans. Paraquat can be used in soybeans to control emerged annual weeds by contact action before the soybeans emerge. When paraquat is tank-mixed with metribuzin, or (three-way) with metribuzin-plus-alachlor or metribuzin-plus-oryzalin, later-emerging annual weeds are also controlled.

PROMETON

Prometon is the common name for 2,4-bis(isopropylamino)-6-methoxy-*s*-triazine. The trade name is Pramitol®. It is a white crystalline solid with a water solubility of 750 ppm at 20°C. It is formulated as a wettable powder, as a liquid that can be applied in water or oil, and as granules in combination with simazine, sodium chlorate, or sodium metaborate. Prometon has an acute oral LD_{50} of 2980 mg/kg for rats.

Uses

Prometon is a nonselective preemergence and postemergence herbicide used to control most annual and broadleaf weeds and certain perennial weeds on noncropland. It is also applied under asphalt pavement and mixed with cutback asphalt for weed control in areas being stabilized. When prometon is combined with simazine, sodium chlorate, or sodium metaborate, a greater variety of perennial weeds is controlled and the period of weed control lasts longer. When prometon is combined with diesel oil, fuel oil, or weed oil, foliar contact activity is increased.

PROMETRYN

Prometryn is the common name for 2,4-bis(isopropylamino)-6-(methylthio)-*s*-triazine. The trade name is Caparol®. It is a white crystalline solid with a water solubility of 48 ppm at 20°C, and is formulated as a wettable powder. Prometryn has an acute oral LD_{50} of 3750 mg/kg for rats and mice.

Uses

Prometryn, a selective herbicide, controls annual grass and broadleaf weeds in celery and cotton. In celery it is used as a postemergence treatment to seedbeds (Florida only) and transplants as well as preemergence and postemergence to direct-seeded celery (California only). In cotton, prometryn is applied preemergence, preplant soil-incorporated (Arizona, California, and New Mexico), and postemergence directed. When combined with MSMA in cotton as a directed postemergence application, the mixture controls a wider range of weeds, especially nutsedges. Also see page 419.

PROPAZINE

Propazine is the common name for 2-chloro-4,6-bis(isopropylamino)-*s*-triazine. The trade name is Milogard®. It is a colorless crystalline solid with a water solubility of 8.6 ppm at 28°C. It is formulated as a wettable powder, flowable liquid, and water dispersible granules. The acute oral LD_{50} is greater than 5000 mg/kg for rats and mice.

Uses

Propazine is widely used to control annual grass and broadleaf weeds in sorghum. It can be applied both before and after planting, but before weeds emerge. Shallow incorporation, not more than 2 in. deep, after application generally gives better weed control, particularly under dry or minimum soil-moisture conditions. Rates of application vary in different locations, and certain soil types are excluded. Also see page 419.

SIMAZINE

Simazine is the common name for 2-chloro-4,6-bis(ethylamino)-*s*-triazine. The trade names are Princep® and Aquazine®. It is a white crystalline solid with a water solubility of 3.5 ppm at 20°C. The herbicide is formulated as a wettable powder, flowable liquid, water-dispersible granules, and granules. Its acute oral LD_{50} is greater than 5000 mg/kg for rats and mice.

Uses

Simazine was the first widely used triazine herbicide. Its largest crop use was in corn, but atrazine has largely replaced it for this purpose. It is primarily used to control annual grass and broadleaf weeds as a preemergence or preplant soil-incorporation treatment; it should be applied before the weeds emerge. Also see page 419.

Simazine is registered for use on more crops than any other triazine herbicide. These include alfalfa, artichokes, asparagus, blueberries, caneberries, corn (see Figure 18-1), cranberries, grapes, numerous orchard-tree crops, and sugarcane. It is also used selectively on alfalfa-orchard grass mix, forage bermudagrass, many species of ornamental woody plants, Christmas tree plantings, tree plantations for timber, shelterbelts, and certain turf grasses for sod. It is widely used as a nonselective herbicide on noncropland. Simazine is also used as a tank mix with atrazine for corn and with ametryn for grapefruit or oranges. Simazine tank-mixed with paraquat controls emerged and later emerging annual weeds in several orchard crops.

Certain of the above uses may not apply generally but require specific formulation, method of application, and/or geographical location; the user should check the labels.

Simazine is used in ponds to control several submerged and floating aquatic weeds and algae.

Figure 18-1. Weed-free plots were treated with simazine, applied preemergence to both weeds and corn. The corn was not injured. (Ciba-Geigy Corporation.)

TERBUTRYN

Terbutryn is the common name for 2-(*tert*-butylamino)-4-(ethylamino)-6-(methylthio)-*s*-triazine. The trade name is Igran®. It is a white crystalline solid with a water solubility of 25 ppm at 20°C. Terbutryn is available as a wettable powder formulation The acute oral LD_{50} of the 80W formulation is 2500 mg/kg for rats.

Uses

Terbutryn is a selective herbicide used to control annual grass and broadleaf weeds in winter wheat and winter barley in Washington, Oregon, and Idaho. It can also be used in sorghum. It can be applied either before or after the stem emerges. Postemergence treatments should be applied before weed rosettes reach 3 in. in diameter or 4 in. high. Also see page 419.

SOIL INFLUENCE ON THE TRIAZINES

Triazine herbicides are reversibly adsorbed by clay and organic colloids. They are not subject to excessive leaching in most soil types. In a study of five triazine herbicides on 25 soil types, adsorption almost always increased in the following order: propazine > atrazine > simazine > prometon > prometryn (Talbert and Fletchall, 1965). Correlation analysis indicated that adsorption of the methylthio-(prometryn) and methoxy-(prometon) triazines was more highly related to clay content, whereas adsorption of chloro-triazines (simazine, atrazine, propazine) was more highly related to organic matter. The following relative leachability in Lakeland fine sand has been reported: atraton > propazine > atrazine > simazine > ipazine > ametryn > prometryn (Rodgers, 1968).

Note that the order of triazine herbicides common to these two studies is the same, indicating that the leachability of triazine herbicides is directly related to their adsorption to soil colloids. These two studies also indicate that adsorption and leachability have little or no relationship to water solubility of the compounds. A reduction in phytotoxicity of triazine herbicides is associated with increasing amounts of clay and organic matter in soil (Weber, 1970).

Triazine herbicides vary widely in their persistence in soils. Soil type and environmental conditions have considerable influence on the actual period of persistence. Methoxytriazines are generally more persistent than methylthio- or chloro-triazines. Prometon, the most persistent, can remain at phytotoxic levels for several years. Atrazine, propazine, and simazine are less persistent but can still injure sensitive plants the next season. Ametryn, cyprazine, prometryn, and terbutryn are usually even less persistent, but may last from six to nine months. Cyanazine and metribuzin appear to be least persistent of the triazines; their half-life is two to four weeks under most conditions. At these rates of disappearance, less than 10% of that applied would remain after two to four months.

A monograph contains several review papers on interaction of triazine herbicides and soil (Gunther, 1970).

MODE OF ACTION OF TRIAZINES

The triazine herbicides inhibit plant growth, but this is considered to be a secondary effect caused by an inhibition of photosynthesis (Ashton and Crafts, 1981). At herbicidal concentration, triazine herbicides cause foliar chlorosis followed by death of the leaf (see Figure 18-2). Other leaf effects include loss of membrane integrity and chloroplast destruction. At sublethal levels, however, increased greening of leaves may occur.

Triazine herbicides are absorbed by leaves, but translocation from them is essentially nil. The amount of foliar absorption varies for various compounds. Propazine and simazine are poorly absorbed by leaves, whereas ametryn and prometryn are readily absorbed. The others appear to be intermediate. All triazine herbicides are rapidly absorbed by roots and readily translocated throughout the plant by the transpiration stream. They are considered to be translocated almost exclusively in the apoplast system (see Figure 18-3).

The rate of degradation of triazine herbicides in higher plants varies greatly with different species. In resistant species they are rapidly degraded, whereas in susceptible species, the herbicides are degraded slowly; thus the rate of degradation appears to be the primary basis of selectivity. This process occurs by hydroxylation, dechlorination, demethoxylation, or demethythiolation, depending on the parent substitution. Conjugation with glutathione and perhaps other peptides is also an important inactivation reaction in certain species. Dealkylation of the alkyl side chains also occurs.

The mechanism of action of triazine herbicides involves a severe inhibition of the Hill reaction of photosynthesis (Gast, 1958; Moreland et al., 1959). Total herbicidal effect, however, must be more complex than this because the plants do not merely starve to death. It has been postulated that the action involves the interaction of light, chlorophyll, and triazine to produce a secondary phytotoxic substance (Ashton et al., 1963).

Figure 18-2. Leaf chlorosis induced by the triazine herbicides. *Upper*: Chlorotriazines (e.g., simazine) almost always show interveinal chlorosis. *Lower*: Methylthiotriazines (e.g., ametryn) usually show veinal chlorosis. (C. L. Elmore, University of California, Davis.)

Certain triazine-resistant weed biotypes appear to differ from the susceptible biotypes of the same species at the adsorption site on the thylakoid membrane of the chloroplast (Pfister et al., 1979). The triazine molecule is adsorbed at this active site, and photosynthesis is inhibited in the susceptible biotype; but in the resistant biotype, adsorption does not occur and photosynthesis is not inhibited.

"Triazines and Physiology of Plants" is a review covering the mode of action of these herbicides (Ebert and Dumford, 1976).

Figure 18-3. Radioactive simazine, simazine degradation products, or both. The pictures show marginal accumulation in (*A*) susceptible cucumber, (*B*) localized distribution in moderately susceptible cotton, and (*C*) general distribution in resistant corn. (Davis et al., 1959.)

SUGGESTED ADDITIONAL READING

Antognini, J., and B. E. Day, 1955, *Proceedings of the 8th Southern Weed Conference,* pp. 92–98.

Ashton, F. M., 1965, *Proceedings of the 18th Southern Weed Control Conference,* pp. 596–602.

Ashton, F. M., and A. S. Crafts, 1981, "Triazines," pp. 328–374, in *Mode of Action of Herbicides,* Wiley, New York.

Ashton, F. M., E. M. Gifford, and T. Bisalputra, 1963, *Bot. Gaz.* **124**, 329.

Davis, D. E., H. H. Funderburk, and N. G. Sansing, 1959, *Weeds* **7**, 300.

Ebert, E., and S. W. Dumford, 1976, *Residue Rev.* **65**, 1.

Esser, H. O., G. Dupuis, E. Ebert, G. Marco, and C. Vogel, 1975, "*s*-Triazines," pp. 129–208, in P. C. Kearney and D. D. Kaufman, Eds., *Herbicides,* Vol. 1, Marcel Dekker, New York.

Gast, A., 1958, *Experientia* **13**, 134.

Gast, A., 1970, *Residue Rev.* **32**, 11.

Gast, A., E. Knüsli, and H. Gysin, 1955, *Experientia* **11**, 107.

Gunther, F. A., Ed., 1970, *Residue Rev.* **32**.

Gysin, H., 1960, *Weeds* **4**, 541.

Gysin, H., 1974, *Weed Sci.* **22**, 523.

Gysin, H., and E. Knüsli, 1960, *Adv. Pest Control Res.* p. 289.

Knüsli, E., 1970, *Residue Rev.* **32**, 1.

Moreland, D. E., W. A. Genter, J. L. Hilton, and K. L. Hill, 1959, *Plant Physiol.* **34**, 432.

Pfister, K., S. R. Radosevich, and C. J. Arntzen, 1979, *Plant Physiol.* **64**, 995.

Rodgers, E. G., 1968, *Weed Sci.* **16**, 117.

Talbert, R. E., and O. H. Fletchall, 1965, *Weeds* **13**, 46.

Weber, J. B., 1970, *Residue Rev.* **32**, 93.

Weed Science Society of America, 1979, *Herbicide Handbook,* 4th Ed., WSSA, Champaign, IL.

For chemical use, see manufacturer's label and follow the directions. Also see the Preface.

19 Ureas and Uracils

Substituted urea and uracil-type herbicides are included in the same chapter because they have several similar uses, and their modes of action have many features in common. These herbicides have been arranged alphabetically by their common names, and substituted ureas are discussed first. Pioneering development of both classes of herbicides was carried out by E. I. duPont de Nemours and Company.

UREAS

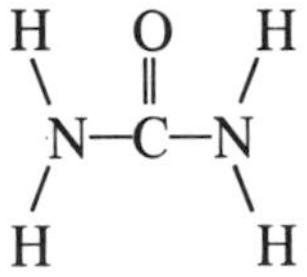

urea

Urea is a common nitrogen fertilizer. By substituting three of the hydrogen atoms of urea with other chemical groups, effective herbicides are produced. Common substitutions include a phenyl, methyl, and/or methoxy group. Other groups occur in certain molecules. The phenyl group may also have chlorine, bromine, or other substitutions (see Table 19-1).

Most urea herbicides are relatively nonselective at high rates of usage, and they are usually applied to the soil. Certain ones, however, have foliar activity, which may be increased by the addition of surfactants. At low rates of application, they may be selective. Selectivity is obtained by both depth protection, provided by little or no leaching characteristics of the herbicide, and by differences in plant tolerance, provided by inherent tolerance.

Diuron

Diuron is the common name for 3-(3,4-dichlorophenyl)-1,1-dimethylurea. The trade name is Karmex®. It is a white crystalline solid with a water solubility of 42 ppm. It is formulated as a wettable powder and as a flowable liquid suspension. Diuron has an acute oral LD_{50} of 3400 mg/kg.

Uses

Herbicidal uses of diuron are numerous on both crops and noncropland, and it is combined with other herbicides. Diuron is used primarily to control annual grass and broadleaf weeds before emergence in at least twenty-three crops: alfalfa, artichokes, asparagus, bananas, barley, bermudagrass pastures, birdsfoot trefoil, blueberries, caneberries, gooseberries, corn, cotton, grapes, perennial grass-seed crops, papayas, peppermint, pineapple, plantains, sorghum, sugarcane, small grains, and several fruit- and nut-tree crops as well as certain ornamentals. For these selective uses, the amount of diuron applied is relatively low, usually 1–4 lb/acre. The rate varies for different crops and different soil types. Also see page 419.

Diuron alone has very little foliar activity on most plants. However, by adding certain surfactants to the spray solution, considerable foliar toxicity is obtained. In this way, emerged annual weeds as well as germinating seedlings may be controlled.

Diuron is also combined with other herbicides to control a wider variety of weeds. Some of these combinations are (1) preplant soil-incorporation treatment of trifluralin followed by a preemergence application of diuron in cotton, (2) diuron and surfactant in combination with DSMA for postemergence weed control in Western irrigated cotton, (3) diuron plus bromoxynil in winter wheat in Washington, Oregon, and Idaho, (4) diuron plus bromacil in citrus and noncropland, and (5) diuron plus terbacil in certain deciduous fruit-tree crops.

Fluometuron

Fluometuron is the common name for 1,1-dimethyl-3-(α,α,α-trifluoro-*m*-tolyl)urea. The trade names are Cotoran® and Lanex®. It is a white crystalline solid with a water solubility of 90 ppm at 20°C. It is formulated as a wettable powder, and the acute oral toxicity of the technical product is 6416 mg/kg for rats.

Table 19-1. Common Name, Chemical Name, and Chemical Structure of the Substituted Urea Herbicides

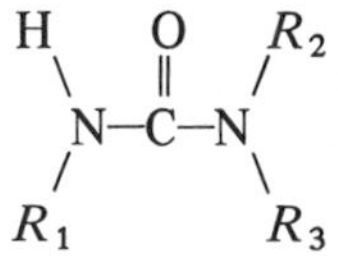

Common Name	Chemical Name	R_1	R_2	R_3
Diuron[1]	3-(3,4-dichlorophenyl)-1,1-dimethylurea	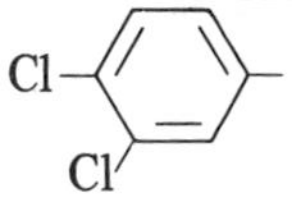	CH_3—	CH_3—
Fluometuron	1,1-dimethyl-3-(α,α,α-trifluoro-*m*-tolyl)urea	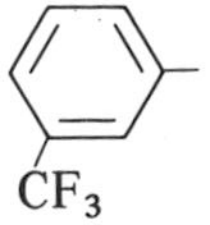	CH_3—	CH_3—
Karbutilate	*tert*-butylcarbamic acid ester with 3-(*m*-hydroxyphenyl)-1,1-dimethylurea	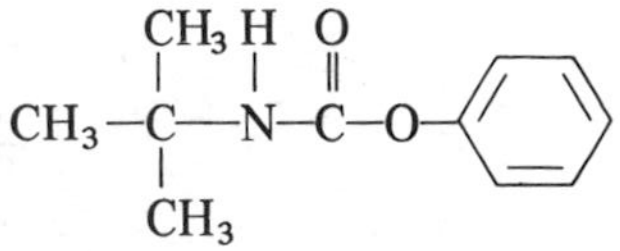	CH_3—	CH_3—

Linuron	3-(3,4-dichlorophenyl)-1-methoxy-1-methylurea	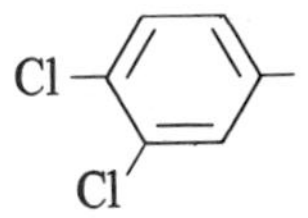	CH_3—	CH_3O—
Monuron[1]	3-(*p*-chlorophenyl)-1,1-dimethylurea	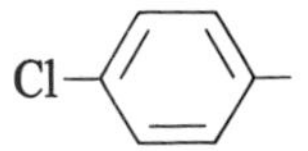	CH_3—	CH_3—
Siduron	1-(2-methylcyclohexyl)-3-phenylurea	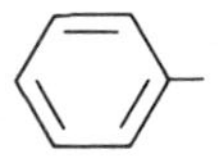	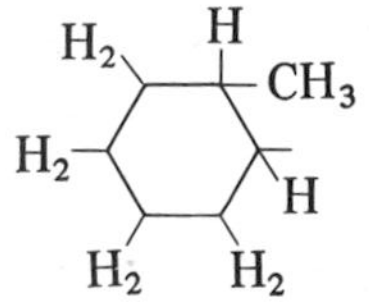	H
Tebuthiuron	*N*-[5-(1,1-dimethylethyl)-1,3,4-thiadiazol-2-yl]-*N,N'*-dimethylurea	CH_3—	CH_3—	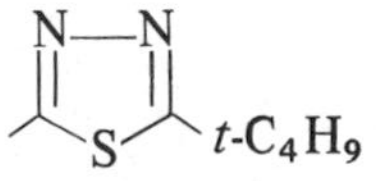

[1] Monuron and diuron were previously designated as CMU and DCMU, respectively; the photosynthetic biologists continue to use the older terminology.

Uses

Fluometuron controls annual grass and broadleaf weeds in cotton and sugarcane. It can be applied preemergence to the crop or as a directed spray after the crop and weeds have emerged. Sequential application may also be used, a preemergence application followed by one or two postemergence-directed applications. DSMA or MSMA may be tank-mixed with fluometuron for a directed postemergence application in cotton to broaden the variety of weeds controlled.

Karbutilate

Karbutilate is the common name for *tert*-butylcarbamic acid ester with 3-(*m*-hydroxyphenyl)-1,1-dimethylurea. It is apparent from the chemical name and the molecular structure that karbutilate could be considered either a carbamate- or urea-type herbicide, but because its herbicidal properties are more similar to those of the urea herbicides, it is included here. The trade name is Tandex®. Karbutilate is a white crystalline solid with a water solubility of 325 ppm. It is formulated as a wettable powder and in granular form. The acute oral LD_{50} is about 3000 mg/kg.

Uses

Karbutilate is a soil-applied, nonselective herbicide used to control annual and perennial broadleaf weeds and grasses plus woody species on noncropland. Emerged weeds are controlled better when karbutilate is combined with amitrole.

Linuron

Linuron is the common name for 3-(3,4-dichlorophenyl)-1-methoxy-1-methylurea. The trade name is Lorox®. It is a white crystalline solid with a water solubility of 75 ppm; it is formulated as a wettable powder. The acute oral LD_{50} of linuron is about 1500 mg/kg.

Uses

Linuron is applied to soil to control germinating annual seedlings and it has certain contact effect when applied to foliage. Best results from foliar applications are obtained when weeds are young and succulent, temperatures are 70°F or higher, and humidity is high. Emerged weeds under drought stress are usually not controlled.

As a preemergence herbicide, linuron is used in asparagus, carrots, corn, parsnips, potatoes, and sorghum. Postemergence, it is used on asparagus, carrots, celery, corn, and soybeans. Linuron is also used preemergence or postemergence in winter wheat in Idaho, Oregon, and Washington as well as postemergence in cotton east of the Rocky Mountains. Postemergence applications in corn, cotton, sorghum, and soybeans should be *directed* to minimize the amount of linuron received by crop plants.

Linuron is often combined with other herbicides as a tank mix when used preemergence: (1) *in corn,* alachlor, atrazine, or propachlor; (2) *in sorghum,* propachlor or propazine (propazine allowed only in Southwest); (3) *in soybeans,* alachlor, chloramben, metolachlor, oryzalin, pendimethalin, or propachlor (propachlor allowed only on seed crop). Linuron is also used preemergence *in soybeans* following a preplant soil-incorporation application of trifluralin or profluralin. In a minimum- or no-tillage cultural practice in *soybeans,* linuron may be applied immediately before, during, or after planting, but before the crop emerges, in combination with paraquat, or three-way mixes with oryzalin-plus-paraquat, metolachlor-plus-paraquat (or glyphosate), or alachlor-plus-paraquat (or glyphosate). In addition, linuron is used alone as a postemergence directed application in *soybeans* or in combination with 2,4-DB or dinoseb.

Linuron also gives short-term control of annual weeds on noncropland. It is relatively nonselective but persists only about four months. Most other herbicides used for total vegetation control last much longer (see Chapter 30). Addition of an appropriate surfactant increases linuron's activity on emerged annual weeds.

Monuron

Monuron is the common name for 3-(*p*-chlorophenyl)-1,1-dimethylurea. It is a white crystalline solid with a water solubility of 230 ppm. It is formulated as a wettable powder and in combination with other materials as special products. The acute oral LD_{50} for rats is 3600 mg/kg.

Uses

The herbicidal potential of monuron was first described by Bucha and Todd (1951). It was the first urea-type herbicide widely used. Monuron has been used to control annual weeds in certain crops, but presently its use is limited. Some specialty products include monuron plus bensulide, diphenamid, or neburon, plus fertilizers, for weed control on established dicondra turf. Certain other products contain atrazine or simazine, with or without fungicides, to control algae in aquariums, ponds, or pools. Monuron also controls weeds on noncropland.

Siduron

Siduron is the common name for 1-(2-methylcyclohexyl)-3-phenylurea. The trade name is Tupersan®. It is a white crystalline solid with a water solubility of 18 ppm at 25°C. It is formulated as a wettable powder. A variety of formulations including fertilizers, insecticides, or both, is available from formulators. The acute oral LD_{50} of siduron is greater than 5000 mg/kg.

Uses

Siduron is a speciality herbicide used almost exclusively to control certain annual grasses in turf (most turf grasses are tolerant even when germinating from seed) and is particularly effective on smooth and hairy crabgrass, foxtail, and barnyardgrass. It will not control annual bluegrass, clover, or most broadleaf weeds. Although most turf grasses are resistant to injury, a limited number of bentgrass strains and bermudagrass turfs may be injured by siduron.

Tebuthiuron

Tebuthiuron is the common name for *N*-[5-(1,1-dimethylethyl)-1,3,4-thiadiazol-2-yl]-*N,N'*-dimethylurea. The trade names are Spike®, Graslan®, and Perflan®. Pure, it is a white, odorless crystalline powder, formulated both as a wettable powder and as pellets. It is light-stable with essentially no volatility.

When tebuthiuron is fed to rats, rabbits, dogs, mallard ducks, and fish, the chemical is rapidly absorbed, metabolized, and excreted through the kidneys.

Feeding 1000 ppm of tebuthiuron to both the rat and dog for three months caused no effect.

Uses

Tebuthiuron is used for total vegetation control at high rates of application. At lower rates it is an effective brush and woody-plant herbicide (see Figure 27-12).

At rates required for total vegetation control, the chemical may persist for more than one year, providing effective year-round weed control on railroads and industrial sites. The chemical kills many woody species; therefore, *application to the rooting area of desirable trees and shrubbery must be avoided.*

Soil Interaction of Urea Herbicides

As a class, urea-type herbicides are relatively persistent in soils. Under favorable moisture and temperature conditions with little or no leaching, most of them can be expected to persist 6 months at selective rates and 24 months or more at higher nonselective rates.

Linuron appears to be the least persistent of this group with no detectable residue after four months. Fluometuron has a half-life of 30 days. At selective rates, diuron, monuron, and siduron last less than one year. Tebuthiuron is very persistent with a half-life of 12 to 15 months in areas receiving 40 to 60 in. of annual rainfall, and it persists longer in low-rainfall areas. Karbutilate also appears to be quite persistent in soils.

Principal factors affecting persistence of substituted ureas in the soil are microorganism decomposition, leaching, adsorption on soil colloids, and photodecomposition. The latter is important only when the herbicide remains on the soil surface for an extended period of time. Researchers believe volatility and chemical decomposition are of minor importance.

Microorganism decomposition is probably the most important factor. Bacteria such as *Pseudomonas, Xanthomonas, Sarcina,* and *Bacillus,* and fungi such as *Penicillium* and *Aspergillus* can use some substituted ureas as a direct source of energy. (Hill and McGalen, 1955). Conditions such as moderate moisture and temperature with adequate aeration, favoring such organisms, would also favor decomposition. Therefore, under dry, cold, or very wet soil conditions (poor aeration), the chemicals normally persist for a long time.

The adsorptive forces between the chemical and the soil colloids directly affect the chemical's rate of leaching; its solubility is a less-important factor.

In a study using four urea-type herbicides, leachability was correlated with adsorption and water solubility (WSSA, 1979; Wolf et al., 1958) (see Table 19-2). Fenuron was leached the most, followed in order by monuron, diuron, and neburon. The water solubilities and available adsorption values for Keyport silt loam of the other urea and uracil herbicides are given in Table 19-2.

Mode of Action of Ureas

Phytotoxic symptoms of urea-type herbicides can be seen largely in the leaves (Figure 19-1). They may show merely a slight chlorosis, which develops slowly with low-rate applications of the herbicide, or a water-

Table 19-2. Water Solubility and Adsorption on Soil of Urea and Uracil Herbicides (WSSA, 1979, Wolf et al., 1958)

Compound	Solubility in Water (ppm)	Adsorption on Keyport Silt Loam
Fenuron	3850	0.3
Tebuthiuron	2300	–
Bromacil	815	1.5
Terbacil	710	1.7
Karbutilate	325	–
Monuron	230	2.6
Fluometuron	90	–
Linuron	75	5.5
Diuron	42	5.2
Siduron	18	2.5
Neburon	5	16.0

[1] Expressed as ppm (active ingredient) present on soil in equilibrium with 1 ppm in soil solution.

soaked appearance, becoming necrotic in a few days at higher rates (Geissbühler et al., 1975; Ashton and Crafts, 1981).

Most of the urea herbicides are readily absorbed by roots and rapidly translocated to upper plant parts via the apoplastic system. Applications to the leaves are also translocated apoplastically with little if any translocated from the treated leaf. However, the actual amount absorbed and translocated from roots to shoot varies greatly with various compounds. Furthermore, differences in absorption and translocation between species have been sufficient with different urea herbicides to allow their selective use in certain crops (Geissbühler et al., 1975; Ashton and Crafts, 1981).

Inhibition of photosynthesis is generally acknowledged to be the primary action of the urea-type herbicides. This inhibition prevents the formation of high-energy compounds (ATP and NADPH) that are required for carbon dioxide fixation and numerous other biochemical reactions. See Moreland and Hilton (1976) for a comprehensive discussion.

However, many workers do not believe that this explains the light-dependent phytotoxic symptoms of the urea herbicides. In other words, plants do not merely starve from lack of photosynthate; rather, it has been postulated that a secondary phytotoxic substance is formed.

The nature of this secondary phytotoxic substance has not been determined (Ashton and Crafts, 1981).

Figure 19-1. Diuron induced chlorosis in peaches, usually veinal chlorosis but sometimes interveinal. *Left to right*: Untreated to increasing rates. (C. L. Elmore, University of California, Davis.)

Urea herbicides are subject to degradation by higher plants, but a given herbicide may be degraded at different rates by various species. This is the basis of certain selective uses of some urea herbicides. See Ashton and Crafts (1981) for detailed information.

URACILS

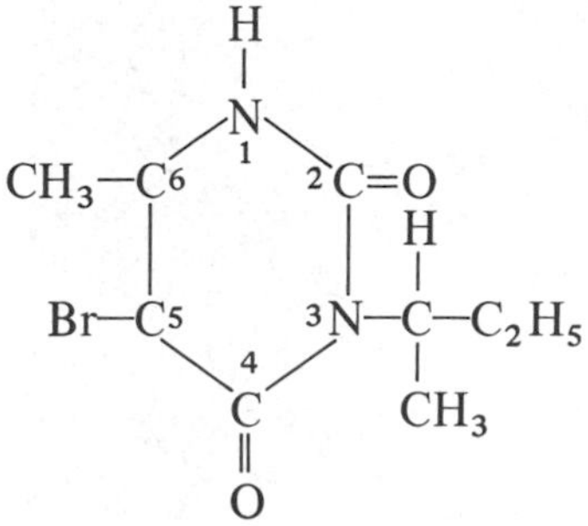

bromacil

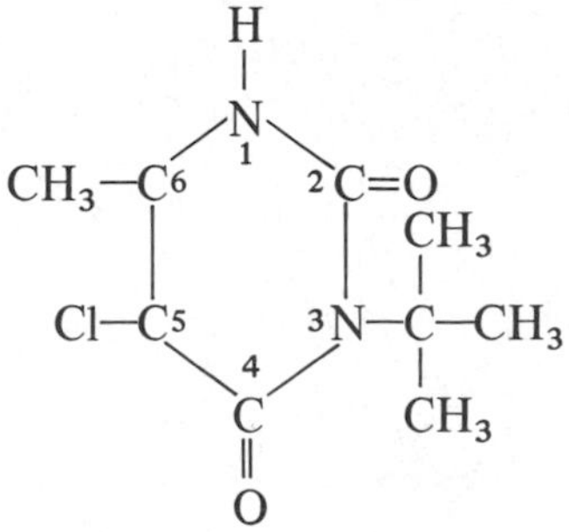

terbacil

Bromacil and terbacil are the two uracil-type herbicides used in the United States. Bromacil has a bromine atom in position 5 and a *sec*-butyl group in position 3; terbacil has a chlorine atom in position 5 and a *tert*-butyl group in position 3. They are both applied to the soil. Lenacil is another uracil used primarily in Europe for weed control in sugar beets and strawberries.

Bromacil

Bromacil is the common name for 5-bromo-3-*sec*-butyl-6-methyluracil. The trade name is Hyvar®. It is a white crystalline solid with a water solubility of 815 ppm. It is formulated as a wettable powder (Hyvar® X) and as the lithium salt in a water-soluble liquid form (Hyvar® X-L). The acute oral LD_{50} is 5200 mg/kg for rats. Wettable powder formulations of bromacil plus diuron are trade-named Krovar®.

Uses

Bromacil is used for selective weed control in citrus and pineapple crops, and for general vegetation control on noncropland. Many annual grass and broadleaf weeds are controlled at low rates. Some perennial weeds and several brush species are controlled at moderate rates, whereas johnsongrass and other somewhat-resistant perennial weeds may require high rates. These higher rates are not selective to citrus or pineapple (see Figure 19-2).

Bromacil plus diuron (Krovar®) is used in citrus and noncrop areas.

Figure 19-2. Chlorosis in walnuts induced by bromacil. (C. L. Elmore, University of California, Davis).

Bromacil plus surfactant can be tank-mixed with the following herbicides to control undesirable vegetation on noncrop areas: amitrole, dicamba, 2,4-D amine, dalapon, MSMA, paraquat, or the sodium salt of cacodylic acid.

Bromacil is also available in granular form as a mixture with certain borate salts for use on noncropland.

Terbacil

Terbacil is the common name for 3-*tert*-butyl-5-chloro-6-methyluracil. The trade name is Sinbar®. It is a white crystalline solid with a water solubility of 710 ppm. It is formulated as a wettable powder. The acute oral LD_{50} is greater than 5000 mg/kg, but less than 7500 mg/kg for fasted rats.

Uses

Terbacil is used at relatively low rates to selectively control many annual grass and broadleaf weeds in alfalfa, apples, blueberries, caneberries, citrus, grass-seed crops, mint, peaches, pecans, strawberries, and sugarcane. At higher rates, 4 to 6 lb/acre, bermudagrass and certain other perennial weeds can be controlled in citrus trees two or more years old.

Soil Interactions of Uracils

Bromacil and terbacil are adsorbed less on soil colloids than the urea-type herbicides monuron, diuron, or neburon, but more tightly than fenuron (see Table 19-2). Therefore, they are leached more readily than monuron, diuron, or neburon, but less readily than fenuron.

Bromacil and terbacil have a half-life of about five to six months when applied at 4 lb/acre, but at sterilant rates they persist for more than one season (see Table 5-1). This loss is apparently a result of microbiological degradation, because volatilization and photodecomposition losses are negligible, and leachability is limited. Soil diphtheroids, *Pseudomonas,* and *Penicillium* species have been shown to be able to degrade bromacil (WSSA, 1979).

Mode of Actions of Uracils

The uracil-type herbicides are similar to the urea-type herbicides in their mode of action (Gardiner, 1975; Ashton and Crafts, 1981). They are readily absorbed by roots and translocated apoplastically to the leaves, where they block photosynthesis. These herbicides cause chlorosis and necrosis in leaves, and bromacil has been shown to inhibit root growth. The structure of leaf chloroplasts is grossly altered and cell-wall development of roots is modified by bromacil.

SUGGESTED ADDITIONAL READING

Ashton, F. M., and A. S. Crafts, 1981, "Ureas," pp. 375–404, and "Uracils," pp. 441–446, in *Mode of Action of Herbicides,* Wiley, New York.

Bucha, H. C., and C. W. Todd, 1951, *Science* **114**, 493.

Gardiner, J. A., 1975, "Substituted uracil herbicides," pp. 293–321, in P. C. Kearney and D. D. Kaufman, Eds., *Herbicides,* Vol 1, Marcel Dekker, New York.

Geissbühler, H., H. Martin, and G. Voss, 1975, "The substituted ureas," pp. 209–291, in P. C. Kearney and D. D. Kaufman, Eds., *Herbicides,* Vol. 1, Marcel Dekker, New York.

Hill, G. D., and J. W. McGalen, 1955, *Proceedings of the 8th Southern Weed Conference,* p. 284.

Moreland, D. E., and J. L. Hilton, 1976, "Action on photosynthetic systems," pp. 493–523, in L. J. Audus, Ed., *Herbicides,* Academic, New York.

Wolf, D. E., R. S. Johnson, G. D. Hill, and R. W. Varner, 1958, *Proceedings of the 15th North Central Weed Conference,* p. 7.

Weed Science Society of America, 1979, *Herbicide Handbook,* WSSA, Champaign, IL.

For chemical use, see the manufacturer's label and follow the directions. Also see the Preface.

20 Other Organic Herbicides

This chapter covers those organic herbicides that do not fall into any chemical class previously discussed.

AMITROLE

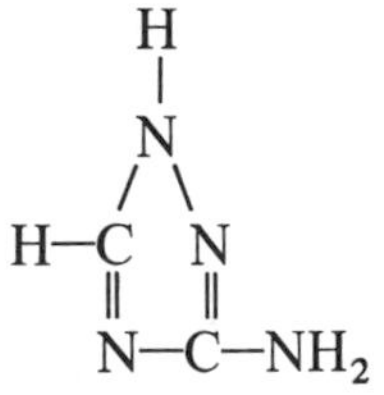

amitrole

Amitrole is the common name for 3-amino-*s*-triazole. It is a white crystalline solid and is very (about 28%) soluble in water. A trade name is Amino Triazole Weed Killer 90®. However, it is commonly used as a formulation containing ammonium thiocyanate, Amitrole-T® or Cytrol® Amitrole-T. Amitrole is also available in formulations containing simazine (Amizine®) or fenac-plus-atrazine (Fenamine®). The acute oral LD_{50} of amitrole is 24,600 mg/kg for rats. Two-year, lifelong studies with rats indicate that amitrole is a goitrogen (causing enlargement of the thyroid gland).

Uses

Amitrole is applied as a foliar spray to control essentially all emerged annual weeds and several perennial weeds on noncropland. Perennial weeds may require the treatment of regrowth to gain adequate control. With certain perennial weeds, the time of year when the initial application is made is critical.

Addition of ammonium thiocyanate to the formulation appears to increase its activity when applied to foliage.

When simazine is applied with amitrole, amitrole controls emerged plants, and simazine controls plants that germinate later. In addition to noncrop use, this mixture is also useful in certain ornamental nursery stock and forest plantations.

When amitrole is applied with fenac plus atrazine on noncropland, amitrole controls emerged plants, atrazine controls plants that germinate later, and fenac is effective applied to the foliage and soil.

Soil Influence

Amitrole disappears rapidly from soils (Carter, 1975). Nonbiological reactions apparently explain most of this loss, but microorganisms may also be involved. The average persistence in soils is about two to four weeks.

Mode of Action

The most striking symptom of amitrole phytotoxicity is the albino appearance of leaves and shoots that develop after application. Amitrole is one of the most readily translocated herbicides. It is translocated in both the symplastic and apoplastic systems, and is therefore classified as a systemic herbicide.

Although amitrole is rapidly degraded in soil (less than a one-month persistence), it appears to be considerably more stable in plants. In plants it forms conjugates with sugars and amino acids which then may be degraded to yield free amitrole. Symptoms of amitrole may be observed in perennial plants a year or more after application.

Amitrole reportedly blocks development of chloroplast ribosomes (Bartels et al., 1967). Such a blockage would interfere with formation of chloroplast membranes, enzymes, and pigments. Accumulation of chloroplast DNA is also inhibited by amitrole (Bartels and Hyde, 1970). These effects on chloroplasts would explain the chlorosis one observes following amitrole treatments. However, it is possible the inhibition of histidine biosynthesis also contributes to the phytotoxicity of amitrole in higher plants (Wiater et al., 1971a, 1971b).

BENSULIDE

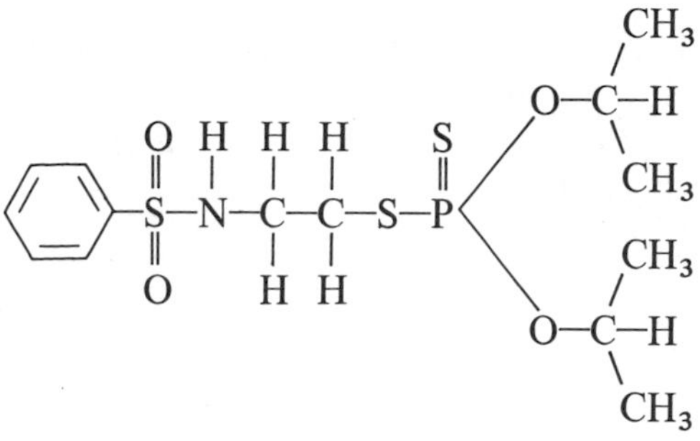

bensulide

Bensulide is the common name for *O,O*-diisopropyl phosphorodithioate *S*-ester with *N*-(2-mercaptoethyl)benzenesulfonamide. The trade names are Betasan®, Prefar®, and Betamex®. Bensulide has a relatively low melting point (34.4°C) and may therefore be a liquid or crystalline solid when pure. It has a water solubility of 25 ppm at 20°C and it is formulated as an emulsifiable concentrate and in granular form. The acute oral LD_{50} is 770 mg/kg for rats.

Uses

Bensulide controls several annual grasses including crabgrass, annual bluegrass, and goosegrass, as well as certain broadleaf weeds in established grass and dichondra lawns. It can also be used to control weeds in many established flowers, ornamentals, and ground covers. Bensulide can also be used in carrots, cole crops, cotton, cucurbits, lettuce, onions, peppers, and tomatoes. A tank mix of bensulide and naptalam is commonly used in cucurbit crops, and a formulation of bensulide plus monuron is applied on established dichondra turf. Also see page 418.

Bensulide is applied as a premergence or preplant soil-incorporated treatment depending on the crop and cultural practice.

Soil Influence

Bensulide is relatively persistent in soils, for 8–12 months. Therefore, it is suggested that crops omitted above should not be planted until 18 months after the last application. Bensulide is inactivated in soils containing high amounts of organic matter. The leaching of bensulide is restricted in all soil types.

Mode of Action

Bensulide inhibits the growth of roots and partially inhibits cell division (Cutter et al., 1968). It is adsorbed on root surfaces and a small amount is absorbed by the root. However, little, if any, is translocated upward to the leaves. It appears to be degraded by higher plants.

DCPA

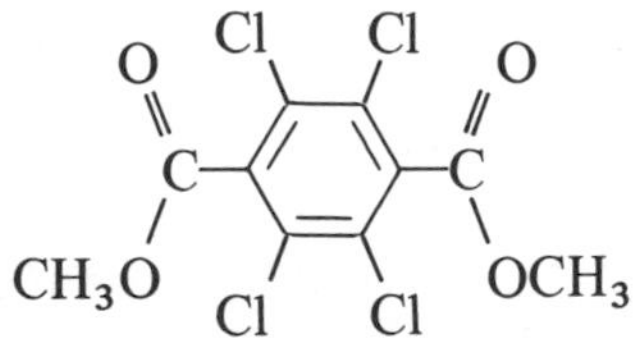

DCPA

DCPA is the common name for dimethyl tetrachloroterephthalate. The trade name is Dacthal®. It is a white crystalline solid with a water solubility of 0.5 ppm at 25°C. It is formulated as a wettable powder and in granular form. The acute oral LD_{50} is greater than 3000 mg/kg for rats.

Uses

DCPA is applied as a preemergence or preplant soil-incorporation treatment to control most annual grasses and many broadleaf weeds in beans, cole crops, collards, cucurbits, eggplant, garlic, kale, lettuce, mustard greens, onions, peppers, potatoes, southern peas, soybeans, strawberries, sweet potatoes, tomatoes, turnips, and yams. It is also used in grass turf, nursery stock, and established ornamentals. When used to control dodder in alfalfa, DCPA prevented the parasite from attaching itself to alfalfa (Bayer et al., 1965). Also see page 418.

Soil Influence

This herbicide is adsorbed to organic matter in the soil and thus is not subject to leaching. It is slowly degraded in soils by microorganisms and chemical hydrolysis. It persists for somewhat more than six months, with an average half-life of 100 days.

Mode of Action

DCPA is absorbed by roots, but scientists think it is not absorbed by leaves. DCPA is absorbed by the coleoptiles of grass seedlings and is readily absorbed by the hypocotyl of cucumber and translocated into the foliage (Nishimoto and Warren, 1971). It is translocated from roots slightly, if at all. DCPA appears to be a general growth inhibitor and especially inhibits germinating seeds and roots. It does not appear to be metabolized by higher plants.

DINOSEB

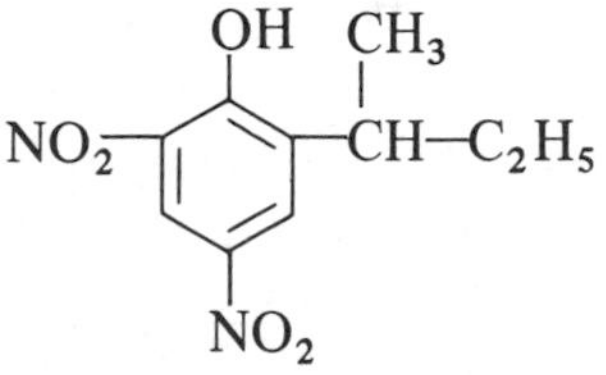

dinoseb

Dinoseb is the common name for 2-*sec*-butyl-4,6-dinitrophenol. It is also often called DNBP. It has several trade names. The phenol form is usually formulated as an emulsifiable concentrate; it is also soluble in oil. It may also be formulated as water-soluble salts; the most common are the ammonium, triethanolamine, and a mixture of ethanol and isopropanolamine salts. Dinoseb is a dark-brown solid or a dark-orange liquid, depending on temperature. As the phenol, it has a water solubility of 52 ppm at 25°C, but the salts are quite soluble in water.

Toxicity

Dinoseb and its salts are *dangerous poisons* if taken internally, if inhaled, or if absorbed through the skin. Symptoms of poisoning are excessive fatigue, sweating, thirst, and fever. With normal precautions, the chemical can be applied routinely with little or no hazard to the applicators. Daily bathing and change of clothing is recommended whether the applicator thinks he is contaminated or not.

The acute oral LD_{50} for rats is 58 mg/kg. It is considered to be quite toxic.

Uses

Dinoseb is very toxic to growing plants, so it is used as a general contact herbicide. It is so toxic to all leaves that it lacks the selectivity of its salt derivatives. Dinoseb is valuable where mowing is impractical; for example, along fencerows, ditch banks, and roadsides. It kills most annual weeds and removes the tops from perennial weeds. Underground parts of perennial plants are not killed except by repeated treatments. Thus dinoseb can be used in dormant alfalfa to kill annual weeds. See page 418.

It is also used for preemergence soil treatments in beans, corn, cucumbers, peanuts, potatoes, and soybeans. In most small-seeded legumes it can be used immediately after cutting and before new growth appears; however, treated areas should not be grazed and treated forage should not be fed to livestock.

Dinoseb is also used as a directed spray to weeds in the fall after harvest or in early spring before bloom in currants, gooseberries, grapes, and raspberries. Directed sprays are also used in many fruit and nut crops. It is also used as a preharvest desiccant in small-seeded legumes, peas, and soybean seed crops, as well as for preharvest vine killing in potatoes.

For simplicity, the uses of all three common salt formulations (ammonium, triethanolamine, and the mixture of ethanol and isopropyl amine salts) are considered together. However, these salts may not be registered for use on all crops mentioned, and it is advisable to check the manufacturer's label, federal registration, or both, before using.

Salts of dinoseb are used for selective postemergence weed control in alfalfa, beans, small grain, small-seeded legumes, corn, garlic, onions, peas, and peanuts (see Figure 20-1). They are also applied as a directed spray to emerged weed ground cover in certain tree and nut crops while the crop is dormant and before bloom. The most favorable conditions for application are dry, sunny weather, temperatures of 70–85°F, and no rain for about 12 hr after treatment. At strong concentrations the salts resemble dinoseb in their toxic contact effects on plants.

Salts are also used for preemergence weed control; effects usually last for three to five weeks. Salts are applied preemergence to beans, corn, cucumbers, mint, peanuts, peas, potatoes, pumpkins, soybeans, and squash.

Light rain after treatment is beneficial to carry the chemical into the soil and to reduce volatility losses.Heavy rains may cause leaching, which may injure the crop or prevent effective weed control by leaching the chemical below the weed seeds in the surface soil. No rain soon after treatment will most likely cause poor preemergence weed control, especially in hot weather. This is a result of volatilization (loss of the chemical as a vapor) under several days' exposure to high soil temperatures.

Figure 20-1. Soybeans treated with dinoseb (amine salt) just after the soybeans emerged (crook stage). This treatment controlled weeds for one month without harm to the soybeans. The nearly weed free plot in the background was treated similarly—but nine days later. Treatment at this stage was much less effective (Kentucky Agricultural Experiment Station.)

Soil Influence

Dinoseb leaches readily, but there is some evidence of partial adsorption in certain organic and clay soils. Persistence of dinoseb in soil depends on many of the same factors discussed in Chapter 5. Under warm, moist conditions, you can expect dinoseb to only last for three to five weeks. No residual carry-over is expected from one season to the next.

Mode of Action

Dinoseb is translocated little, if at all, within the plant. It acts almost completely as a contact herbicide; therefore, uniform coverage of all foliage is important.

Dinoseb stimulates respiration in low concentration and powerfully inhibits respiration at high concentrations (Kelly and Avery, 1949; Loomis and Lyman, 1948); it also inhibits the coupling during phosphorylation and oxidation of pyruvate. Experimental evidence indicates that dinoseb acts on a basic mechanism in the cell by which the phosphate bond is coupled to oxidative reaction (Loomis and Lyman, 1948). Dinitrophenols act as protein coagulants (Crafts and Harvey, 1950), as well as causing the respiratory response.

ENDOTHALL

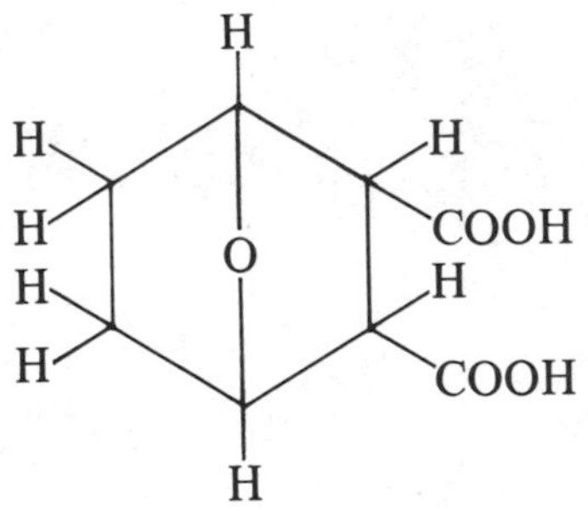

endothall

Endothall is the common name for 7-oxabicyclo[2.2.1]heptane-2,3-dicarboxylic acid. There are are several trade names. It is a white crystalline solid with a water solubility of about 10%, formulated as water-soluble liquids and in granular forms. Various endothall salts are available, such as disodium, dipotassium, or amine. Endothall is also formulated in combination with silvex. The acute oral LD_{50} of the acid, sodium, or amine salts ranges from 38 to 206 mg/kg for rats.

Uses

In the past endothall has controlled annual weeds in several crops. However, its current major use is control of aquatic weeds in irrigation and drainage canals, lakes, and ponds. It has a relatively low level of toxicity to fish (Walker, 1963; Yeo, 1970). It is also used in established turf or as a preharvest desiccant. In sugar beets, it is combined with pyrazon, with or without dalapon, to control emerged annual weeds.

Mode of Action

Endothall is absorbed readily by leaves and roots. It is translocated to a limited extent from roots to shoots of plants via the xylem, but it is not phloem-mobile and is thus not translocated from leaves to other plant parts. Its action appears to be contact in nature, causing rapid desiccation to germinating seedlings, browning of the foliage, or both.

Endothall is subject to considerable leaching in soils. It is rapidly degraded in both soil and water.

FENAC

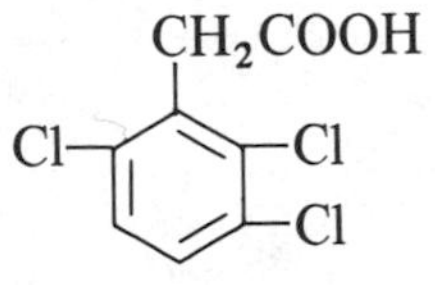

fenac

Fenac is the common name for (2,3,6-trichlorophenyl) acetic acid. It is formulated as the sodium salt (Fenatrol®), the dimethylamine salt plus the dimethylamine salt of 2,4-D (Fenatrol® Plus), and the ammonium salt plus amitrole and atrazine (Fenamine®). Fenac is a white crystalline solid, only slightly soluble in water as the acid; however, the salt forms listed above are relatively soluble in water. The acute oral LD_{50} of fenac is 1780 mg/kg for rats.

Uses

The sodium salt formulation of fenac is used alone or combined with silvex as a preemergence treatment in sugarcane. The dimethylamine salts of fenac and 2,4-D are also used in sugarcane as a preemergence treatment or an early postemergence treatment. The ammonium salt of fenac formulated in combination with amitrole and atrazine will control a broad range of weed species on noncrop areas and under asphalt or cement paving. Also see page 419.

Mode of Action

Fenac has growth-regulating properties causing epinasty, bud inhibition, and bud necrosis (see Figure 20-2). However, unlike auxin-type herbicides, its translocation appears to be primarily restricted to the apoplast; therefore, it is usually applied to the soil and absorbed by the roots.

Fenac is strongly adsorbed by soil colloids and resists leaching. It is degraded slowly in soil, remaining phytotoxic from one to two years.

Figure 20-2. Modification of leaf structure of cotton induced by fenac. (W. B. McHenry, University of California, Davis.)

FLURIDONE

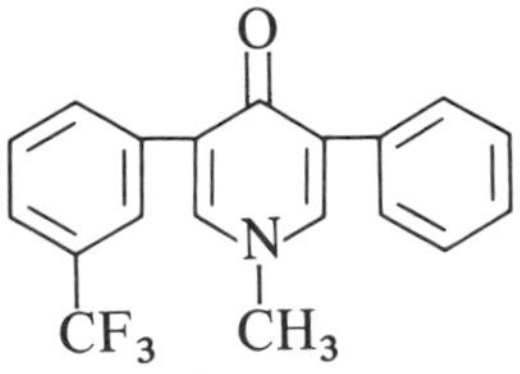

fluridone

Fluridone is the common name for 1-methyl-3-phenyl-5-[3-(trifluoromethyl)phenyl]-4(1*H*)-pyridinone. The trade name is Sonar®. It is a white crystalline solid with a water solubility of 12 ppm. It is available as a wettable powder and an aqueous suspension. The acute oral LD_{50} for rats is greater than 10,000 mg/kg.

Uses

Fluridone controls most annual grasses and many broadleaf weeds when applied preemergence or mixed into the soil. At relatively high rates, it also controls certain perennial grasses and sedges. Cotton is the only annual crop that has been shown to be tolerant to fluridone. It is also being evaluated for aquatic weed control. Also see page 389.

Fluridone is strongly adsorbed to organic matter in soils and is not subject to leaching.

Mode of Action

Phytotoxic symptoms include slowing of growth and leaf chlorosis or albinism followed by necrosis. It is absorbed by both roots and emerging shoots of seedlings; however, there are quantitative differences in the rates of absorption among various species (Albritton and Parka, 1978; Rafii and Ashton, 1979a). It appears to be translocated primarily in the apoplast, but the rate of transport also varies among species (Bernard et al., 1978).

Fluridone has been reported to inhibit photosynthesis as well as RNA, protein, and carotenoid synthesis (Rafii and Ashton, 1979b; Bartels and Watson, 1978). Other scientists suggest that the tolerance of cotton to fluridone is related to the differential site of uptake, translocation, and/or effect on metabolic processes (Albritton and Parka, 1978; Bernard et al., 1978; Rafii and Ashton, 1979a, 1979b). Radioactive studies indicate no significant metabolism of fluridone in higher plants.

METHAZOLE

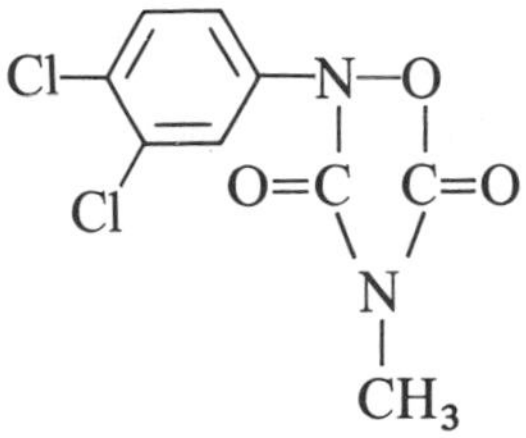

methazole

Methazole is the common name for 2-(3,4-dichlorophenyl)-4-methyl-1,2,4-oxadiazolidine-3,5-dione. The trade name is Probe®. It is a tan solid with a water solubility of 1.5 ppm at 25°C. It is formulated as a wettable powder. The acute oral LD_{50} of technical methazole is 2501 mg/kg for rats. In early literature, the common name for this herbicide was bioxone.

Uses

Methazole is primarily a cotton herbicide. It is applied both preemergence and postemergence-directed at lay-by time. Preemergence sprays control many annual grasses and broadleaf weeds.Postemergence treatments control mainly broadleaf weeds. A tank mix of methazole with MSMA for postemergence treatments expands the control of weeds, especially annual grasses and nutsedges.

Soil Influence

Methazole is adsorbed by soil colloids, so is not subject to excessive leaching in most nonsandy soils. It is subject to degradation by soil microorganisms and has a half-life of less than 30 days.

Mode of Action

Methazole induces foliar chlorosis but also delays emergence of seedlings and inhibits both shoot and root growth (Keeley et al., 1972). It is readily absorbed by roots and rapidly translocated to the shoot; accumulation in older leaves indicates apoplastic transport (Jones and Foy, 1972a; Wills, 1976). Methazole has been shown to be degraded in cotton, beans, wheat, and onions (Jones and Foy, 1972b; Dorough et al., 1973; Dorough, 1974). Certain metabolites of methazole form conjugates.

Although methazole does not inhibit photosynthesis, its first metabolite [3,4-(dichlorophenyl)-3-methylurea] is a strong inhibitor of the Hill reaction (Good, 1961). Researchers also have suggested that the effect of methazole on plant metabolism may be similar to the effect of carbamate-type herbicides.

NORFLURAZON

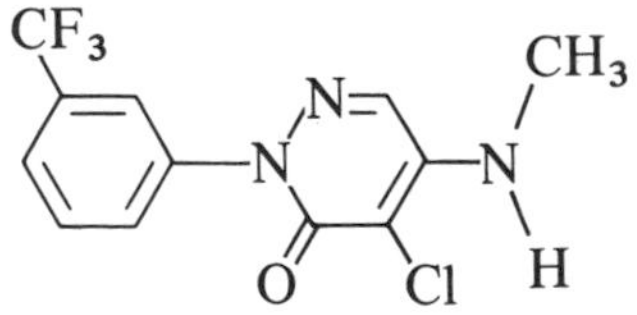

norflurazon

Norflurazon is the common name for 4-chloro-5-(methylamino)-2-(α,α,α-trifluoro-*m*-tolyl)-3(2*H*)-pyridazinone. The trade names are Evital®, Solicam®, and Zorial®. It is a white to brownish-gray solid with a water solubility of 28 ppm at 25°C. Both wettable powder and granule formulations are available. The acute oral LD_{50} for rats is greater than 8000 mg/kg.

Uses

Norflurazon is applied as a preemergence herbicide to control many annual broadleaf and grass weeds. It is registered for use in cotton, cranberries, and several tree fruit and nut crops including apricots, cherries, filberts, nectarines, peaches, plums, and walnuts. Also see page 419.

Soil Influence

Norflurazon is adsorbed by soil colloids and is not subject to leaching. It is degraded by soil microorganisms and has an average half-life of 45–180 days in soils of the Mississippi Delta and the Southeast.

Mode of Action

Norflurazon causes light-grown seedlings to emerge with chlorotic leaves and die following exhaustion of food reserves (Bartels and Watson, 1978). It inhibits the biosynthesis of carotenoids, and since carotenoid pigments protect chlorophyll from photodegradation, chlorophyll is destroyed. See Ashton and Crafts (1981) for a review of original research on this phenomenon.

Norflurazon is absorbed by roots and translocated to shoots, presumably via the apoplastic system. It is metabolized by higher plants. In cotton, tolerance appears to be associated with limited transport rather than degradation (Strang and Rogers, 1974). However, in cranberries, degradation rather than limited transport may play the dominant role in tolerance (Yaklich et al., 1974).

PERFLUIDONE

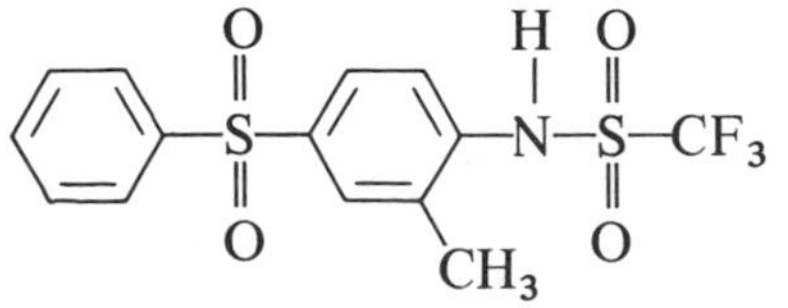

perfluidone

Perfluidone is the common name for 1,1,1-trifluoro-*N*-[2-methyl-4-(phenylsulfonyl)phenyl] methanesulfonamide. The trade name is Destun®. It is a white solid with a water solubility of 60 ppm at 22°C. It is available in wettable powder and liquid forms. The acute oral LD_{50} for rats is 633 mg/kg.

Uses

The major value of perfluidone is control of yellow and purple nutsedge in cotton. It also controls most annual grasses and certain broadleaf weeds. It is applied as a preemergence treatment. It is not recommended for cotton grown in Arizona, California, New Mexico, Oklahoma, or Texas.

Soil Influence

Although perfluidone appears to be adsorbed to some extent by mineral and organic colloids, it is subject to leaching. It is leached more readily in neutral and alkaline soils than in acid soils. It is subject to microbial degradation, and some herbicide disappears due to photodecomposition and volatilization. The half-life is about one month (Weed Science Society of America, 1979).

Mode of Action

Perfluidone induces foliar necrosis, inhibits shoot and root growth, and interferes with cell division (Davis and Dusbabek, 1975). Some responses in nutsedge resemble those caused by cytokinins, including breaking of bud dormancy and apical dominance, inducing basal bulb formation, and dwarfing shoot growth (Bendixen, 1975). Absorption of perfluidone by roots is slow, but translocation is moderate (Davis and Dusbabek, 1975; Lamoureux and Stafford, 1977). It appears to undergo hydroxylation and later combination with glucose.

PICLORAM

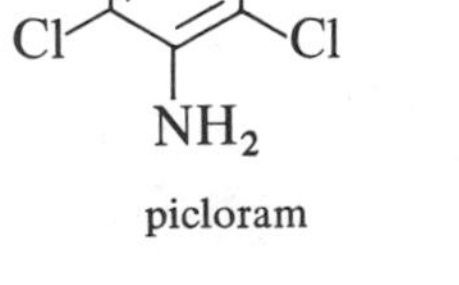

picloram

Picloram is the common name for 4-amino-3,5,6-trichloropicolinic acid. The trade names are Amdon® and Tordon®. It is a white solid with a water solubility of 430 ppm at 25°C. It is available in four basic forms: potassium salt, triethylamine salt, triisopropanolamine salt, and isooctyl ester. The major manufacturer produces 13 formulations of picloram.

These include liquid, pellet, and bead forms alone or in combination with 2,4-D, 2,4,5-T, or sodium borates. The acute oral LD_{50} of picloram for rats is 8200 mg/kg.

Uses

Picloram is effective on most perennial-broadleaf-herbaceous weeds and many woody species; however, most grasses are resistant. It will also control many annual broadleaf weeds. Rates as low as $\frac{1}{50}$ lb/acre will control many annual broadleaf weeds. On some species it is 10 times as potent as 2,4-D. When applied to the foliage of woody plants and broadleaf perennial weeds as a liquid spray, rates from $\frac{1}{2}$ to $1\frac{1}{2}$ lb/acre are used. When applied to thick stands of brush as pellets, rates of 6–8 lb/acre may be needed. It may also be applied as bark or cut-surface treatment to control woody plants, or to soil in noncrop areas for general weed control, (see Chapter 28, and Figure 20-3).

Soil Influence

Picloram is adsorbed by soil colloids, and the degree of leaching is inversely correlated to this adsorption and water-holding capacity of the soil. It is most easily leached through sandy, montmorillonitic soils low in organic matter.

Figure 20-3. Picloram-induced injury to wheat. *Left*: Untreated. *Right*: Treated. (C. L. Elmore, University of California, Davis.)

In contrast, it is leached with greatest difficulty through soils high in organic matter and lateritic clay soils. Picloram is very persistent in soils, but is subject to slow degradation by microorganisms. Phytotoxicity may often be detected well over one year after application.

Mode of Action

Picloram has high phytotoxicity; it is easily absorbed by roots and foliage, is truly systemic, and it is degraded slowly. It is absorbed by leaves, stems, and roots and is translocated both in the symplast and apoplast. Toxicity symptoms of picloram are quite similar to those of 2,4-D and other auxin-type herbicides–epinasty, cuplike leaves, and tissue proliferation. The mechanism of action is probably associated with modification of nucleic acid metabolism, and certain species may be resistant because of their high levels of nucleases (Molhotra and Hanson, 1970).

PYRAZON

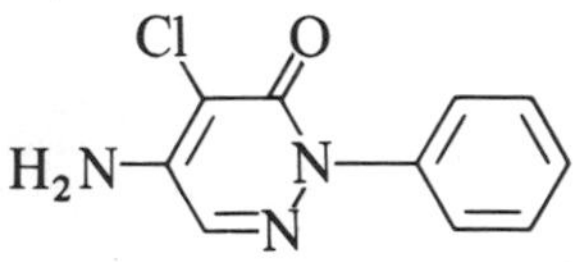

pyrazon

Pyrazon is the common name for 5-amino-4-chloro-2-phenyl-3(2*H*)-pyridazinone. The trade name is Pyramin®. It is a tan-to-brown solid with a water solubility of 400 ppm at 20°C, formulated as a wettable powder and as a flowable liquid. Pyrazon has an acute oral LD_{50} of 3000 mg/kg for rats.

Uses

Pyrazon is used for annual, broadleaf-weed control in sugar beets and red beets. Control of grasses has been variable. Also see page 419.

In sugar beets, it is applied preemergence, preplant soil-incorporated, or early postemergence. In red beets, pyrazon is applied as a preemergence or early postemergence treatment.

To control annual grasses and other tolerant weeds, pyrazon is often combined with other herbicides. A tank mix of pyrazon and endothall, as a preemergence application, will control additional weeds including green foxtail, koçhia, and wild buckwheat. This use is restricted to Colorado, Idaho, Nebraska, and Wyoming. Pyrazon plus ethofumesate is used preemergence or preplant soil-incorporated for wild oat and volunteer-cereal control.

Postemergence applications of pyrazon plus dalapon, phenmedipham, or nonphytotoxic emulsifiable oil are also used. The time of a postemergence application of pyrazon to both sugar beets and red beets is critical: It should be applied after beets have at least two expanded true leaves but before the weeds have two to four true leaves. Young beets in the cotyledonary stage may be injured, and larger weeds are not controlled.

Soil Influence

Soil type seems to have considerable influence on effectiveness and selectivity of pyrazon. Soil applications are not recommended on sands or loamy sands because of leaching and possible crop injury. On the other hand, adsorption on soils containing more than 5% organic matter precludes adequate weed control. Pyrazon is degraded fairly rapidly in warm, moist soils, persisting from one to three months.

Mode of Action

Pyrazon induces wilting, chlorosis and necrosis in leaves as well as inhibiting growth in susceptible species (Frank and Switzer, 1969; Rodebush and Anderson, 1970). Microscopic studies have shown that chloroplast structure is also altered (Anderson and Schaelling, 1970). Although pyrazon is absorbed by leaves, it is not translocated from them to a significant degree. It is readily absorbed by roots and distributed throughout the plant; thus pyrazon is primarily translocated in the apoplast. Although pyrazon is degraded in plants, the primary basis of the resistance of sugar beets appears to be associated with the conjugation of pyrazon with glucose to form a nontoxic molecule. Pyrazon has been shown to inhibit photosynthesis (Eshel, 1969; Hilton et al., 1969).

PETROLEUM OILS

Petroleum oils are hydrocarbons. Oils have gained commercial use as weed killers mostly since about 1940. However, use has decreased since 1973 because of the dramatic increase in the price of all petroleum products. Oils act as contact herbicides and may be used either as selective or as general weed killers. As selective weed killers, the varsol or Stoddard solvent-type oils are used to control both grass and broadleaf weeds in carrots, celery, parsnips, parsley, and in certain conifer seedlings. As a directed spray they effectively control weeds in cotton and woody ornamental plantings. General or nonselective oils are used to kill all plant life. The oils have been popular on railway roadbeds, canal banks, roadsides, barnyards, and in orchards where nontillage is practiced.

Size and structure of the oil molecule influences toxicity of oils to plants. Range in size and structure of molecules is partially reflected by the number of atoms and the degree of saturation of the compounds. Boiling point, viscosity, and specific gravity also influence the phytotoxicity of oils. Most oils wet vegetation more rapidly and more thoroughly than water, because of their low surface tension.

Classification

Petroleum products can be classified into *saturated* and *unsaturated* hydrocarbons.

Saturated Hydrocarbons

Saturated hydrocarbons have *single bonds between the carbon atoms.* Chemically, they are relatively inert. Saturated hydrocarbons can be further divided into straight-chain or the normal paraffin hydrocarbons, branched-chain paraffins, and ring or cyclic saturated hydrocarbons, also known as naphthenes and as cycloparaffins.

Straight-Chain–Saturated

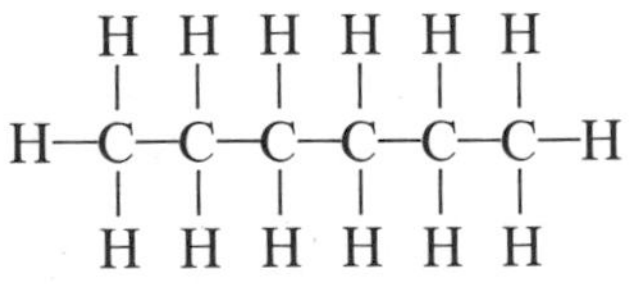

n-hexane (C_6H_{14})

Branched-Chain–Saturated

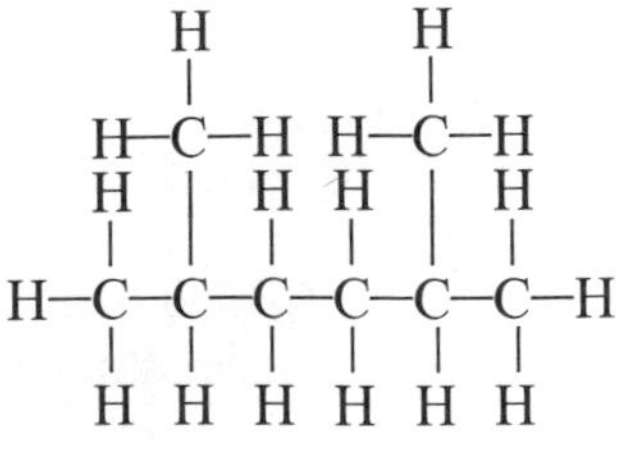

2,5-dimethylhexane (C_8H_{18})

Ring Structure–Saturated

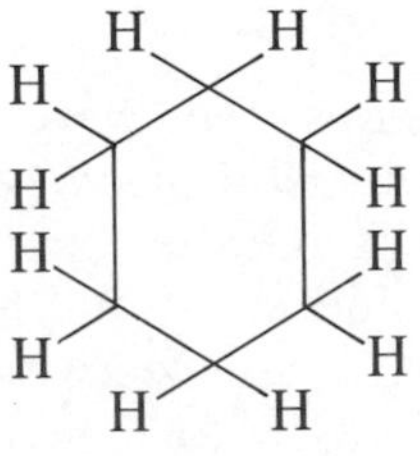

cyclohexane (C_6H_{12})

Unsaturated Hydrocarbons

Unsaturated hydrocarbons have *double or triple bonds between the carbon atoms.* Each molecule may have from one to many such bonds. Unsaturated hydrocarbons can be further classified as olefins (ethylene series), which include both straight-chain and branched-chain unsaturated hydrocarbons, and ring, or cyclic, unsaturated hydrocarbons, also referred to as aromatic hydrocarbons. The basic structure of this group is the unsaturated benzene ring. Two or more rings may condense; thus the naphthalene molecule may be represented as two benzene rings joined together, as illustrated. Compared with saturated hydrocarbons, the unsaturated hydrocarbons are chemically more reactive.

Olefin group (Straight-chain, unsaturated)

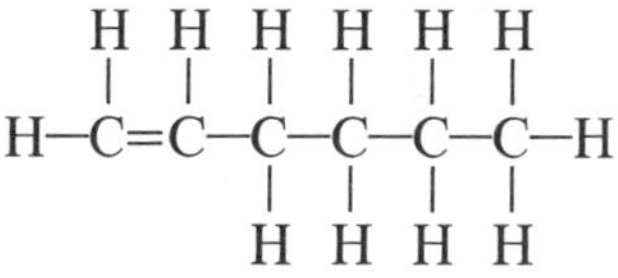

n-hexene-1 (C_6H_{12})

Ring structure–unsaturated (aromatic)

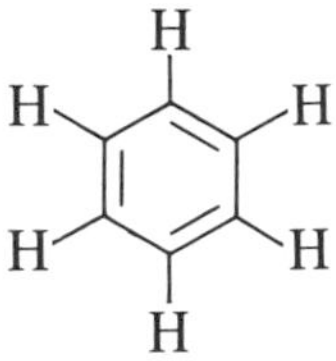

benzene (C_4H_6)

Double-ring structure-unsaturated (aromatic)

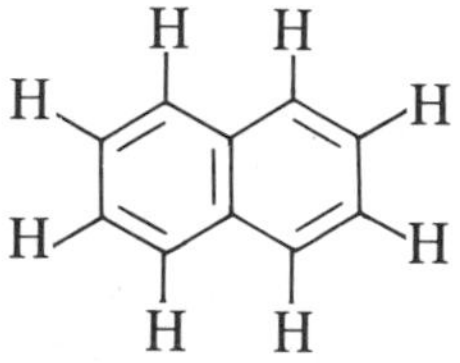

naphthalene ($C_{10}H_8$)

The aromatic molecule appears to be more toxic to plants than the olefin; and the saturated, straight-chain (paraffin) is the least toxic of the three (Havis, 1948) (see Figure 20-4). As one or more side chains are added to the aromatic group, toxicity to plant tissue increases; it increases until the molecular weight of the side chain equals that of the aromatic portion. Above this point the molecule becomes more like its paraffinic component (Bell and Norem, 1950).

Forms Sold on Market

Stoddard Solvents or Varsol®

Stoddard solvents or Varsol®, petroleum spirits, and mineral spirits are petroleum distillates of low flammability often used for dry-cleaning clothes or as paint thinners. They are normally clear and smell like re-

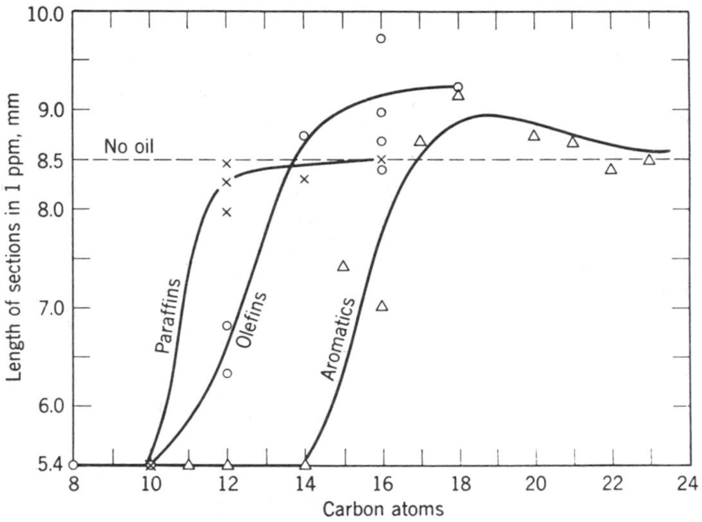

Figure 20-4. The effect of a 2-min exposure of various hydrocarbons on the growth of corn coleoptiles. The smaller the molecule, the more toxic the oil was to the plants. Also, as a class, the aromatics were most toxic, the olefins intermediate, and the paraffins least toxic to the plant. (J. van Overbeek and R. Blondeau, Shell Development Company, Modesto, CA.)

fined naphtha. Most Stoddard solvents have a gravity API (American Petroleum Institute) of 42° or above, a minimum flash point of 100°F, and an aromatic content of 8–20%.

These products are valuable for selective weed control in vegetable crops such as carrots, parsnips, celery, and parsley (Sweet et al., 1946). They are also used effectively in conifer-tree nurseries.

Gasoline and Kerosene

Gasoline and kerosene are somewhat phytotoxic but are not usually used for weed control for two reasons. (1) *Gasoline* presents considerable fire and explosion hazards to the applicator. Although certain susceptible plants may be killed very rapidly, others may not be injured because of the high volatility and low concentration of phytotoxic components of gasoline. (2) *Kerosene* is highly refined today and contains a high share of straight-chain compounds and a low share of unsaturated and ring-structure compounds; therefore, its toxicity to plants is relatively low.

Fuel Oils

Fuel oil is sold by grades, depending on the degree of refining, flash point, and distillation temperatures. Number 1 is the most highly refined; it is a relatively volatile, light oil, generally with a low content of unsaturated compounds. It is similar to, and may be the same as, kerosene. Number 6 is usually the lowest grade, having considerable quantities of impurities, a high viscosity, and usually many unsaturated and ring-structure compounds. Usually the lower the grade, the greater the contact toxicity to plants.

Diesel Oils

Diesel oils are usually moderately heavy, low-grade oils designed for burning in diesel motors. The gravity API commonly ranges from 35 to 43° and the flash point from 150 to 180°F. The product appears to be quite variable, with the aromatic and unsaturated content differing with the source and degree of refinement. In general, diesel fuel oils are too phytotoxic to use for selective weed control. They are effective as general contact herbicides and are often made even more toxic by fortification with dinoseb.

Weed Oils

Some petroleum oils are actually labeled as weed oils. These are similar to the products described above. Selective weed oils are essentially the same as Stoddard solvents. Nonselective weed oils are similar to low-grade fuel oils or diesel oils; however, in some cases distillate fractions high in unsaturated and ring-structure compounds are added to these two products to increase their phytotoxicity.

Factors Affecting Toxicity

Storage

Some oils apparently increase in toxicity with storage (Crafts and Reiber, 1948). Therefore, any selective weed oil that has been stored for any appreciable time should be tested on a small area before using it for field-scale weed control.

Boiling Point and Range of Distillation

The boiling point of oil is related to molecular weight, volatility, and toxicity to plants. Crude petroleum oils are composed of fractions with differing boiling points. If the oil has a very low boiling point, being volatile, it may evaporate before it affects the plant (Havis, 1950). High-boiling materials are viscous liquids or solids with molecules so large that they cannot penetrate the tissues (Bell and Norem, 1950). In general, hydrocarbons within the boiling range of 150–275°C (302–527°F) are most toxic to plants (Havis, 1950).

Flash Point

Flash point is of little importance in determining herbicidal toxicity, but it is especially important in determining explosion and fire hazards. Flash point is the temperature at which an oil vaporizes so rapidly that it forms an ignitable mixture with air in a container of given dimensions. Most contact oils have flash points above 180°F; most selective oils (Stoddard solvents) have flash points above 100°F.

Viscosity

Viscosity describes the flowing quality of an oil: It is the resistance of the fluid to motion. Light oils flow more rapidly, or pump more easily, than heavy oils. Viscosity of oils is usually determined experimentally as the time required for a given volume of the oil at a given temperature to drain through a specified opening. Viscosity influences the rate at which an oil will spread over and penetrate into a plant.

Specific Gravity

Specific gravity indicates the density or weight of an oil. Specific gravity of most oils lies between 0.73 and 0.95, compared with 1 for water. Specific gravity is measured by a hydrometer; the petroleum industry uses a Baumé scale (hydrometer) that has been adopted by the American Petroleum Institute (API). Therefore the "gravity API" will usually be given. The heavier the oil, the lower the degree reading. With other characteristics constant, the lower the reading, the higher the percentage of aromatics, and, therefore, the greater the toxicity.

General contact herbicidal oils usually have a gravity API maximum of 32° (heavy); selective oils have a gravity API of approximately 42° (light) (Bell and Norem, 1950).

How Oils Affect the Plant

Oils vary considerably in their toxicity to plants. As with most herbicides, the basic cause for the herbicidal action of oils is only partially understood; the effects of such an action are much better understood.

Plants sprayed with oil usually first show a darkening of the youngest leaf tips, presumably a result of a leakage of cell sap into intercellular spaces. This gives the plant a water-soaked appearance. There is loss of turgidity and drooping of stems and leaves. The plants have an odor of macerated tissue or freshly mown hay (Currier, 1951).

Theory of Oil Action

Scientists have depicted the cell membrane or plasma membrane as a double layer of lipoids (fat-like substances) held in place by one layer of protein on each side.

Other researchers have proposed that once oil reaches the inside of the leaf, it solubilizes the lipoids of the cell membrane. This process makes the semipermeable membrane more permeable, cell sap leaks into intercellular spaces, and the cell collapses (VanOverbeek and Blondeau, 1954) (see Figure 20-5). Because of cell-sap leakage, the plant appears water soaked.

Selective Action of Oils

The plasma membrane of some plants seems to resist the solubilizing action of Stoddard-solvent types of oils. Thus carrots, celery, parsnips, parsley, and some conifer-tree seedlings are not seriously injured by treatments that kill most annual weeds.

Penetration Into Plants

Oils may penetrate into plants through the stomates, thin cuticle, epidermis, bark, and even through injured roots. Plant leaves with a heavy cuticle and a small number of stomates permit little penetration of the oil. This explains why some desert plants with their heavy cuticle and few stomates absorb oils slowly. All oils exhibit low interfacial tension with plant cuticle, and, if viscosity permits, the oil rapidly spreads over the waxy surface.

Penetration through stomates has been demonstrated on citrus leaves (Knight et al., 1929; Turrell, 1947). Oil may also be absorbed directly through the cuticle (Ginsbury, 1931; Kendall, 1932; Knight et al., 1929).

Stomatal penetration can be demonstrated readily. If an emulsion of a light oil is sprayed on young plants in light when the stomates are open, the plant is killed. However, when the same emulsion is applied at night when the stomates are closed, the plant is not harmed (Van Overbeek and Blondeau, 1954).

Oils penetrate the cuticle, especially if it is thin. However, unfortified oils almost never penetrate through a tough, continuous, waxy cuticle such as the upper-leaf surface of the apricot (Van Overbeek and Blondeau, 1954).

Oils wet plant surfaces readily and tend to spread as a thin film. When oils containing a dye were placed on one-year-old apple twigs, lower-viscosity oils easily penetrated the bark, and the oil was slowly translocated through the twigs. The rate of penetration is related to the viscosity of the oil (Ginsbury, 1931) and to its surface tension (Knight and Cleveland, 1934). Light oils have lower viscosity and lower surface tension; they enter plant tissue more readily than heavy oils.

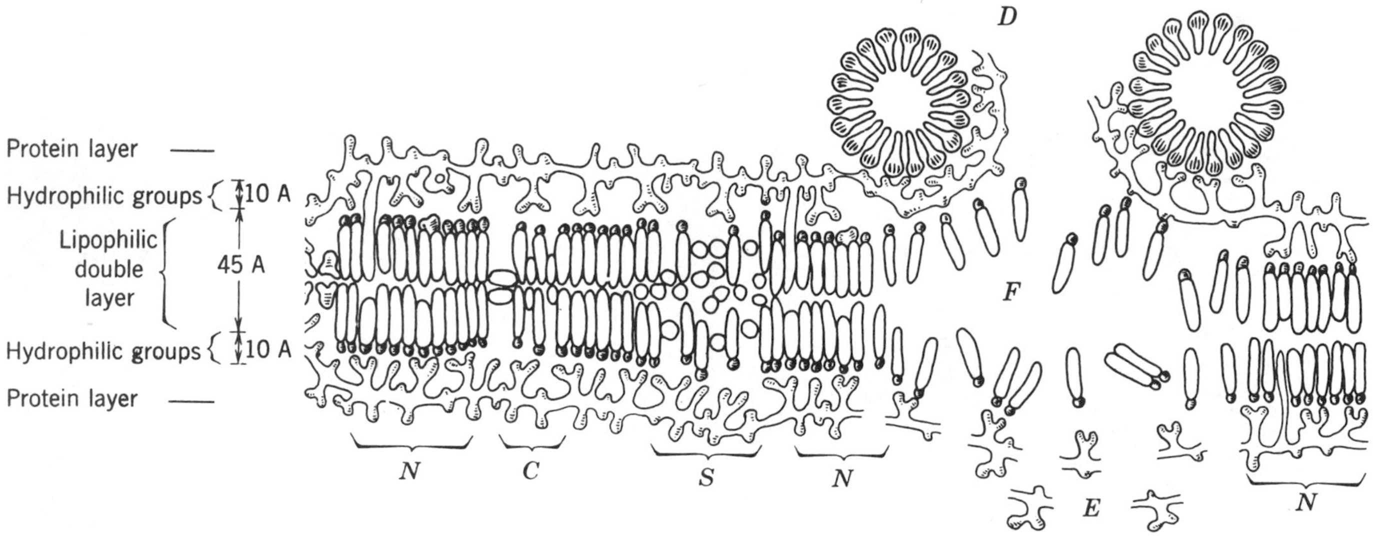

Figure 20-5. The plasma membrane as affected by various toxic molecules (*N*) Normal membrane consists of a double layer of fatty molecules stabilized by protein layers. Cell sap is kept inside the cell by fatty portion of the plasma membrane. (*C*) Fatty molecules being pushed apart by polycyclic hydrocarbons. The large molecules penetrate slowly. (*S*) Xylene solubilized into the fatty layer, penetration is rapid. (*D*) Detergent micelles are pulling away the protein layer, rendering the fatty layer *F* unstable. (*E*) Disruption of the protein layer is brought about by agents that liquify the protein. Solubilization of the plasma membrane causes leaks in the membrane, permitting the cell sap to leak into the intercellular spaces. (Van Overbeek and Blondeau, 1954.)

Translocation

After penetrating the leaf, the oil moves into the intercellular spaces (Kendall, 1932; Rohrbaugh and Rice, 1949). Although oils move from one part of the plant to other parts, scientists still do not fully know how they move. Most research workers believe that oils move principally through the intercellular spaces (Minshall and Helson, 1949; Rohrbaugh, 1934; Young, 1934, 1935) with little or no movement through the vascular system. Intercellular translocation greatly lessens the amount, rate, and distance of movement. Oil may move in any direction—up, down, radial, or tangential.

Scientists have studied movement of kerosene-like oils in dandelions, carrots, and parsnips. Oil applied to cut roots moved up into the leaves, and oil applied to leaves moved down to the roots. Movement was confined to intercellular spaces. In a large, turgid dandelion root, oil moved 4–5 cm (2 in.)/hr (Minshall and Helson, 1949). When oils were applied to dandelion tops, respiration rates increased in the roots (Rasmussen, 1947). Kerosene sprayed on bluegrass at proper rates will effectively remove dandelions; the dandelion is killed and the bluegrass is injured little or not at all (Loomis, 1938).

SUGGESTED ADDITIONAL READING

Albritton, R., and S. J. Parka, 1978, *Proc. Southern Weed Sci. Soc.* **31**, 253.

Anderson, J. L., and J. P. Schaelling, 1970, *Weed Sci.* **18**, 455.

Ashton, F. M., and A. S. Crafts, 1981, *Mode of Action of Herbicides,* Wiley, New York.

Bartels, P. G., and A. Hyde, 1970, *Plant Physiol.* **46**, 825.

Bartels, P. G., G. K. Matsuda, A. Siegel, and T. E. Weier, 1967, *Plant Physiol.* **42**, 736.

Bartels, P. G., and C. W. Watson, 1978, *Weed Sci.* **26**, 198.

Bayer, D. E., E. C. Hoffman, and C. L. Foy, 1965, *Weeds* **13**, 92.

Bell, J. M., and W. L. Norem, 1950, *Agr. Chem.* **5**, 31.

Bendixen, L. E., 1975, *Weed Sci.* **23**, 445.

Bernard, D. F., D. P. Rainey, and C. C. Lin, 1978, *Weed Sci.* **26**, 252.

Bonner, J., 1949, *Amer. J. Bot.* **36**, 429.

Carter, M. C., 1975, "Amitrole," pp. 377–398 in P. C. Kearney and D. D. Kaufman, Eds., *Herbicides,* Vol. 1, Marcel Dekker, New York.

Crafts, A. S., and W. A. Harvey, 1950, *Agr. Chem.* **5**, 28.

Crafts, A. S., and H. G. Reiber, 1948, *Hilgardia* **18**, 77.

Currier, H. B., 1951, *Hilgardia* **20**, 383.

Cutter, E. G., F. M. Ashton, and D. Huffstutter, 1968, *Weed Res.* **8**, 346.

Davis, D. G., and K. K. Dusbabek, 1975, *Weed Sci.* **23**, 81.

Dorough, H. W., 1974, *Bull. Environ. Toxicol.* **12**, 473.

Dorough, H. W., D. M. Whitacre, and R. A. Cardona, 1973, *J. Agr. Food Chem.* **21**, 797.

Eshel, Y., 1969, *Weed Res.* **10**, 196.
Frank, R., and C. M. Switzer, 1969, *Weed Sci.* **17**, 344.
Ginsbury, J. M., 1931, *J. Agr. Res.* **43**, 469.
Good, N. E., 1961, *Plant Physiol.* **36**, 788.
Havis, J. R., 1948, *Proc. Amer. Soc. Hort. Sci.* **51**, 545.
Havis, J. R., 1950, Cornell Agric. Expt. Sta. Mem. No. 298.
Hilton, J. L., A. L. Scharen, J. B. St. John, D. E. Moremald, and K. H. Norris, 1969, *Weed Sci.* **17**, 541.
Jones, D. W., and C. L. Foy, 1972a, *Weed Sci.* **20**, 116.
Jones, D. W., and C. L. Foy, 1972b, *Weed Sci.* **20**, 8.
Keeley, P. G., C. H. Carter, and J. H. Miller, 1972, *Weed Sci.* **20**, 71.
Kelly, S., and G. S. Avery, Jr., 1949, *Amer. J. Bot.* **36**, 421.
Kendall, J. C., 1932, N.H. Agric. Expt. Sta. Bull. No. 262.
Knight., H., J. C. Chamberlin, and C. D. Samuels, 1929, *Plant Physiol.* **4**, 299.
Knight, H., and C. R. Cleveland, 1934, *J. Econ. Entomol.* **27**, 269.
Lamoureux, G. L., and L. E. Stafford, 1977, *J. Agr. Food Chem.* **25**, 512.
Loomis, W. E., 1938, *J. Agr. Res.* **56**, 855.
Loomis, W. E., and F. Lyman, 1948, *J. Biol. Chem.* **173**, 807.
Minshall, W. H., and V. A. Helson, 1949, *Proceedings of the 3rd Northeastern Weed Control Conference,* p. 8.
Molhotra, S. S., and J. B. Hanson, 1970, *Weed Sci.* **18**, 1.
Nishimoto, R. K., and G. F. Warren, 1971, *Weed Sci.* **19**, 156.
Rafii, Z. E., and F. M. Ashton, 1979a, *Weed Sci.* **27**, 321.
Rafii, Z. E., and F. M. Ashton, 1979b, *Weed Sci.* **27**, 422.
Rasmussen, L. W., 1947, *Plant Physiol.* **22**, 377.
Rodebush, J. E., and J. L. Anderson, 1970, *Weed Sci.* **18**, 443.
Rohrbaugh, P. W., 1934, *Plant Physiol.* **9**, 699.
Rohrbaugh, L. M., and E. L. Rice, 1949, *Botanical Gaz.* **11**, 85.
Strang, R. H., and R. L. Rogers, 1974, *J. Agr. Food Chem.* **22**, 1119.
Sweet, R. D., R. Kunkel, and G. J. Raleigh, 1946, *Proc. Amer. Soc. Hort. Sci.* **48**, 475.
Turrell, F. M., 1947, *Botanical Gaz.* **108**, 476.
Van Overbeek, J., and R. Blondeau, 1954, *Weeds* **3**, 55.
Walker, C. R., 1963, *Weeds* **11**, 226.
Wiater, A., T. Klopotonski, and G. Bagdasarian., 1971a, *Acta Biochim. Pol.* **18**, 309.
Wiater, A., K. Krajewska-Grynkiewicz, and T. Klopotonski, 1971b, *Acta Biochim. Pol.* **18**, 299.
Weed Science Society of America, 1979, *Herbicide Handbook,* 4th ed., Champaign, IL.
Wills, G. D., 1976, *Weed Sci.* **24**, 370.
Yaklich, R. W., S. J. Karczmarczyk, and R. M. Devlin, 1974, *Weed Sci.* **22**, 595.
Yeo, R. R., 1970, *Weed Sci.* **18**, 283.
Young, P. A., 1934, *J. Agr. Res.* **49**, 559.
Young, R. A., 1935, *J. Agr. Res.* **51**, 925.

For chemical use, see the manufacturer's label and follow the directions. Also see the Preface.

21 Inorganic Herbicides

Inorganic herbicides are those weed-control chemicals that contain no carbon. The principal ones are the borates and chlorates, but the cyanates, calcium cyanamide, and ammonium sulfamate (AMS) also fall into this group.

Most inorganic herbicides were used before the modern era of organic herbicides began with 2,4-D in the mid-1940s. Although various organic herbicides have replaced these inorganic herbicides for many uses, they are still used.

SODIUM CHLORATE

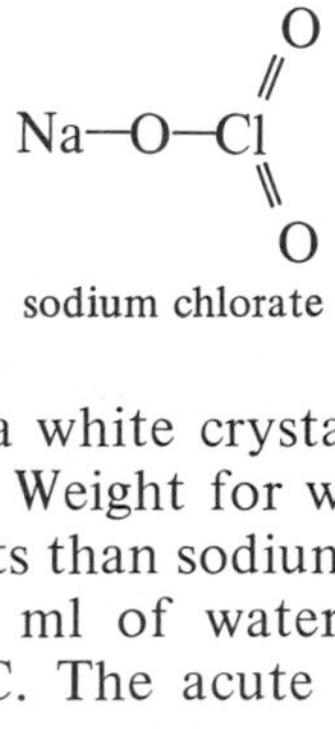

sodium chlorate

Sodium chlorate ($NaClO_3$), a white crystalline salt, looks like common table salt (sodium chloride). Weight for weight, sodium chlorate is 30–50 times more toxic to plants than sodium chloride. Sodium chlorate is very soluble in water; 100 ml of water at 0°C will dissolve 79 g—increasing to 230g at 100°C. The acute oral LD_{50} is 5000 mg/kg for rats.

Sodium chlorate has a salty taste. "Salt-hungry" animals may eat enough to be poisoned; 1 lb of this chemical per 1000 lb of animal weight is considered lethal. Also, after spraying, some poisonous plants ordinarily avoided by livestock become palatable.

Fire Danger

Sodium chlorate, with three atoms of oxygen, is highly flammable. This is especially true when it is mixed with organic materials such as clothing, wood, leather, or plant parts and allowed to dry. It has been ignited by the sun's rays, clothing friction, or shoes scraping a rock. The fire cannot be smothered, since the chemical provides the needed oxygen.

Sodium chlorate can be applied as a spray or as dry crystals. Spraying gives more-even coverage and a faster kill, but the fire hazard may be serious when the foliage dries. Dry crystals applied to *dry* vegetation usually bring no serious fire hazard. The chemical is leached into the soil with the first rain. But if crystals are applied to *damp* vegetation, they will adhere. When vegetation dries, it will be flammable (almost as if it was sprayed).

The person applying sodium chlorate must follow these precautions: (1) Wear rubber shoes or overshoes (not leather), as the crystals tend to shed from rubber; (2) wear trousers without cuffs; (3) change contaminated clothing before it dries, and *wash it immediately*. If allowed to dry, these clothes will be a serious fire hazard.

Uses

Sodium chlorate is generally used as a sterilant to kill all vegetation. It is widely used on cropland for spot treatment to control the spread of serious perennial weeds, and is broadcast on rights-of-way of highways and railroads. Other chemicals have proved more effective for selective weed control.

In practice sodium chlorate is often combined with sodium borates. See the section on borates later in this chapter.

Persistence in the Soil

Leaching quickly removes sodium chlorate from the soil (Seely et al., 1948). Also, soil microorganisms decompose chlorates to chlorides. Decomposition usually moves fastest in moist soils above 70°F. With low rainfall, chlorate may remain toxic for five years or longer. In humid Southeastern states, toxicity may disappear in 12 months on heavy soils and in 6 months on sandy soils.

Ease of leaching may be a disadvantage—heavy rains or irrigation soon after application may remove the chemical from the upper 2–3 in. of soil. This absence would allow shallow-rooted weeds to keep on growing.

Mode of Action

Absorption

The plant absorbs sodium chlorate rapidly through both roots and leaves. Dormant seeds in the soil usually survive the rates commonly used.

When sodium chlorate is sprayed on leaves, chlorate ions penetrate the cuticle and come into direct contact with living cells (Loomis et al., 1933). The stomates need not be open for the chemical to enter the leaf (Meadly, 1933).

Translocation

Chlorate moves rapidly from the roots upward through the xylem tissues. Since xylem tissue is composed mostly of dead cells, chlorate moves freely regardless of its toxicity.

Since the chlorate kills living cells rapidly, it probably moves downward through the phloem very slowly, if at all. This is because only *living* phloem cells are active in translocation.

Effect on Metabolism

Sodium chlorate does three things: (1) It depletes the plant's food reserves (Bakke et al., 1939; Crafts, 1935; Latshaw and Zahnley, 1927); (2) it temporarily increases the rate of respiration; (3) it decreases catalase activity (Neeler, 1931). This decrease can cause an increase in hydrogen peroxide, which is toxic to plants.

BORATES

Boron is an essential minor element for plant growth. In some areas, approximately 30 lb/acre are applied to improve plant growth. However, in large quantities boron is toxic to plants; it acts as a soil sterilant.

Borates used as herbicides are boron salts containing sodium and oxygen plus water. Herbicidal salts of boron are sodium metaborate tetrahydrate ($Na_2B_2O_4 \cdot 4H_2O$), sodium metaborate hexahydrate ($Na_2B_2O_4 \cdot 6H_2O$), disodium octaborate tetrahydrate ($Na_2B_8O_{13} \cdot 4H_2O$), and sodium tetraborate-anhydrous ($Na_2B_4O_7$). The latter is also available in hydrated forms, $Na_2B_4O_7 \cdot 5H_2O$ (pentahydrate) and $Na_2B_4O_7 \cdot 10H_2O$ (decahydrate).

Boron, a nonselective herbicide, has been used effectively by railroads, farms, government agencies, and industry to destroy fire-prone and unsightly vegetation. The paving industry uses it under asphalt to prevent weeds from growing up through the asphalt.

These sodium borates are nonflammable; noncorrosive to ferrous metals, rubber, and plastics; nonvolatile; only slowly leached; and can be stored with little or no risk to fungicides, insecticides, and fertilizers. They are not deactivated by light. When used in sufficiently large quantities to be an effective herbicide, they are toxic to most soil microorganisms.

The toxicity of sodium borate to man and animals is low when normally handled and applied. The acute oral LD_{50} of these compounds ranges from 2000 to 5560 mg/kg for rats.

Uses

In the past, borates have been used alone as herbicides; however, now they are usually combined with other herbicides such as sodium chlorate, bromacil, diuron, and monuron. Combinations can all be applied with granular-application equipment, and several can also be dissolved or suspended in water and applied with standard spray equipment.

The combination of borates and sodium chlorate reduces the fire hazards of sodium chlorate discussed previously in this chapter. This combination also gives more-effective, long-term, nonselective weed control, because sodium chlorate is leached more readily than the borates. Diuron or monuron is added to some formulations to give better control of certain annual weeds and seedlings of most perennial weeds. Bromacil is added to some formulations to control certain perennial grasses more effectively; also, because boron is toxic to most soil microorganisms, it greatly retards microbial degradation of diuron, monuron, and bromacil.

Mode of Action

Sodium borates are absorbed principally by roots, translocated through the xylem to all parts of the plant, and accumulated in the leaves. In herbicidal quantities borates cause plant desiccation, beginning with burning and necrosis of leaf margins.

The chemical is usually carried into the soil by rainfall. The herbicide is most effective on young and tender plants. Therefore, treatment should be timed so that the borate will have reached the root-absorption zone by the time plant growth is just starting. The quantity required will vary with soil type, rainfall, and the weed species to be controlled. Recommended rates vary between 3 and 12 lb/100 ft^2.

In warm, moist soils the chemical usually remains effective as a soil sterilant for about one year, but applications can usually be reduced in succeeding years. In dry or frozen soils, sterilant effects continue for several years.

AMS (AMMONIUM SULFAMATE)

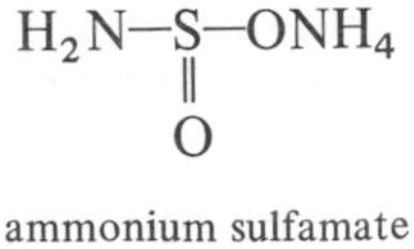

ammonium sulfamate

Ammonium sulfamate, sold as "Ammate," is an effective woody-plant killer. It is nonflammable, nonvolatile, and as normally used, nonpoisonous to man or livestock. It can be applied safely where 2,4-D and related products are hazardous to apply. That is why it is used principally on rights-of-way, roadsides, and ditches that adjoin crops such as cotton, tomatoes, grapes, and tobacco (see Figure 21-1).

It is also applied to stumps to prevent sprouting and in frills or notches to kill undesirable hardwood trees without cutting. Undesirable woody plants mixed in with a desirable woody hedge can be killed by the jar method. This involves placing and leaving the tops of undesirable plants in a jar or bucket of ammonium sulfamate solution.

The chemical acts as both a contact and translocated herbicide. It will also give temporary soil sterility. Because of its nitrogen and sulfur content, it may have a fertilizing effect after breakdown in the soil.

Ammonium sulfamate as sold commercially is a yellow crystalline substance. The pure chemical is a colorless crystalline substance. It is very soluble in water and absorbs moisture when exposed to the air.

In solution the chemical corrodes some metals, especially brass and copper; it also affects steel surfaces exposed to air such as the pump, the outside of tanks, and truck or tractor parts. Stainless steel, aluminum, and bronze are resistant. Metal parts covered by the spray solution inside the tank, pump, and lines corrode much more slowly.

Exposed areas should be coated with protective paints or covered with oil. Rinsing exposed surfaces with water after each day's use is important. Thorough cleaning inside and out at the end of the season, fol-

Figure 21-1. Ammonium sulfamate used as a spray to control the brush on the left side of the road. The tobacco on the right side of the road was not injured. (E. I. du Pont de Nemours and Company.)

lowed by a coating of oil, will preserve the equipment. Corrosive effects are considered negligible on fences, guy wires, and telephone wires.

Ammonium sulfamate has a very low order of toxicity to humans and livestock. When the chemical was included in the daily feed of rats, and even when reasonably large quantities were injected into their bloodstreams, no serious ill effects were noted (Ambrose, 1943). Sheep have been fed up to 0.5 lb/day without apparent injury. The acute oral LD_{50} for rats is 3900 mg/kg.

Ammonium sulfamate was tested on 194 human subjects using the procedure described by the Office of Dermatoses Investigation, U.S. Public Health Service. No contact irritation or evidence of sensitization was found.

OTHER SALTS

Many salts are toxic to plant tissues if applied in high concentrations. This is especially true of many fertilizer salts such as ammonium nitrate, urea, and potassium chloride. In sufficient concentration these provide a contact burning effect. Dissolved in water and added as a wetting spray, many annual weeds, especially broadleaf weeds, are easily killed. Addition of a wetting agent may double the contact killing effect (Klingman and Davis, 1954) (see Figure 6-6).

SUGGESTED ADDITIONAL READING

Ambrose, A. M., 1943, *J. Ind. Hyg. Toxicol.* **25**, 26.

Bakke, A. L., W. G. Gaessler, and W. E. Loomis, 1939, Iowa Agric. Expt. Sta. Bull. No. 254.

Crafts, A. S., 1935, *Plant Physiol.* **10**, 699.

Klingman, G. C., and J. C. Davis, 1954, *Proceedings of the 7th Southern Weed Conference* p. 167; 174.

Latshaw, W. L., and J. W. Zahnley, 1927, *J. Agr. Res.* **35**, 757.

Loomis, W. E., V. E. Smith, R. Bissey, and L. E. Arnold, 1933, *J. Amer. Soc. Agron.* **25**, 724.

Meadly, G. R. W., 1933, *J. Dept. Agric. W. Australia* **10**, 481.

Neeler, J. R., 1931, *J. Agr. Res.* **43**, 183.

Seely, C. E., K. H. Klages, and E. G. Schafer, 1948, Washington Agric. Expt. Sta. Bull. No. 505.

Weed Science Society of America, 1979, *Herbicide Handbook,* 4th ed., WSSA, Champaign, IL.

For chemical use, see the manufacturer's label and follow directions. Also see the Preface.

22 Small Grains and Flax

Small grains discussed here include wheat, oats, barley, rye, and rice. Flax is included because its cultural practices and weed problems are similar to those of small grains.

Winter varieties are planted in the fall, live through winter, and are harvested the following summer. Spring varieties are planted in early spring and harvested in mid to late summer. As an average, tolerance to cold is in this order: rye, wheat, barley, oats, and rice. Therefore, winter rye is grown in far northern areas. Rice in the United States is normally planted in the spring.

In general, winter varieties are most often infested with winter-annual weeds and to a lesser extent by summer annuals that germinate in early spring. Spring varieties are primarily infested by summer annuals that germinate in the early spring. Perennial weeds are also troublesome in certain areas.

EFFECT OF WEEDS ON YIELD

Weeds compete directly with the grain crop for light, moisture, carbon dioxide, and soil nutrients. Also, many weeds have an allelopathic effect on the crop (see Chapter 2). Grain yield reductions range from crop failure to losses so slight that they are not measurable (see Figure 22-1 and Tables 22-1, 22-3 and 22-4).

Fifty fields were selected at random in 1956 and another 50 in 1957 near Winnipeg, Canada, to clearly establish actual losses in small grain and flax resulting from weeds. These fields included a complete range of weed infestations from those with satisfactory weed-control programs to those badly infested with weeds. In each field 10 paired plots were staked out. One plot of each pair was kept weed free by hand weeding. Principal weed species included wild mustard, wild oat, Canada thistle, sowthistle, and green foxtail. Average losses in yields caused by weeds are shown in Table 22-1.

Figure 22-1. Oats sprayed with 2,4-D to remove wild mustard. (New Jersey Agricultural Experiment Station.)

Lowest losses were recorded where the crop-management program in previous years had provided effective weed control. Highest losses occurred when there was neither herbicide-spraying program nor crop rotation. Farmers with no weed-control programs experienced much heavier losses than indicated in Table 22-1.

Weed competition early in the season reduces yields more than late-season competition. Although yields are not greatly reduced, late-season weeds may cause difficulty in harvesting. As well as causing smaller yields and harvesting difficulties, weeds lower crop quality and may reduce the protein content of grain.

Table 22-1. Yield Reductions Expressed as a Percentage From Weed Competition (Friesen, 1957)

Year	Wheat (%)	Barley (%)	Oats (%)	Flax (%)	Total Crop Average (%)
1956	19.5	–	10.6	26.6	15.9
1957	14.3	18.6	6.5	27.6	17.3

WEED-CONTROL METHODS

Weed-control methods in small grains and flax include clean seed, crop rotation, good seedbed preparation, full use of competition, and application of herbicides.

Clean Seed

The importance of clean seed can hardly be overemphasized. This is discussed in Chapter 3.

Crop Rotation

With continuous production of small grains or flax, weeds that grow during the same season and are favored by the crop-management programs build up rapidly.

Farmers learned that small grain or flax could usually be grown for several years on "new land" before weeds became a serious problem. After this time, they were forced to use more-effective methods of weed control. Crop rotation proved effective against many species.

Crop rotation is a strong link in the chain of improved weed-control practices. Herbicides can be combined effectively into a crop-rotation program. In some areas, the rotation of cotton, corn, and small grain with lespedeza illustrates this point. As a result of variations in the time of seedbed preparation, time of cultivations, and period of growth, plus the use of different herbicides in each of the four crops, the buildup of many weed species is prevented. Here "crop rotation" and "chemical rotation" are combined for maximum weed control.

Seedbed Preparation

Weed control is one of the principal purposes of seedbed preparation. Because small grains cannot be effectively cultivated after sowing, the importance of controlling weeds before sowing is obvious. In some areas where weed seedlings appear before sowing, weed competition can be reduced by a presowing cultivation. Where practical, summer fallow is especially effective because weed-seed production can be prevented during an entire summer season, and the growth of perennial weeds can be checked.

Competition

Thickly planted, fast-growing small grains offer considerable competition to weeds. In such grains few or no weeds may be found at harvest time. Yet had the grain not crowded them out, the area would have been solidly covered with weeds (see Figure 22-2).

Well-adapted, disease-resistant varieties, planted at the proper time with adequate soil fertility and moisture may be able to grow slightly faster than weeds. If the crop emerges ahead of the weeds, it may compete effectively with many troublesome weed species.

Flax is only partially effective in competing with weeds. With its slow early growth and small leaf surface, the flax plant offers little competition to the weeds. Nevertheless, a thick stand of flax is important for weed control, as shown in Table 22-4. Even with good stands of flax, the amount of weeds was still large.

Figure 22-2. Competition from the crop crowds out many weeds. (North Carolina State University.)

Chemical Control

Weeds are often a serious problem in small grains and flax, even with good cultural practices. Some weeds are favored by the same management programs that favor small grains.

Chemicals can often be used in small grains in such a way that crop competition, as discussed above, is increased. The herbicide may stunt weeds with little or no reduction in the growth of crop plants. With this slight competitive advantage, small grain may "crowd out" the weeds (see Figure 22-2).

Information given here and in following chapters on chemical weed control should be considered only as a guide. The performance of herbicides can and does vary under different climatic and soil conditions as well as the stage and rate of plant growth (see Chapter 4). Therefore, state or county extension agents and local company field representatives should be consulted. *It is essential to follow directions on the product label.* These may vary for different geographical areas. For a listing of herbicides used in wheat, barley, oats, rice, and flax, see Table 22-2.

BARLEY, OATS, AND WHEAT

2,4-D controlled weeds on 71% of U.S. wheat acreage in 1976, according to the U.S. Department of Agriculture. Dicamba was used on 6.8% and MCPA on 5.5% of the wheat acreage (USDA, 1980). Only these three herbicides will be discussed in detail.

2,4-D

2,4-D controls a wide spectrum of weeds in small grains. It is relatively low-cost and can be applied by the usual farm spray equipment. More acres of small grain are treated with 2,4-D than any other crop with any other herbicide.

Time of Treatment

Good weed control without crop injury usually depends on proper timing in the application of herbicides. This is particularly true for 2,4-D in small grains. For our discussion, growth of small grains is divided into four stages:

1. Zero- to four-leaf stage.
2. Four-leaf to boot stage.
3. Boot stage to flowering.
4. Soft-dough-grain stage to maturity.

Table 22-2. Wheat, Barley, Oats, Rice, and Flax Herbicides[1]

Herbicide[2]	Wheat	Barley	Oats	Rice	Flax
Barban	X	X			X
Bentazon				X	
Benthiocarb				X	
Bifenox				X	
Bromoxynil	X	X	X		X
Butachlor				X	
Chlorbromuron	X				
2,4-D	X	X	X	X	
2,4,5-T				X	
Dalapon					X
Diallate		X			X
Dicamba	X	X	X		
Dichlofop-methyl	X	X			
Difenzoquat	X	X			
Dinoseb	X	X	X		
Diuron	X	X	X		
EPTC					X
Glyphosate	X	X	X		
Linuron	X				
MCPA	X	X	X	X	X
Metribuzin	X	X			
Molinate				X	
Oxadiazon				X	
Paraquat	X	X			
Picloram	X	X	X		
Propanil	X			X	
Propham				X	X
Silvex				X	
Terbutryn	X	X	X		
Triallate	X	X			
Trifluralin	X				

[1] For rates, instructions, combinations with other herbicides, and restrictions, see the label.

[2] For weeds controlled, see page 418. For trade names see page 422.

Periods of greatest susceptibility to 2,4-D are during the periods of rapid growth. It appears that the rate of meristematic development is closely related to the plant's susceptibility to 2,4-D.

The four-leaf up to just before the boot stage is the recommended stage for the application of 2,4-D for reasons discussed below.

Zero- to four-leaf stage. Small grains, including rice, are very sensitive to 2,4-D during germination and seedling stages. Therefore, treatment during this time will usually cause many malformations of the head and onion leaves, general stunting of the plant, and reduced yields (see Figures 22-3 and 22-4).

Four-leaf to boot stage. Four-leaf up to just before the boot stage includes the fully tillered stage and is considered to be the *most desirable time* to apply 2,4-D (see Figure 22-3). In addition to the grain's tolerance at this time, weeds are usually small and easily killed. Also, weeds have not caused serious damage as far as competition is concerned, and ground-spray equipment causes but slight mechanical injury to the grain.

Most varieties will withstand up to $\frac{1}{2}$ lb of 2,4-D (acid equivalent) per acre at this time without yield reductions; tolerant varieties will withstand 2 lb/acre (Friesen, 1957). In general, wheat varieties are the most tolerant, barley is intermediate, and oats are the least tolerant. Rates over $\frac{1}{2}$ lb/acre are suggested only where resistant weeds require this amount for control. The above high rates are for amine forms of 2,4-D. Ester forms are used at about half the above rates ($\frac{1}{4}$-$\frac{1}{2}$ lb/acre).

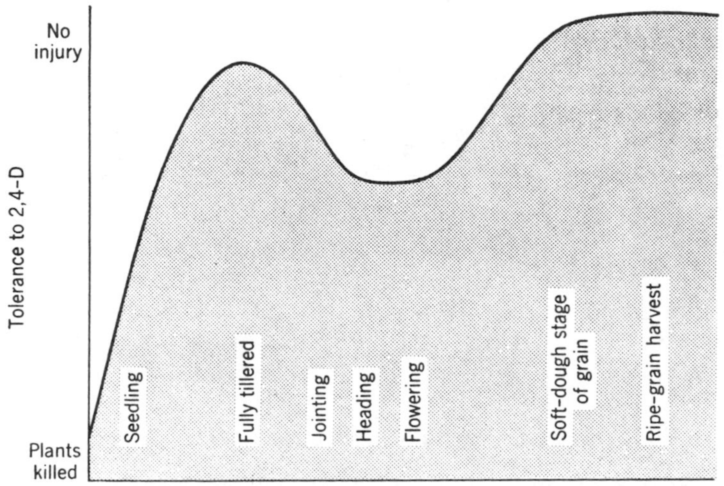

Figure 22-3. Stages of small-grain growth and the degree of tolerance to excessive rates of 2,4-D, *with most tolerant stage listed first.* (1) Soft dough of grain stage to maturity. (2) Fully tillered; four leaves or more per plant; 5–8 in. tall. (3) Jointing stage through flowering. This includes the boot stage to the flowering stage. (4) Germination to four-leaf stage. (North Carolina State University.)

Figure 22-4. Thatcher wheat injured by premature treatment with 2,4-D. The wheat was about 3 in. tall when treated, and from 60 to 70% of the heads were abnormal. (J. B. Harrington, University of Saskatchewan.)

Boot stage through flowering. During the boot stage the internodes elongate rapidly. This stage is also known as the jointing stage. From the start of the boot stage through flowering, small grains are susceptible to injury. 2,4-D sprays at this time may reduce yields seriously.

Soft-dough-grain stage to maturity. Small grains are very tolerant to 2,4-D at this time. However, this treatment is not usually recommended because ground-spray equipment will crush much of the grain, weed competition has already done its damage to crop yield, and there are possible residues of 2,4-D in harvested grain (see Figure 22-4).

Effect of 2,4-D on Germination

Germination studies indicate little or no damage to germination of grain from plants treated with 2,4-D when the grain is treated at the proper time and rate. When treated during a susceptible stage (boot stage), wheat showed an 11% drop in germination (Helgeson, et al., 1948).

Effect of Soil Fertility on Crop Tolerance to 2,4-D

Oats were grown under three levels of soil fertility in New Jersey. All plots were treated at 1 lb/acre of 2,4-D. A 10-10-10 fertilizer was applied to provide medium- and high-fertility plots. Yields were not reduced on low-fertility plots compared with low-fertility check plots, even though several 2,4-D sprays were applied at stages when grain is normally susceptible to the rate of 2,4-D used.

Yields were reduced on medium- and high-fertility plots when treated during a period of susceptibility. Treatment in the fully tillered stage (tolerant stage) did not reduce yields. Therefore, it is concluded that high soil fertility can be expected to increase the crop's susceptibility to injury.

If, as previously discussed, plants are most susceptible during periods of rapid growth or rapid development of meristematic tissues, the above facts are to be expected. Higher rates of fertility will have a tendency to stimulate growth and thus increase the rate of meristematic tissue development.

Variety Tolerance to 2,4-D

Many studies have been made of variety tolerance to 2,4-D. Each study has shown considerable variation in tolerance among varieties.

Due to widespread use of the chemical, 2,4-D tolerance might well be one point of selection in an improved-variety breeding program.

Dicamba

In small grains, dicamba, an auxin-type herbicide, controls some broadleaf weeds not readily controlled with 2,4-D; for example, Canada thistle, field bindweed, silverleaf nightshade, smartweed, velvetleaf, and wild buckwheat. It is applied in the spring before the joint stage of fall-seeded grain, or to spring-seeded grain prior to the five-leaf stage.

MCPA

MCPA is less injurious to small grains, flax, and some legumes than 2,4-D (this is especially so for flax and oats), and it is more effective on a few broadleaf weeds, for example, hempnettle and Canada thistle. However, it is less effective than 2,4-D on most broadleaf weeds, and it usually costs more in the United States.

MCPA is less injurious than 2,4-D to most legumes, especially red clover. There is no difference in alfalfa. Vetch and sweetclover seedlings may be severely damaged.

MCPA is especially suggested for weed control in flax, in those varieties of oats known to be susceptible to 2,4-D, and with tolerant clovers such as red clover.

TWO SPECIAL TOPICS

Two aspects of small grains exert major effects on these crops. Both deserve special discussion. These are wild oat and ecofallow.

Wild Oat

Wild oat is a serious problem weed in spring wheat, spring barley, spring oats, and flax. It is less of a problem in winter cereals. Yield losses in North America have been estimated as high as $1 billion per year. Crop yield losses are due to both allelopathy and competition. More than 50 different strains of wild oat give the weed a wide area of adaptation. Many of the seeds are dormant, making eradication impossible. Most seeds germinate with soil temperatures of 50°–60°F.

Crop rotation and summer fallow provide only limited control. Delayed spring seeding, with effective spring cultivation just before seeding, is reasonably effective. However, seeding delayed later than the recommended date of seeding may result in a serious drop in crop yield.

Herbicides such as triallate, trifluralin, and diallate can be soil-applied to partially control wild oat. To selectively control wild oat through foliar-spray application, the following herbicides are available: barban, dichlofop-methyl, difenzoquat, and triallate.

Ecofallow for Small Grains

Where moisture regularly limits production of small grain, crops are often produced every second year or in other cases with two crops in three years. During the no-crop (fallow) year, the soil is kept weed-free for the main purpose of storing moisture in the soil. *Ecofallow* means controlling weeds with herbicides during a fallow period with reduced or no tillage.

The soil loses moisture principally through weedy plants and from drying of the soil. Tillage of the soil usually increases the drying process–to the depth of tillage. Without herbicides farmers usually must till the soil five to seven times to control weeds during a fallow season.

Herbicides help control weeds and thereby reduce the number of tillage operations needed. Herbicides used to control emerged fallow weeds include 2,4-D, dicamba, glyphosate, paraquat, and terbutryn. Other herbicides (alachlor, atrazine, cyanazine, metolachlor, and propham) provide preemergence residual weed control and therefore provide a longer time interval between tillages.

Advantages to this program are the following:

1. *More moisture stored* through improved weed control, less soil tillage, and more surface mulch.
2. *Less energy required* because of reduced soil tillage.
3. *Less soil erosion* because of a greater soil mulch cover with plant residues–such as straw, sorghum residues, or corn stalks.

4. *Larger crop yields.* There is direct relationship between crop yields and moisture built up in the soil—in areas where fallow is practiced.

Herbicides used for this program must provide long periods of weed control, but must be decomposed to harmless soil residues by planting time for the next crop. Further large-scale development of this program depends upon more-effective, safe herbicides that farmers can afford to use.

RICE

Weeds reduce yields of rice in the United States an average of about 15%, even with present weed-control methods. In Arkansas, with 10 rice plants/ft^2, one barnyardgrass/ft^2 reduced rice yields by 40% and five barnyardgrass plants reduced rice yields by 66% (see Table 22-3).

Rice responds to essentially the same weed-control practices as wheat, oats, and barley, and its response to phenoxy herbicides is also similar. Rice is grown under "paddy" conditions (flooded with water) or "upland" conditions (dryland conditions). In the U.S., rice is grown almost entirely under paddy conditions. These cultural differences influence weed-control practices. For example, in paddy rice, barnyardgrass can be partially controlled by seeding the rice in water and by keeping 4–8 in. of water over the soil surface for the first four weeks. The water lowers the oxygen content of the soil, thus reducing germination of many weed seeds. However, if bluegreen algae (scum) appears, draining the water may be necessary. Algae can also be controlled in rice with organic copper formulations.

Principal weeds of upland rice are coffeeweed, curly indigo, and grassy weeds, especially barnyardgrass. Main weeds of paddy rice are barnyardgrass and aquatic species, especially arrowhead, ducksalad, waterplantain, sedges, spikerush, and bulrush.

Table 22-3. Effect of Barnyardgrass on the Yield of Rice in Stuttgart, Arkansas (Smith, Flinchum, and Seaman, 1977)

Plants per Square Foot			Yield Loss per Acre	
Rice	Grass	Yield of Rice (lb/acre)	Percent	Dollar Value[1]
10	0	5160	None	None
10	1	3110	40	102
10	5	1760	66	170
10	25	580	89	229

[1] Rough rice valued at $5 per hundredweight.

Crop rotation with related herbicide rotation effectively holds back hard-to-control weeds such as red rice, some sedges, and some broadleaf weeds. A crop–herbicide rotation with two years of soybeans with herbicides such as alachlor, trifluralin, profluralin, and metribuzin will effectively control red rice and many other weeds. Paraquat can be applied as a directed postemergence treatment to remove red rice from soybeans. When rice is planted the third year, red rice causes little loss in yield (Smith, Flinchum and Seaman, 1977).

Herbicides used in rice are listed in Table 22-2.

Phenoxy-Type Herbicides

2,4-D has been used widely to control many broadleaf weeds in rice at rates of $\frac{1}{2}$–$1\frac{1}{2}$ lb/acre. If weeds are covered with water, the water is lowered to expose the weeds to spray. 2,4-D is applied between the stages of late tillering and early jointing. MCPA is more selective to rice and is a safer material to control some broadleaf weeds. Other phenoxy herbicides used in rice are 2,4,5-T and silvex.

Precautions should be taken to avoid drift to susceptible plants.

FLAX

Weed control in flax is much like that in small grains. As discussed in the section on competition, flax has a small leaf surface. It offers little competition to weeds for light, so naturally weeds are often a serious problem in flax (see Table 22-4). Herbicides used in flax are listed in Table 22-2.

Flax is usually sprayed with MCPA for broadleaf-weed control when the flax has at least three leaves and before weed seedlings are 2 in. tall (see Figure 22-5). MCPA (amine form) is usually applied at $\frac{1}{4}$-lb/acre for

Table 22-4. Effect of Stand of Flax on Yields of Seed, Straw, and Weeds (Hay, 1970)

Stand of Flax on July 25 (%)	Flax Seed (bushels/acre)	Flax Straw (lb/acre)	Weed Plants (lb/acre)
130	15.8	2068	291
100	17.3	2073	312
55	14.2	1537	576
17	11.1	1083	1329
LSD (0.05)	3.5	329	350

Figure 22-5. Flax weeds that should have been killed earlier in the season. (L. A. Derscheid, South Dakota State University.)

weeds such as wild mustard, lambsquarters, ragweed, fanweed, or cocklebur; and up to $\frac{1}{2}$-lb/acre for smartweed.

Rates of herbicide given on the label are not expected to affect germination of flaxseed, oil content, or iodine number. A number of herbicide treatments are specifically used to control wild oats in flax.

SUGGESTED ADDITIONAL READING

Weed Science published by WSSA, 309 West Clark St., Champaign, IL., 61820. See index at end of each volume.

Weeds Today. Same availability as above.

FAO International Conference on Weed Control, 1970. Same availability as above.

Regional Weed Science research reports and proceedings. Available in the libraries of most land grant universities. Published annually.

Friesen, G., 1957, *North Central Weed Control Conference Proceedings* **14**, 40.

Hay, J. R., 1970, *FAO International Conference on Weed Control* p. 38, WSSA, Urbana, IL.

Helgeson, E. A., K. L. Blanchard, and S. D. Sibbitt, 1948, *North Dakota Experiment Station. Bimonthly Bulletin* **10**(5), 166.

Smith, R. J., W. T. Flinchum, and D. E. Seaman, 1977, USDA-ARS Agric. Handbook No. 497.

USDA Agric. Economic Rpt. No. 461, 1980.

Obtain weed-control bulletins, leaflets, and so forth, from your state university or college of agriculture on weed control in specific crops.

For chemical use, see the manufacturer's label for method and time of application, rates to be used, weeds controlled, and special precautions. *Label recommendations must be followed–regardless of statements in this book.* Also, see the Preface, Inclusion or omission of a product name does not constitute recommendation or non-recommendation of a product.

23 Small-Seeded Legumes

Weed control is a major problem in growing legumes. Alfalfa, white clover, and ladino clover are perennial crops usually grown for hay and pasture. Red clover, alsike clover, crimson clover, sweet clover, and lespedeza are hay and soil-improving crops. They are often interplanted in small grain. With either group there is little opportunity after seeding to control weeds by cultural methods, unless the legume is planted in rows.

Weed control of large-seeded legumes planted in rows, such as soybeans and peanuts, is discussed in Chapter 24.

Weeds in small-seeded legumes can decrease yields, lower quality, and increase disease and insect problems, as well as cause premature loss of stand, harvesting problems, and irritation in animals when a spiny weed is eaten. Some weeds, such as fiddleneck, spurge, and common groundsel, are toxic to livestock when heavily infested hay is fed.

Most weeds in small-seeded legumes can be controlled by one or more of the following methods:

1. Clean seed.
2. Weed control before seeding.
3. Proper date of seeding.
4. Competitive nature of crop.
5. Mowing.
6. Flaming.
7. Cultivation.
8. Chemicals.

CLEAN SEED

The use of certified seed is of prime importance to avoid sowing weed seeds along with the crop. "Bargains" in low-cost seed almost always mean costly weed infestations for many years. This is especially true for dodder.

The seeds of many serious weeds are nearly the same size, shape, and weight as seeds of small-seeded legumes, and once the legume seeds are contaminated, they can be only partially cleaned by present methods. In recent years several mechanical devices have been used that clean seed well enough to meet most state seed-law requirements. Many of these devices are quite ingenious and take advantage of small physical differences between the seeds to be separated. Shape, size, weight, and differences in seed coat, hairiness, and appendages have all proved useful in separations.

One example is the dodder cleaner. Seed coats of most legumes are smooth and waxy, but seed coats of dodder are rough and pitted and stick to felt cloth; legume seeds do not. Adjoining felt-covered rollers rotate in opposite directions and seeds pass between them. Dodder seeds are carried to the outside while legume seeds remain between the rollers. Rollers are slanted so that legume seeds slide out at the lower end.

WEED CONTROL BEFORE SEEDING

Small-seeded legume seedlings are not vigorous growers and offer little competition to aggressive weeds. Where the area is infested with serious perennial weeds that will persist after the legume crop is seeded, control *before seeding* may be necessary. Control methods must be appropriate for the weed species concerned.

Annual weeds are often brought under control by two methods. The first is crop rotation before seeding. For example, a rotation of row crops and small grain *in which weeds are effectively controlled for two years or more* usually reduces weed-seed populations in the soil. The second method is preparation of the seedbed well before seeding. This timing permits killing one or more crops of weeds. Cultural implements are preferred that *do not* bring deeply buried seed to the surface. Deeply buried seed usually remain dormant, so they should give no problem if undisturbed.

DATE OF SEEDING

Date of seeding may determine weediness of the crop (D. Klingman, 1970). Most small-seeded legume crops can be either fall-planted or spring-planted. Weediness will depend on whether winter- or summer-annual weeds are more serious. Summer-annual weeds are often more serious. Thus alfalfa is usually fall planted. With fall planting, the legume crop is well established by spring and competes well with summer-annual weeds. However, there are the hazards of fall drought and winter injury in some areas and excessive rainfall in other areas.

Spring planting may be preferred if winter-annual weeds are the major problem. Spring planting avoids the flush of winter-annual-weed growth. However, with spring planting, summer-annual weeds may crowd out legume seedlings before they become established. More-effective herbicides may make it possible to take advantage of more-ideal seeding conditions usually found in the spring.

If legume crops are seeded when soil temperature and moisture especially favor them, they may germinate and grow fast enough to crowd out most weeds.

COMPETITION

One effective control for many weeds is to maintain a thick stand of the legume. Therefore, conditions that make the legume more vigorous usually reduce weed growth. Conversely, less competition from the legume means more weeds. Proper fertilization, drainage where needed, moisture conservation where moisture is limited, disease and insect control, mowing time, use of adapted varieties, and other proper management practices help maintain a thick stand of legumes.

Irrigating a crop just before mowing may allow the crop to continue growth during the harvest period. Rapid regrowth of the crop may crowd out many weeds.

MOWING

Few weeds can survive competition with vigorously growing alfalfa in addition to being mowed two or more times per growing season. This double-barreled attack weakens and may kill most annual weeds and

Table 23-1. Effect of Cutting for Two Years on Yield and Stand of Alfalfa and on Number of Weeds in Wisconsin (Nelson, 1925)

Stage of Cutting Alfalfa	Number of Cuttings per Season	Yield of Weed-Free Alfalfa (tons)	Number of Alfalfa Plants at End of Experiment (per ft^2)	Weeds in Hay (%)
Full bloom	2	3.8	12	17.2
Early bud	3	2.1	3	45.1
Succulent	4	0.7	0	72.8

many serious upright growing perennial weeds. Prostrate weeds escape control by mowing. Mowing controls weeds especially well when a thick legume stand is also maintained.

If alfalfa is consistently cut when it is too immature, its vigor will be reduced, thus giving weeds a chance to grow (see Table 23-1).

Mowing weeds that outgrow legume seedlings is an effective method of control.

FLAMING

Propane or diesel burners can be used to control weeds in established alfalfa. Winter-annual broadleaf weeds are controlled with flaming just before they resume growth in the spring. This treatment also suppresses alfalfa weevil larval populations.

Flaming just after a cutting has controlled established dodder plants. However, it usually results in a few day's suppression of alfalfa growth. This treatment is usually applied only in the dodder-infested areas. Because of the high price of petroleum, flaming is expensive.

CULTIVATION

Tillage of alfalfa fields to control annual weeds has been frequently recommended and practiced, even though little experimental evidence supports the practice.

In Nebraska, cultivation of alfalfa did not increase yields; however, neither did it reduce yields (see Table 23-2).

Therefore, where a good stand of alfalfa persists, one cultivation may help control weeds subject to such cultural methods. The spring-tooth harrow kills many annual weeds without serious injury to alfalfa crowns. The disc harrow may cause considerable damage to alfalfa by cutting the crowns. The spike-tooth harrow is effective only on very small weeds.

Table 23-2. Yield (4-Yr Average) of Alfalfa Receiving Various Types of Cultivation (Kiesselback and Anderson, 1927)

Treatment	Alfalfa Yield (tons/acre)
No cultivation	3.39
Three disk-harrow treatments	3.22
Three spring-tooth-harrow treatments	3.31
Spike-tooth-harrow treatments	3.37

Table 23-3. Herbicides and Method of Application to Small-Seeded Legumes[1]

Herbicide[2]	Alfalfa	Alsike Clover	Crimson Clover	Ladino, White Clover	Red Clover	Sweet Clover	Birdsfoot Trefoil
Benefin	1	1		1	1		1
Chlorpropham	3	3		3	3	3	3
2,4-DB	3	3		3	3		3
Diallate	1	1			1	1	
Dinoseb	3	3	3	3	3	3	3
Dichlobenil	3, 4			3, 4			
Diuron	3				3		3
EPTC	1	1	1	1	1	1	1
MCPA	3		3		3		
Metribuzin	3						
Paraquat	4						
Pronamide	3	3	3		3	3	3
Propham	1, 2, 3	1, 2, 3	1, 2, 3	1, 2, 3	1, 2, 3	1, 2, 3	
Simazine	3						
Terbacil	3						
Trifluralin	3, 4						

[1] 1 = Preplant incorporated. 2 = Preemergence to crop. 3 = Postemergence to crop. 4 = Special, see label.
[2] For herbicide rates, instructions, combinations, and restrictions, see the herbicide label. Follow the label instructions.

Weed control is vital in legume-seed production. When grown mainly for seed, alfalfa and certain other legumes are occasionally planted in rows. Row planting permits weed control by cultivation. It also permits directed and shielded application of herbicides.

CHEMICAL WEED CONTROL

Small-seeded legumes vary widely in their tolerance to various herbicides. This tolerance plus susceptiblity of weeds is important in choosing an appropriate herbicide treatment.

Chemicals are applied before planting to eliminate troublesome species that cannot be controlled efficiently after planting. For example it may be desirable to treat bermudagrass, johnsongrass, horsenettle, or wild garlic to control them before seeding alfalfa. The length of residual herbicide toxicity in the soil and its effect on the newly seeded crop must also be considered (see Table 5-1).

Table 23-3 lists chemicals labeled for use in small-seeded legumes, and the time or method of application. Herbicides are available for application at the following periods of crop growth: (1) preplant incorporation, (2) preemergence to the crop, and (3) postemergence to the crop. *See the herbicide label for specific details of application.*

See page 418 for weeds controlled by various herbicides. Herbicide labels also list the weeds controlled.

SUGGESTED ADDITIONAL READING

Kiesselback, T. A., and A. Anderson, 1927, Nebraska Experiment Station Bulletin 222.

Klingman, D. L., 1970, *FAO International Conference in Weed Control,* pp. 401–424, WSSA, Urbana, IL.

Nelson, N. T., 1925, *J. Amer. Soc. Agron.* **17**, 100.

Weed Science published by WSSA, 309 West Clark St, Champaign, IL, 61820. See index at end of each volume.

Weeds Today. Same availability as above.

FAO International Conference on Weed Control, 1970. Same availability as above.

Regional weed science research reports and proceedings. Available in the libraries of most land grant universities. Published annually.

Obtain weed control bulletins, leaflets, and so forth, from your state university or college of agriculture on weed control in small-seeded legumes.

For chemical use, see the manufacturer's label for method and time of application, rates to be used, weeds controlled, and special precautions. *Label recommendations must be followed–regardless of statements in this book.* Also, see the Preface. Inclusion or omission of a product name does not constitute recommendation or nonrecommendation of a product.

24 Field Crops Grown in Rows

Why are crops planted in rows? What determines the width between rows? Just why are row crops cultivated? What effect does cultivation have on crop yields? These are important questions to any farmer planning his future operations. Unfortunately we can only partially answer all of these questions.

While man depended on the hoe for weed control, he was also undoubtedly looking for easier ways to do the job. He found one great labor-saver in dragging a heavy hoe behind a horse. Early agricultural writings refer to "horse-hoeing." Planting crops in rows made horse-hoeing easier.

Thousands of research reports and articles have been written on weed control in various crops, especially in row crops. These reports are available in *Weed Science* and *Weed Research,* two scientific publications; Regional Weed Conference proceedings and research reports; College of Agriculture bulletins, leaflets, and circulars; commercial company publications; and in various trade journals and farm magazines. These sources give detailed information, but no attempt will be made to review all this literature here. Only those practices that have proven their usefulness or that appear especially promising are discussed below.

Several review papers on weed control in these crops were prepared for the FAO International Conference on Weed Control 1970. Crop yield losses caused by weeds were discussed in Chapter 2.

Weed control in row crops includes all methods that were discussed in Chapter 2–mechanical, competition, crop-rotation, fire, and chemical methods. It could include biological control through predators and diseases; however, this refinement has not been perfected for weed control in cultivated crops.

Top-yielding crops usually provide maximum competition to weeds (see Figure 22-2). Planning for and using such practices are the first steps in an effective weed-control program.

Farmers cultivate principally *to control weeds* (see Chapter 2, subsection "Tillage"). Some crops may benefit from one early cultivation to loosen the soil if it becomes hard and packed when dry. However, on other soils, research data indicate that cultivation is of no value if weeds are controlled.

Cultivation has disadvantages. It requires considerable costly fuel. Also, large tractors are an expensive investment for the farmer. Wet weather or other work may prevent timely cultivation, and weeds may get ahead of the crop. Some crops grow slowly, and weeds may get a head start. Some weeds are very difficult to control in the row; cultivation may injure roots and reduce yields, and heavy weed growth may develop after the last cultivation. Also, repeated cultivation, especially in wet soils, injures the physical condition of the soil.

Weed control in most annual crops can be divided into *early-season* and *late-season* phases. Early-season weeds usually have a greater effect on crop yields than late-season weeds. Late-season weeds make harvesting difficult, may contaminate the harvested crop (grass in cotton, weed seeds in grain), and reinfest the soil with weed seeds.

For most row crops, a crop canopy covers the row middles within seven to eight weeks after planting. A herbicide that provides preemergence weed control is important in keeping the crop weed-free until harvest. If the crop is 100% free of weeds when the canopy is formed, few weeds can *start* and cause serious crop yield losses. However, weeds that escape control may cause serious yield losses.

See page 422 for weeds controlled by most herbicides discussed in this chapter.

WEED CONTROL IN SPECIFIC ROW CROPS

These methods have proven effective in various parts of the United States, but not necessarily in all areas. In managing a row crop, herbicides are applied at four different times:

1. *Preplant* means before planting.
2. *Preemergence* means after planting but before crop emergence.
3. *Postemergence* means after emergence of the crop.
4. *Minimum-till* or *no-till.* Minimum-till is the production of a crop with a small amount of tillage; no-till is crop production with no tillage.

Preplant applications usually require *soil incorporation* of a selective, soil-applied herbicide. The seedbed is prepared as usual, and the herbicide is applied and evenly mixed into the soil to a depth of 2–3 in. This is deep enough to control germination of susceptible weed seeds. Crop seeds must be tolerant to the herbicide, or the seeds must be placed below the herbicide-treated layer. See Figure 5-1.

Incorporation equipment includes power-driven rotary hoes with L-shaped knives, ordinary rotary hoes, discs, sweeps, drag-harrows, and other such soil-tillage implements. To be effective the equipment must break up large clods and mix the herbicide evenly in the soil. Rain is not needed to activate the herbicide; therefore, the herbicide seldom fails. Other preplant herbicide treatments are discussed under the section on minimum-till or no-till.

Preemergence treatments are applied to the soil surface after planting, but before the crop emerges. Rain or sprinkle-irrigation water is required to leach the chemical into the soil. Without water the treatment fails. With too much water the herbicide may be leached below the weed-seed germination area and also cause failure. Crop seeds must be tolerant to the herbicide.

Postemergence treatments are made any time after the crop emerges. With highly selective herbicides the chemical may be applied broadcast as a spray overtop the crop and weeds. The crop is not injured and the weeds are killed. Crops can be protected more by directing the spray to the small weeds below the taller crops; this technique minimizes the amount of spray hitting the crop plant. *Drop-nozzles* are often used to make this type of application. If weeds are taller than the crop, a recirculating sprayer (Figure 7-4) or a rope-wick applicator (Figure 7-5) can be used to apply a nonselective herbicide. In this case, selectivity is attained through application. Glyphosate is the principal herbicide applied in this manner. See pages 126 and 127.

Minimum-till or *no-till* is based on the fact that farmers till the soil primarily to control weeds (see Chapter 2, subsection "*Tillage*"). Before planting or at that time, a broad-spectrum herbicide or combination of herbicides is applied. Ideally this treatment will control all existing vegetation, leaving a weed-free mulch on the soil surface. Another, more persistent herbicide, may be added to control weeds until harvest.

The most successful program using this technique is small grain followed by soybeans. The growing season must be long enough for both crops. After small-grain harvest the area is treated with a herbicide such as glyphosate or paraquat. Soybeans are planted through the straw mulch. Then another herbicide such as alachlor, metolachlor, or oryzalin is applied to provide selective residual weed control until soybean harvest. In some practices, the soybeans are planted through the straw mulch, followed immediately by a combination spray applying the contact and residual herbicides. In this case the herbicides must be applied before soybean emergence to avoid crop injury.

Corn is also successfully grown with minimum-till and no-till. The soil is not plowed. The new crop is planted through the previous crop's residue. Glyphosate and paraquat are broad-spectrum herbicides used in combination with other herbicides like atrazine, alachlor, cyanazine, or metolachlor.

The small-grain–soybean program provides two crops in one year. Minimum- or no-till programs provide soil and water conservation plus reduced tillage costs. They do require special planting equipment and extra herbicide costs. Also, insect control and fertilizer costs may increase. As herbicides become more selective, minimum- and no-till programs will become far more important. For additional information on No-Till see Lewis (Editor). 1981.

Corn

Modern weed control in corn, depending upon soil type (see Chapter 2), may include use of both herbicides and cultivation. Location of corn roots should determine depth and placement of cultivator shovels. Corn should be cultivated only deep enough to remove the weeds or deep enough to cover them, to minimize root pruning. When corn is $2\frac{1}{2}$ ft tall, cultivation within 6 in. of the stem to a depth of 6 in. will cut off much of the root system (see Figure 24-1). In dry weather such plants may seriously wilt following deep cultivation. Deep, late-season cultivation will nearly always reduce crop yields. Therefore, if late cultivation is done, it should be with shallow sweeps. Roots dragging on the cultivator shank are a sign that you are cultivating too deeply.

Farmers have many methods of cultivation. Rotary weeders, finger weeders, and harrows are often used when the corn is small. By the

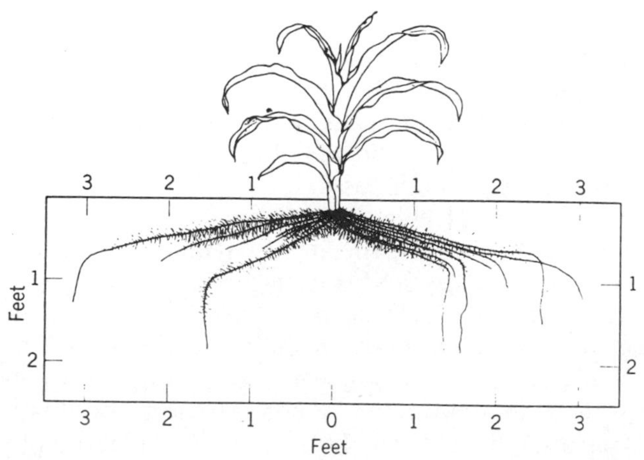

Figure 24-1. Cultivation should be shallow to avoid destroying corn roots, especially after the plant is 15 in. tall. (Adapted from the *Nebraska Research Bulletin* No. 161, 1949.)

time the corn is 3–4 in. tall, shovel or sweep cultivation can be started, throwing soil into the row to bury small weeds. Special shields may control the amount of soil thrown to the corn to avoid burying the corn. Cultivation is repeated as often as needed until the corn is 20–24 in. tall. The last cultivation is often called "lay-by" cultivation. Where cultivation is the only method of weed control, farmers usually do so three to four times.

Many farmers now combine a reduced-cultivation program with a herbicide program and get better weed control than by cultivation alone.

Very effective herbicides are available for weed control in corn. They can be applied during the entire life span of the corn plant. However, because of application difficulties, most herbicides are applied before the corn reaches 30 in. tall. After this, high-clearance sprayers are needed to prevent breaking the corn.

Chemical Weed Control in Corn

Since 2,4-D was introduced in the mid-1940s to control weeds in corn, many herbicides have been developed for this purpose (see Table 24-1). According to USDA (1980), of total 1976 corn acreage in the United States, 49% received atrazine (Figure 24-2), 28% received alachlor, 12% received butylate, and 4% received EPTC.

Most of these herbicides are used to control annual weeds, both grasses and broadleaf weeds; however, a few may also give temporary control of certain perennial weeds. Combinations of two corn herbicides are often used to increase the spectrum of weed species controlled.

Figure 24-2. Atrazine applied as a band treatment over the corn row at planting time. Usually, the weeds in the middle should be controlled by cultivation while the weeds are much smaller. The rows to the left were not treated. (Ciba-Geigy Corporation.)

We give details here on the use of 2,4-D as a postemergence, directed spray to control broadleaf weeds in corn; this is because the treatment is relatively inexpensive and readily available. These factors are particularly important for developing countries, where some of the herbicides listed in Table 24-1 may not be available.

When corn is 3–4 in. tall, 2,4-D applied at the rate of $\frac{1}{4}$–$\frac{1}{3}$ lb/acre will control most broadleaf annual weeds; corn that is 4–12 in. tall will withstand double this rate. Postemergence treatments with 2,4-D will not control established grasses.

Corn 12–30 in. tall will withstand $\frac{1}{3}$–$\frac{1}{2}$ lb of 2,4-D per acre for broadleaf-weed control. The spray is preferably applied with nozzles dropped between the corn rows to avoid spraying most of the corn plant. Also, spray nozzles that produce uniform coarse spray droplets should be used (300 μm or larger) to avoid spray drift (Chapter 6, 7).

Usually 2,4-D esters are applied at the lower rates suggested above, and 2,4-D amines at the higher rate. If there are 2,4-D-susceptible crops in the area, the amine form of 2,4-D should be used to avoid the volatility hazard of 2,4-D esters.

Corn becomes more susceptible to 2,4-D after excessively high temperatures and high soil-moisture content. After such periods the rate of 2,4-D should be lowered to reduce corn injury.

Corn varieties differ considerably in tolerance to 2,4-D. Some varieties will withstand rates of treatment higher than rates suggested above.

Table 24-1. Field Corn Herbicides[1]

Minimum or No-till	Preplant Incorporated	Preemergence to Crop	Postemergence to Crop
Alachlor	Alachlor	Alachlor	Alachlor
Atrazine	Atrazine	Atrazine	Ametryn
Cyanazine	Butylate	Chloramben	Atrazine
Glyphosate	EPTC[2]	Cyanazine	Bentazon
Paraquat	Metolachlor	2,4-D	Butylate
Pendimethalin	Simazine	Dicamba	Cyanazine
		Diuron	2,4-D
		Glyphosate	Dicamba
		Linuron	Diuron
		Metolachlor	Linuron
		Paraquat	Paraquat
		Pendimethalin	Pendimethalin
		Propachlor	Trifluralin
		Simazine	

[1] *Always* read—and follow—manufacturer's label exactly, also see page 418.
[2] Special formulation of EPTC with antidote.

Table 24-2. Sorghum Herbicides, Including Milo and Grain Sorghum[1]

Minimum or No-till	Preplant Incorporated	Preemergence to Crop	Postemergence to Crop
Atrazine	Atrazine	Atrazine	Atrazine
Glyphosate	Metolachlor	Bifenox	2,4-D
Paraquat	Propazine	Cyanazine	Dicamba
	Terbutryn	Glyphosate	Diuron
		Linuron	Glyphosate
		Metolachlor	Linuron
		Paraquat	MCPA
		Propachlor	Paraquat
		Propazine	
		Terbutryn	

[1] *Always* read—and follow—manufacturer's label exactly.

Sorghum, Including Milo and Grain Sorghum

Weed control in sorghum is similar to that in corn. In general, sorghum is less tolerant to herbicides than corn. With development of suitable herbicides to give seasonlong weed control in sorghum, the rows can be planted close together (see Table 24-2).

Sorghum and corn have similar development stages of susceptibility and tolerance to 2,4-D. Grain sorghum 4–12 in. tall is reasonably tolerant to 2,4-D. In the five- to eight-leaf stage, most varieties will tolerate $\frac{1}{3}$ lb/acre of a 2,4-D ester or $\frac{1}{2}$ lb of an amine salt of 2,4-D. At about 12 in. height, when the head starts to develop rapidly, the plant becomes susceptible to 2,4-D. When the grain is half-formed, the plant again becomes tolerant to 2,4-D (see Figure 24-3).

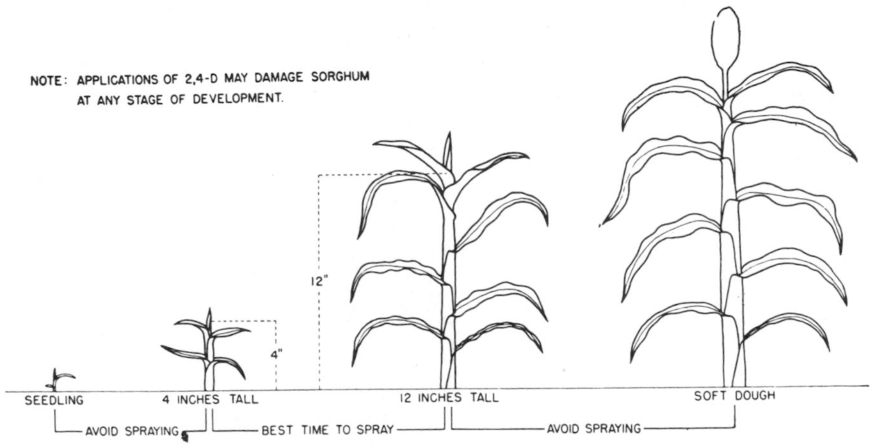

Figure 24-3. Sorghum tolerance to 2,4-D at various stages of development. (W. M. Phillips, Kansas Agricultural Experiment Station and USDA.)

Soybean

Soybeans are not planted until the soil is thoroughly warmed; thus several crops of weed seedlings can be destroyed before planting. Management practices such as thorough seedbed preparation, adequate soil fertility, choice of a well-adapted variety, and use of good seed all contribute to a soybean crop that will compete with weeds.

The rotary hoe is an effective and economical weed killer in soybeans. For best results, it should be used when the ground is slightly crusted and when weeds are just emerging, and not more than $\frac{1}{4}$ in. high. This treatment is most effective when the ground is dry. However, if weeds are not more than $\frac{1}{4}$ in. tall, research shows that rotary hoeing is effective even though the soil is moist. Application of oryzalin followed by timely rotary hoeing increases control of weeds more than usually expected from either used alone (see Figures 24-4 and 24-5).

Chemical Weed Control in Soybean

U. S. Department of Agriculture (1980) reports that in 1976, of the total soybean acreage in the United States, 37% received alachlor, 26% trifluralin, 8% linuron, and 6% metribuzin. Almost as many herbicides have been developed to control annual weeds in soybeans as in corn (see Table 24-3). For Weeds Controlled, see page 418.

Table 24-3. Soybean Herbicides[1]

Minimum or No-till	Preplant Incorporated	Preemergence to Crop	Postemergence to Crop
Alachlor	Alachlor	Alachlor	Acifluorfen
Bifenox	Bifenox	Bifenox	Barban
Glyphosate	Chloramben	Chloramben	Bentazon
Linuron	Chlorpropham	Chlorbromuron	2,4-DB
Metolachlor	DCPA	Chlorpropham	Dinoseb
Metribuzin	Dinitramine	DCPA	Dichlofop-methyl
Oryzalin	Fluchloralin	Dinoseb	Glyphosate
Oxyfluorfen	Metolachlor	Diphenamid	Metribuzin
Paraquat	Metribuzin	Glyphosate	Oxyfluorfen
	Oryzalin	Linuron	
	Pendimethalin	Metolachlor	
	Trifluralin	Metribuzin	
	Vernolate	Naptalam	
		Oryzalin	
		Oxyfluorfen	
		Paraquat	
		Propachlor	

[1] *Always* read—and follow—manufacturer's label exactly.

Figure 24-4. *Left*: Before rotary hoeing; a heavy infestation of green foxtail in soybeans. Fully 90% of the weeds have not emerged, but are germinated (in the white). This is the time to rotary hoe. *Right*: After timely rotary hoeing; note the lack of both weeds and soybean injury. (Iowa State University, Ames, IA.)

Most farmers start their weed-control program in soybeans with either a preplant soil-incorporated herbicide treatment or a surface-applied preemergence treatment following planting. This may be followed with a rotary hoeing and one or two cultivations. The use of postemergence treatment usually depends on the number of weeds that have survived earlier treatments.

Figure 24-5. *Right*: Soybeans with oryzalin plus metribuzin surface applied after planting, plus one early cultivation. *Left*: Cultivated plot, no herbicide treatment. (Elanco Products Company, a division of Eli Lilly and Company.)

Peanuts

The rotary weeder, flexible-spike weeder, and cultivator remain efficient tools for weed control in peanuts. The rotary weeder is often used just before and again just after peanuts emerge to control early, small weed seedlings. The timely use of the rotary weeder may reduce cultivating time by 25% and hand-hoeing up to 50%. Cultivating two to four times is common.

Chemical Weed Control in Peanuts

The peanut root grows very fast after planting, even though leaves emerge slowly. Under favorable conditions, the root grows downward in about 24 hr and may reach 2 in. long in 4 days, 6 in. long in 6 days, and 15 in. long in 12 days. In contrast, the first leaves usually do not appear above the soil until 7–12 days after planting.

Therefore, when leaves emerge, the root is deep in the soil. Root tips are well protected from preemergence herbicides applied about six days after planting. A delayed preemergence spray usually reduces the risk of injuring peanuts. A delayed application, however, may reduce the effectiveness of weed control because germinating weed seedlings may also have passed the optimal control period. Several preplant soil-incorporated herbicides are selective enough to be used before planting. Preemergence herbicides are usually applied immediately after planting.

Most herbicides used for annual-weed control in peanuts are applied preplant soil-incorporated or preemergence (see Table 24-4). Preplant soil-incorporated treatments are usually applied as part of seedbed preparation. Preemergence herbicides are applied either at planting time or at cracking time. Planting-time treatment is usually applied during actual planting with application equipment attached to the tractor

Table 24-4. Peanut Herbicides[1]

Preplant Incorporated	Preemergence		Postemergence
	At Planting Time	At Cracking Time	
Alachlor	Alachlor	Alachlor	Bentazon
Benefin	Chloramben	Dinoseb	Dinoseb
Metolachlor	Dinoseb	Diphenamid	2,4-DB
Trifluralin	Diphenamid		Metolachlor
Vernolate	Metolachlor		
	Naptalam		
	Trifluralin		
	Vernolate		

[1] *Always* read–and follow–manufacturer's label exactly.

doing the planting. Cracking-time treatment is a delayed preemergence treatment applied when the emerging peanut shoot cracks the surface soil, but has not yet emerged. This treatment usually comes 6–10 days after planting; it may be postemergence to early germinating weeds and, if so, must kill those weeds hit by the spray.

Cotton

Weed control in cotton in previous years required considerable hand labor plus an equal amount of cultivation. Even so, cotton was often still weedy at harvest time. With mechanical harvesting, weed control became increasingly important because weedy trash gets mixed with cotton lint.

Production methods are needed that provide for the best possible growth and yields of cotton. Among these are a well-adapted variety, adequate fertility, and proper control of diseases, insects, and weeds. A vigorous crop of cotton helps control weeds, especially late-season weeds, through competition.

Flaming

A flame cultivator may be used first when cotton stems are $\frac{1}{4}$ in. in diameter. Best results are obtained if the preceding weed-control program has kept weeds small. The flat-type burner that uses volatile gases such as propane or butane is suggested. The continued use of flaming depends on the price of fuel.

Chemical Weed Control in Cotton

Cotton was once planted thick, and the stand was thinned during the first hoeing. Herbicides eliminated much of the need for hand-hoeing, and the desired cotton stand has been obtained by *planting to a stand.* Preparing the seedbed well, using high-quality seed, sideplacing fertilizer properly, and delaying planting until the soil is warmed to at least 60°F have provided the desired stand without chopping. In many areas growers want a stand of 45,000 plants/acre.

Herbicides used in cotton are given in Table 24-5. According to the U. S. Department of Agriculture (1980), of the total 1976 cotton acreage in the United States, 38% received trifluralin, 29% fluometuron, and 18% DSMA or MSMA. Certain preplant soil-incorporated treatments, such as trifluralin, can be done in the fall preceding spring planting, or at any time up to the date of planting. Usually the herbicide is incorporated as a part of seedbed preparation. (See Figure 24-6).

Most preemergence herbicides are applied just after planting, often with application equipment attached to the tractor doing the planting.

Table 24-5. Cotton Herbicides[1]

Preplant, General Weed Control	Preplant Incorporated	Preemergence to Crop	Postemergence to Crop
Glyphosate	Bensulide	Bensulide	Cyanazine
Paraquat	Fluchloralin	Cyanazine	DCPA
	Fluometuron	DCPA	Dinoseb
	Pendimethalin	Diphenamid	Diuron
	Prometryn	Dipropetryn	DSMA; MSMA
	Trifluralin	Diuron	EPTC
		Fluometuron	Fluometuron
		Methazole	Glyphosate
		Norflurazon	Linuron
		Oryzalin	Methazole
		Pendimethalin	Oils, herbicidal
		Perfluidone	Prometryn
		Prometryn	Trifluralin

[1] *Always* read—and follow—manufacturer's label exactly.

Figure 24-6. Diuron applied in a band over the cotton row controlled weeds for six weeks after planting. Areas between rows have been cultivated. There is no need for weed chopping in the row here. (E. I. du Pont de Nemours and Company.)

Herbicidal oils like Varsol or Stoddard solvent are effective postemergence sprays on cotton. Oils are applied when cotton is small; for effective control, oils must be applied while weed seedlings also are small. Oil is aimed into an 8–10 in. band in the cotton row so as not to hit the cotton leaves. The usual rate is 35–50 gal/acre; however, because it is applied in a band, only 7–10 gal are required per acre planted. Treatments are repeated every five to seven days (see Figure 2-6). Usually two or three treatments are made, and the treatment is stopped when cracks appear in the bark of the cotton stalk; if not, the cotton may be injured. Because of the increased cost of petroleum products, herbicidal-oil usage is on the decrease.

Sugarbeets

The major weed problem in sugarbeets is annual weeds, both grasses and broadleaf weeds. They are particularly troublesome at emergence through thinning time and then later after lay-by. Before selective preplant and preemergence herbicides were developed for sugarbeets, these annual weeds were serious competitors with the crop. Use of herbicides now permits sugarbeets to stay essentially weed-free until thinning time. Moreover, it allows use of precision planting and mechanical thinning.

Herbicides now used preplant or preemergence in sugarbeets are degraded fairly rapidly in the soil. Therefore, an additional herbicide is usually required later to give seasonlong weed control.

Herbicides used in sugarbeets are shown in Table 24-6. Three of these

Table 24-6. Sugarbeet Herbicides[1]

Preplant Incorporated	Preemergence to Crop	Postemergence to Crop
Cycloate	Endothall	Barban
Diallate	Ethofumesate	Dalapon
Ethofumesate	Nitrofen	Desmedipham
EPTC	Paraquat	Endothall
Pebulate	Propham	EPTC
Propham	Pyrazon	Ethofumesate
Pyrazon	TCA	Phenmedipham
		Propham
		Pyrazon
		Trifluralin

[1] *Always* read—and follow—manufacturer's label exactly.

–EPTC, pebulate, and cycloate–are thiocarbamates. They were developed in the order listed. With each successive compound, the safety to sugarbeets has increased with little, if any, loss of weed-killing ability.

Sugarcane

Sugarcane is a perennial crop. The crop is started by planting pieces of the stalk in deep drills or furrows. New plants are vegetatively produced from buds located at the nodes of the stem; this crop is known as *plant cane.* After harvest, preparation is made for succeeding crops known as ratoon or stubble cane. Depending on weed control, disease, soil fertility and overall productivity, two to four ratoon crops are produced.

Sugarcane is normaly grown on fertile soils, in areas with high rainfall, irrigation, or both, and usually in tropical and semitropical climates with a long growing season. These conditions, plus the perennial nature of the crop, are favorable to weeds of all kinds–annuals, biennials, and perennials.

The control of weeds in sugarcane is a serious and costly problem. Weeds must be effectively controlled until the cane "closes in" and provides sufficient shade to prevent weed growth.

Plant cane usually closes in the row middles in four to seven months, and ratoon cane in two to five months. Residual herbicides applied before "close-in" may effectively extend weed control through to harvest. Cultivation and herbicides are the two principal methods of weed control. Herbicides are applied with backpack, animal pack, tractor, and aircraft sprayers.

Historically, the crop was heavily cultivated, including hand-hoeing. With the coming of selective herbicides, however, sugarcane producers quickly adapted them into their crop-management program.

The following herbicides are used on sugarcane: ametryn, asulam, atrazine, CDAA, 2,4-D, dalapon, diquat, diuron, fenac, fluometuron, glyphosate, metribuzin, paraquat, silvex, TCA, terbacil, and trifluralin. Each of these controls a specific group of weeds and requires special instructions for application. The manufacturer's label should be consulted for these instructions. See page 418 for weeds controlled by each herbicide.

Tobacco

Tobacco has very small seeds, and at first the seedlings grow slowly; thus the plants are started from seed in plant beds. These plants are later transplanted to the field. These facts lead to a serious weed problem in the plant bed and also a weed problem in the field.

In the plant bed, weeds, soil-borne diseases, and insects and nematodes may create a problem, and a highly effective general-purpose soil fumigant is often used. Three materials are applied: methyl bromide, dazomet, and metham.

Methyl bromide is applied as a gas under a gas-proof cover at least two days before seeding. Metham is sodium methyldithiocarbamate. The seedbed is prepared and the chemical applied three or more weeks before planting. When the treated area is covered, it requires half as much chemical as when no cover is used. Dazomet is tetrahydro-3,5-dimethyl-2*H*-thiadiazine-2-thione. The seedbed is prepared and the chemical applied four weeks or more before planting. The chemical is watered down and no gas-proof cover is needed.

In the field, benefin, isopropalin, diphenamid, pebulate and pendimethalin are applied as herbicides. Diphenamid can be applied up to seven days before transplanting and incorporated into the soil. Also, diphenamid can be applied immediately after transplanting as an over-the-top spray.

Benefin, isopropalin, pebulate and pendimethalin are incorporated into the soil. Tobacco plants can be set directly into the treated soil, providing that the tobacco is not made into beds. If beds are made following an original flat incorporation, the chemical is placed twice as deep in the soil. The roots of tobacco plants set on top of the bed have difficulty penetrating through the deeply placed herbicide. Isopropalin can be used even though the herbicide is incorporated deeply into the soil.

In North Carolina, tobacco on sandy soils showed no increase in yields as a result of cultivation if weeds were otherwise controlled. On loam soils and on clay-loam soils, increased tobacco yields resulted with one to two cultivations. In no case were yields increased by more than two cultivations.

TREATMENT OF SERIOUS PERENNIAL WEEDS

Serious perennial weeds in row crops should be treated while confined to a small area, before they spread. Ideally they should be eradicated by using a soil sterilant. Methyl bromide, a temporary soil sterilant, can be used on many species that do not recover from depths greater than this gaseous chemical penetrates. It is a relatively expensive and laborious treatment. With methyl bromide, crops can be seeded into the treated soil within a few days after treatment without injury.

More-persistent soil sterilants can also be used. Although they are less expensive than methyl bromide, treated soil may remain toxic to crops for several years after treatment. (See Chapter 30 for details on soil sterilants.)

Perennial grasses such as quackgrass, johnsongrass, and bermudagrass can be controlled by dalapon before planting beans, corn, cotton, potatoes, milo, soybeans, and sugar beets. It is applied as a preplant foliar spray at 6 lb/acre when johnsongrass is 8–12 in. tall, and quackgrass 4–6 in. tall. This is followed by plowing or deep discing three days or more after application. Additional delays are required before planting some crops. Follow label instructions.

Trifluralin applied at double the normal rate for two years provides effective control of both seedling and rhizome johnsongrass in soybeans and cotton. The soil should be thoroughly tilled, to work the trifluralin in and around each rhizome. Later, cultivation is needed to cut off or bury occasional surviving plants. This program permits the production of soybeans or cotton without loss of a cropping year. Similar control of bermudagrass has been observed.

Glyphosate, a new herbicide, is particularly effective on perennial grasses, but also controls many broadleaf weeds. It is applied to foliage of the plant to be controlled. Relatively nonselective, it can be applied to tall weeds (6 in. taller than crop) in crops such as cotton, soybeans, and peanuts to control weeds such as volunteer corn, shattercane, johnsongrass, milkweed, cocklebur, pigweed, and sunflower. In noncrop areas, glyphosate controls bermudagrass, Canada thistle, field bindweed, silverleaf nightshade, Texas blueweed, dogbane, kudzu, multiflora rose, trumpet creeper, and many others.

SUGGESTED ADDITIONAL READING

Lewis, W. M. (Editor) 1981. N. Carolina Agric. Ext. Ser. Ag. -273. No-Till Crop Production Systems in Nor. Car.

Weed Science published by WSSA, 309 West Clark St., Champaign, IL 61820. See index at end of each volume.

Weeds Today. Same availability as above.

FAO International Conference on Weed Control, 1970. Same availability as above.

Regional weed science research reports and proceedings. Available in the libraries of most land grant universities.

USDA Agricultural Economic Report No. 461, 1980.

Obtain weed control bulletins, leaflets, and so forth, from your state university or college of agriculture on weed control in specific crops.

Obtain commercial herbicide labels from herbicide manufacturers.

For chemical use, see the manufacturer's label for method and time of application, rates to be used, weeds controlled, and special precautions. *Label recommendations must be followed–regardless of statements in this book.* Also, see the Preface. Inclusion or omission of a product name does not constitute recommendation or nonrecommendation of a product.

25 Vegetable Crops

Losses caused by weeds in vegetable crops can easily mean the difference between profit and loss. To ascertain more clearly the size of these losses, Wisconsin scientists carried out a special research program (Shadbolt and Holm, 1956). They allowed a 15% weed stand in carrots to remain for the first $5\frac{1}{2}$ weeks before removing the weeds; this treatment reduced the yield of carrot roots by 78%. A 50% weed stand reduced carrot yield by 91%.

In onions, leaving a 15% weed stand for the first six weeks before removal reduced onion-bulb weight by 86%. A 50% weed stand reduced yields by 91%.

The conclusion–the first four weeks of crop growth are the most critical in affecting crop yields. Zimdahl (1980) gives many other examples of serious weed losses in vegetables.

In vegetable crops it is especially important that the weed program approach 100% control. A few scattered weeds allowed to survive may grow so rapidly and so large that results may be nearly as serious as a thick weed infestation.

Past methods of weed control in vegetable crops have centered about the cultivator and hoe. But hoeing is expensive and labor is in short supply. With increased labor costs, more-efficient methods of weed control are necessary if the grower is to earn a profit. Tillage (cultivation) is discussed in Chapter 2. Experienced growers usually follow these practices.

The first five references listed at the end of this chapter ("Suggested Additional Reading") are reviews of weed control in vegetables. All were published in 1970.

With selective herbicides, growers can control weeds far more efficiently, with less human effort, than in the past. Savings in labor are important to the grower as well as to the consumer; both reap the benefits.

Before growers may use, or the manufacturer sell, any chemical, the chemical must be proven safe for use on the food crops concerned. This is required before the U.S. government will approve a label. Label clearance means data have been furnished showing that the chemical is useful and safe. The label gives recommendations for use. These recommendations must be followed carefully to assure that the crop will be free of herbicide residues.

Vegetable-crop production is a very specialized business with crops grown intensively under varying conditions of soil and climate. A herbicide may produce excellent results under one set of conditions and either injure the crop or fail to control weeds under other conditions. The label should be checked for specific instructions.

Anyone not experienced in spraying herbicides on a given crop should proceed cautiously. With experience the farmer can expand the control program.

MINOR-CROP HERBICIDES

Development of an agricultural pesticide is expensive: currently, \$10–\$50 million per product. Most of the available herbicides have been developed in the United States; however, many have come from Western Europe and a few are now coming from Japan. Preliminary development of herbicides is done almost exclusively in the private sector. Chemical companies do the synthesis and primary screening, and patent the products of their research. Costs of synthesis, screening, development, registration, and service to the product must be paid ultimately through sales of the product. These costs in turn must be passed to the consumer through the cost of the product. In addition, federal, state university, and county researchers make substantial contributions to the ultimate use of herbicides.

Acreage of any one vegetable crop is small when compared with that of corn, soybeans, or cotton. Therefore, vegetables are considered minor crops, and it is generally not economically feasible for a chemical company to develop a herbicide for a minor crop if it is not suitable for a major crop as well. As a result, most herbicides used in vegetable crops were previously developed for a major crop.

Therefore, minor crops may be somewhat less tolerant to herbicides than major crops; in vegetables, herbicides must be used with greater care. Normally at least a twofold safety factor is desired (twice as much herbicide is required to induce crop injury as is needed for weed control). Also, many minor crops are expensive high-cost crops. The liability or risk of lawsuits further reduces a company's interest in adding such crops to the label.

An additional problem with developing herbicides for minor crops is that of obtaining federal registration for a particular use. Although a considerable amount of data used for registration of the herbicide for use on a major crop also applies to its use on minor crops (e.g., toxicology of the compound, environmental impact, etc.), a considerable amount of specific information about the chemical's effect on the minor crop must also be developed. The cost of this research may exceed potential profit to the chemical company, but this has been somewhat alleviated by the formation of an Interstate Research Project (IR-4) supported by the federal government. This project coordinates and supports federal and state research directed toward the development of the information required for registration of a pesticide for use in a minor crop or other minor uses by the Environmental Protection Agency (EPA).

EPA regulations also permit limited use of a pesticide under sections 18-C (Emergency Need) and 24-C (Special Local Need). These two permit procedures are usually begun at state level through the state department of agriculture.

ARTICHOKES

Artichokes are perennials, and once established they are not replanted for many years. In addition to cultivation, two herbicides are used in this crop. These herbicides are applied to the soil as a directed spray to control annual weeds. Diuron is used in late fall or early spring after the last cultivation. Simazine is used after the last fall cultivation. Both compounds are relatively persistent in soils. Sensitive crops should not be planted in a diuron-treated area within two years after application. Simazine should not be applied to sandy soils.

ASPARAGUS

Asparagus is a perennial crop that may remain in production 10–20 years before replanting or rotating to another crop; therefore, year-round weed control must be provided without injuring the crop.

Because asparagus is dormant in the winter and crowns are planted relatively deep (8–12 in.), growers can control weeds by overall discing during this period. During the rest of the year, herbicides are effectively used.

Diuron or simazine is applied after discing in the fall or spring, but before the spears and weeds appear. These herbicides will control most annual weeds throughout the cutting season. Diuron may also be applied after harvest but before fern development. Under most conditions these treatments give essentially year-round control of most annual weeds. Metribuzin is another soil-applied herbicide used in established asparagus for annual-weed control, but it is less persistent.

Salts of 2,4-D also control annual weeds in established asparagus.

They are applied in the spring after discing but before spear emergence, during the cutting season immediately after cutting, or postharvest as a directed spray after the fern is well developed. Some twisting of the spears may occur if 2,4-D is applied directly to emerged spears. Sodium and alkanolamine salts are used; certain others have been suspected of giving off-flavors to the crop.

2,4-D may also be used to control certain perennial broadleaf weeds, especially field bindweed, in established asparagus. It is usually applied as a directed spray after harvest and after the fern is well developed. Scientists at Washington State University have used 2,4-D safely to control broadleaf weeds. They sprayed asparagus after a normal cutting or harvesting with 2 lb of amine 2,4-D per acre to control Canada thistle and field bindweed. They repeated the experiment for five years. 2,4-D did not reduce the number, size, or total weight of spears produced.

Dicamba or dicamba plus 2,4-D also controls annual and perennial broadleaf weeds in asparagus; only one application per season is permitted.

Dalapon is used to control quackgrass and bermudagrass in asparagus. The product used is Dalapon® M, sodium and magnesium salts. It is applied to weed foliage before, during, and/or after harvest; the user should avoid spraying the crop.

Glyphosate controls many emerged annual and perennial weeds. It is applied to weed foliage after the last harvest until one week prior to emergence of the first asparagus spear. Spray contact with fern, stem, or spears should be avoided.

Paraquat can be used to kill emerged annual weeds and the foliage of perennial weeds. It should be applied to weed foliage before, during, or after planting but before crop emergence.

The above herbicide uses are for *established* asparagus, one year or more after planting crowns. To establish this crop from crown plantings, linuron, paraquat, and diuron may be used preemergence to the crop. Diuron is for use only in the San Joaquin delta of California. Linuron may also be used postemergence in crown plantings.

In asparagus *seedbeds* (planting of true seed for crown production) chloramben controls many annual broadleaf and grass weeds as a preemergence treatment. Since seedlings of asparagus emerge slowly, two to three weeks after planting, paraquat is useful for killing emerged annual weeds on preformed beds before planting the crop and/or after planting. This treatment is preemergence to the crop and postemergence to the weeds. Linuron can also be used preemergence and postemergence to the crop in direct-seeded asparagus.

BEANS

Beans tolerate weed competition more successfully than many other vegetable crops because of their rapid emergence and early growth

Table 25-1. Herbicides Used in Beans[1,2]

Herbicide	Application[3]
Alachlor	PP, PE, PPI
Bentazon	P
Chloramben	PE
Chlorpropham	PE, PPI
Dalapon	PP
DCPA	PE
Dinoseb	PE, P
EPTC	PP, PPI, LBI
Glyphosate	PE (P to weeds)
Profluralin	PPI
Trifluralin	PPI

[1] Every herbicide listed is not suitable for use on all types of beans; *check labels.*

[2] Specific combinations of these herbicides are used for beans; *check labels.*

[3] Relative to crop: PP = preplant; PE = preemergence; PPI = preplant soil incorporated; P = postemergence; LBI = layby soil incorporated.

(Zimdahl, 1980). Most harmful are (1) annual weeds emerging soon after planting and not removed, and (2) tall weeds that compete for light. Nutsedge and quackgrass are also serious competitors. Preirrigation and planting into moisture is used in arid areas of the country to minimize early-annual-weed competition. However, herbicides are widely used in bean culture throughout the country (see Table 25-1). The several types of beans vary considerably in their tolerance to various herbicides; the manufacturer's label should be checked before using.

BEETS, RED

Red or table beets and sugarbeets have a similar tolerance to herbicides and the same application methods are used. However, all sugarbeet herbicides are not registered for use in red beets. Cycloate, EPTC, phenmedipham, and pyrazon are registered for use on red beets; see Chapter 24 for application methods.

CARROT FAMILY (CARROTS, DILL, PARSLEY, PARSNIPS)

One of the oldest selective weed-control methods in vegetables is the highly refined petroleum solvents used to control annual weeds in members of the carrot family. These solvents are like paint solvent or dry-cleaning fluid. They are sold under several names including Stoddard solvent, Varsol, 350° thinner, selective weed oil, and carrot oil. Kerosene or No. 1 fuel oil has been used in the past, but they may cause a kerosene-like flavor and should be avoided. Even more refined products must be applied before the carrot root has reached $\frac{1}{4}$ in. in diameter to avoid off-flavors. The solvent used should be fresh from the refinery. These solvents control most emerged annual weeds and can be applied anytime after the carrot has two true leaves and before the root is $\frac{1}{4}$ in. in diameter.

Oilings may be repeated two to three times if necessary if done early in the season. Weeds are easiest to kill when less than 2 in. tall. Wetting sprays are applied.

High temperatures, but not over 85°F, usually provide best results. Carrots are more likely to be injured if the temperature is high, humidity is high, or plants are wet when they are sprayed. Increased plant toxicity when plants are damp can be used to the grower's advantage. This will permit use of less oil, or provide more-effective weed control. Therefore, some producers have a standard practice of spraying between dusk and dawn.

The following summary of herbicidal treatments for annual-weed control is *restricted to carrots;* however, several herbicides may also be safe to use on other members of the carrot family. Federal registration and the manufacturer's label should be checked before using these on crops other than carrots.

Trifluralin or bensulide may be used as a preplant soil-incorporation treatment. Chloroxuron, linuron, or nitrofen may be applied as a preemergence or postemergence treatment. The time of postemergence treatment for all three herbicides is somewhat different; chloroxuron—after true leaves form; linuron—when the crop is at least 3 in. tall; and nitrofen—2 weeks after crop emerges. Chlorpropham can be used as a preemergence treatment. There are geographical limitations for some of these treatments.

CELERY

Celery seed is usually planted in seedbeds, and then young plants are transplanted into the field. It is also possible to grow celery by direct seeding into the field and thinning to stand.

Seedbeds

Celery seedbeds may be fumigated with a temporary soil sterilant such as allyl alcohol or metham before planting the seed. When fumigation is practiced, additional herbicides usually are not required before removing transplants from the seedbeds.

In seedbeds, trifluralin can be applied as a preplant soil-incorporation treatment. Prometryn or nitrofen are preemergence herbicides for this crop. Prometryn can also be applied postemergence after celery has two to five true leaves. Stoddard solvent is safe to use postemergence on celery seedlings after they have one true leaf but before crown-leaf cups are formed.

Transplants

Basically two types of herbicides are used for annual-weed control in transplanted celery. CDAA, CDEC, and trifluralin control germinating weed seeds but not emerged seedlings. However, chloroxuron, linuron, nitrofen, and prometryn control both small, emerged weed seedlings and germinating weed seeds. Therefore, CDAA, CDEC, and trifluralin must be applied before emergence of weed seedlings, whereas the others can be applied later.

Trifluralin can be applied as a pretransplant soil-incorporation treatment. CDAA can be applied two to three days after transplanting and again after three weeks–in Florida only. CDEC can be applied immediately after transplanting and again three weeks later. Combinations of CDAA and CDEC can be used at similar times–again in Florida only.

The other four herbicides have somewhat different times of application. Those four are applied to small, emerged weeds and also control germinating weed seedlings. Nitrofen must be applied within two weeks after transplanting. Chloroxuron is applied between two and six weeks after transplanting and is most effective when 1–2% of a nonphytotoxic oil is added to the spray solution. Prometryn is also applied two to six weeks after transplanting and before the weeds are 2 in. high. Linuron is applied before the crop is 8 in. tall.

Direct Seeded in the Field

Trifluralin may be used as a preplant soil-incorporation treatment to control annual weeds in direct-seeded celery. Nitrofen is applied preemergence or postemergence within two weeks after crop emergence. Chloroxuron and prometryn may be applied after celery seedlings have two true leaves. Stoddard solvent or other refined petroleum oils (see carrots) can be used at the two- to four-leaf stage, but before crown-leaf cups are formed.

COLE CROPS (BROCCOLI, BRUSSELS SPROUTS, CABBAGE, CAULIFLOWER)

All cole crops appear to respond similarly to herbicides; however, one must check the manufacturer's label for the registered use for each before using.

Cole crops may be directly seeded in the field or transplanted. Since germinating seeds are usually more sensitive to herbicides than older plants, the herbicides used in these two types of culture are somewhat different. However, once the direct-seeded seedling becomes established, it is as tolerant to herbicides as the transplant.

Direct-Seeded

Bensulide, CDAA, CDEC, DCPA, nitrofen, and trifluralin (see Figure 25-1) are used to control annual weeds in direct-seeded cole crops.

The first four of these herbicides may be incorporated into the soil or applied preemergence immediately after seeding. With preemergence treatment, sprinkle irrigation should be used if rain does not fall within a few days after application. Trifluralin is applied preplant and must be incorporated into the soil to be effective. Nitrofen is applied as a preemergence herbicide, and its activity is reduced if it is incorporated into the soil.

Transplants

The following treatments can often be applied to established, direct-seeded cole crops as well as transplants; however, if a preemergence

Figure 25-1. Annual-weed control in broccoli with trifluralin. *Center*: Treated band. *Left and right*: Untreated.

treatment was applied, a second postemergence treatment may not be required.

Trifluralin may be applied as a soil-incorporation treatment before transplanting cole crops to control annual weeds.

CDEC, CDAA, or DCPA may be applied after transplanting cole crops, but before weeds emerge. Nitrofen may be applied after transplants are established or two weeks after direct-seeded plants have emerged. These treatments control annual weeds.

CORN, SWEET

Weed-control methods described for field corn in Chapter 24 also generally apply to sweet corn. Some varieties of sweet corn may be somewhat more sensitive to certain herbicides than field corn.

CUCURBIT FAMILY (CUCUMBERS, MELONS, PUMPKINS, AND SQUASH)

Members of the cucurbit family are primarily warm-weather crops; thus their major weed problems concern summer-annual grasses and broadleaf weeds. Some of these are barnyardgrass, pigweed, and lambsquarter.

Table 25-2. Herbicides Used in Cucurbits

Herbicide	Cucumber	Melon	Pumpkin	Squash
Bensulide[6]	PPI[7]	PPI[1]	–	PPI
CDEC	PE	PE[2]	–	–
Chloramben	–	–	PE, PPI	PE, PPI
DCPA	Post(1)	Post(1)[3]	–	Post(1)
Dinoseb, amine	PE, PT	PE, PT[4]	PE, PT	PE, PT
Napthalam[6]	PE, PPI	PE, PPI[5]	–	–
Napthalam	Post(2)	Post(2)[5]	–	–
Paraquat	–	PE	–	–
Trifluralin	Post(3)	Post(3)[2]	–	–

[1] Cantaloupe, crenshaw, muskmelon, Persian, watermelon.
[2] Cantaloupe, watermelon.
[3] Cantaloupe, honeydew, watermelon.
[4] Muskmelon, watermelon.
[5] Cantaloupe, muskmelon, watermelon.
[6] A combination of bensulide and napthalam can be used PE or PPI; cantaloupe, cucumber, muskmelon, watermelon.
[7] Abbreviations: PPI = preplant soil incorporated; PE = preemergence; PT = pretransplant, lay plastic immediately, delay planting at least 2 weeks; Post(1) = 4–5 true-leaf stage; Post(2) = immediately after transplanting or just before vining (about 1 month after planting); Post(3) = directed spray at 3–4 true-leaf stage, soil incorporated.

Herbicides to control these weeds are listed in Table 25-2. Note that none of these can be used on all four types of cucurbit crops; in fact, there are even further restrictions within the melons (see footnotes of Table 25-2). Restrictions are usually based on the lack of selectivity to a given herbicide in particular varieties, but in some cases it is merely a lack of sufficient information for registration.

LETTUCE AND GREENS

Unlike cucurbit crops, lettuce and greens are cool-weather plants. Their different annual-weed complex may include annual bluegrass, crabgrass, chickweed, henbit, common groundsel and many others. Herbicides to control these weeds in commercial leaf crops, except lettuce, are given in Table 25-3.

Lettuce

Although lettuce is usually a cool-weather crop, it may be subjected to high temperatures. In Arizona and Southern California, lettuce is planted in late August and September when air and soil daytime temperatures may exceed 100°F. Lettuce will not germinate above 90°F. However, the beds are kept wet, so they are cooled by the constant evaporation of water from the bed surface in this arid climate. This cooling allows shallow-planted lettuce seeds to germinate. Under these conditions, the annual-weed complex tends to be more summerlike than those weeds named above.

Table 25-3. Herbicides Used in Greens

	Collards, Kale, Mustard Greens, Turnip Greens	Spinach
Herbicide	Application	Application
CDEC[1]	PE[2]	PE or Post
Chlorpropham	–	PE
Cycloate	–	PPI
DCPA	PE, PPI	–
Trifluralin	PPI	–

[1] CDEC may also be used as a preemergence treatment in endive, escarole, and Hanover salad.

[2] Abbreviations: PE–preemergence; PPI–preplant soil incorporation; Post–postemergence.

Figure 25-2. Annual-weed control in lettuce with benefin. *Center*: Treated band. *Left and right*: Untreated.

Many herbicides that control annual weeds in lettuce are applied as a preplant soil-incorporation treatment. These include benefin (see Figure 25-2), bensulide, CDEC, pronamide, propham, and a combination of benefin plus propham. Propham plus pronamide is also used this way.

Preemergence applications are also used: CDEC, pronamide, and propham. Propham can also be used postemergence when lettuce has four or more leaves.

Paraquat can be used before the crop emerges as a contact spray to control emerged annual weeds; crop plants emerged at the time of application will be killed.

Greens

Greens included here are collards, kale, mustard greens, turnip greens, and spinach. Tolerances of these first four crops to several herbicides are quite similar; however, spinach reacts differently (see Table 25-3).

ONIONS

Although the home gardener often grows onions from bulbs (sets) or transplants, most commercial growers use seed. Weeds are particularly serious in onions produced from seeds because onions germinate and emerge relatively slowly, and their cylindrical-upright leaves make them poor competitors. However, these same leaves are an advantage when using certain contact herbicides whose spray droplets bounce off onion leaves, but remain on leaves of many broadleaf weeds.

Several herbicides used as early postemergence treatments for annual-weed control in onions must be applied only at certain stages of growth to avoid injury to the crop. These stages are classified as loop (crook),

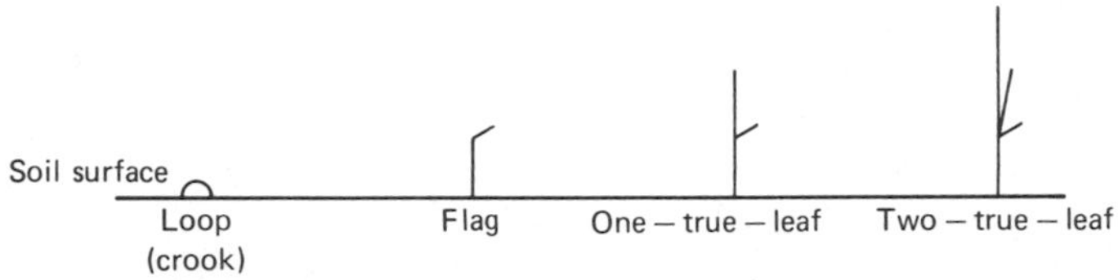

Figure 25-3. Stages of onion growth.

flag, one-true-leaf, and two-true-leaf (Figure 25-3). The following discussion applies primarily to *direct-seeded onions.*

Highly refined petroleum solvents as an oil emulsion have been used to control annual weeds that emerge before the onions. This mixture is applied just before the onions emerge. It has no residual soil activity, hence a second flush of weeds may appear, particularly if the soil has been cultivated. If needed to control this "second round" of weeds, this light aromatic emulsion is applied as a directed spray after onions emerge to the crook stage.

DCPA, chlorpropham, bensulide, and nitrofen control many broadleaf weeds and grasses in onions when applied preemergence. CDAA can also be used preemergence to control annual weeds in onions, but application is delayed until just before onions emerge or in the loop stage.

DCPA, chlorpropham, and CDAA can also be used postemergence to onions to control germinating weed seeds. They usually will not control emerged weeds. However, nitrofen not only controls germinating weed seeds but also small emerged weeds of several species. It should not be applied after the two- to three-leaf stage of onions.

Chlorpropham and CDAA should not be used on sandy soils because the crop may be injured. Bensulide can be used only in Texas and New Mexico.

Figure 25-4. Barnyardgrass control in onions with DCPA. *Center*: Treated band. *Left and right*: Untreated.

The ammonium-salt formulation of dinoseb is registered for use in onions as a postemergence treatment in three California counties from the two-leaf stage to 6 in. tall.

PEAS

Because peas are a cool-weather crop, the primary weed problem is annual cool-weather weeds such as chickweed, henbit, shepherdspurse, wild oat, and annual bluegrass. The perennial weed Canada thistle is also a problem in some areas.

Several herbicides have been developed to control annual weeds in peas. Diallate, EPTC, propham, triallate or trifluralin are used as a preplant soil-incorporation treatment. Alachlor, dinoseb, and propachlor can be used preemergence.

Several forms of dinoseb have been used for many years to control annual weeds in peas at various stages of crop growth. For preemergence applications, the alkanolamine salts of the ethanol series (I) or of the ethanol series plus the isopropanol series (II) are used. For postemergence treatment, the above formulations or the ammonium salt are applied when the peas are 2–8 in. high and the weeds are small. Dosages of formulation I and II depend on air temperature expected within 24 hr after application. These should be between 70 and 85°F, with lower dosages at higher temperatures. They should not be applied within six weeks of harvest. Treated foliage should not be grazed or fed within six weeks after application.

Other postemergence applications for peas include barban, bentazon, dalapon (sodium plus magnesium salt), MCPA (sodium or amine salt), MCPB (sodium salt), or glyphosate.

Peas show varietal differences in their tolerance to some herbicides listed above. There are also geographical limits and special application methods for use of certain herbicides. The labels must be checked before use.

PEPPERS

The main weed problem in peppers is annual broadleaf weeds and grasses. All herbicides registered for use in this crop, except paraquat, control only annual weeds. Paraquat controls emerged annual weeds and gives topkill and suppression of perennials.

In direct-seeded peppers the following herbicides are used as a preemergence treatment: bensulide, DCPA, or diphenamid. Bensulide and napropamide are applied as a preplant soil-incorporation treatment.

In transplants, diphenamid is applied at transplanting or within one month after transplanting. Chloramben and DCPA are applied four to six weeks after transplanting. These herbicides are applied to the soil surface. Trifluralin is used as a pretransplant soil-incorporation treatment.

The above herbicides used in direct-seeded or transplanted peppers do not control emerged weeds.

Paraquat is used as a contact herbicide for emerged weeds on preformed beds prior to planting, after planting but before crop emergence, or postemergence of the crop with shielded sprays.

POTATOES

The major weed problem in potatoes is annual broadleaf and grass weeds; in some areas quackgrass or nutsedge can also be serious.

Dalapon controls quackgrass as a preemergence treatment, fall or spring, or a preplant treatment to emerged quackgrass. Preplant treatments are usually plowed before planting potatoes. Rates of dalapon used vary depending on time of application. *Dichloropropene* also controls quackgrass in potatoes as a soil fumigant; it is injected into the soil 6–8 in. deep, but as a preplant treatment. *EPTC* is incorporated into the soil preplant and/or lay-by to suppress quackgrass and nutsedge growth.

Many herbicides are applied to the soil to control annual weeds in potatoes (see Table 25-4). Dalapon, paraquat, or dinoseb amine are

Table 25-4. Soil-Applied Herbicides Used in Potatoes for Annual Weed Control[1,2]

Herbicide	Application[3]
Alachlor	PE
DCPA	PE, LB
Diallate	PPI
Dinoseb, amine	PE
Diphenamid	PE, LB
EPTC	PPI, LB, I
Linuron	PE
Metribuzin	PE, I, P
Trifluralin	PPI

[1] Some of these treatments have specific limits such as geographic areas, soil type, application time relative to harvest, and potato variety.

[2] Certain herbicide combinations are also used: alachlor + dinoseb, alachlor + linuron, CDAA + CDEC (New York only), EPTC + trifluralin.

[3] Relative to crop: PE = preemergence, PPI = preplant soil incorporation, P = post, LB = lay-by, I = irrigation water (flood, sprinkle, or furrow).

applied to the foliage of emerged weeds prior to potato emergence. At the low rate used for this treatment, dalapon controls only annual grasses.

Ametryn, dinoseb, endothall, and paraquat are also used for preharvest vine killing in potatoes.

TOMATOES

Tomatoes are grown commercially from transplants or are direct-seeded in the field. Although annual weeds are the major problem by either planting practice, they are particularly serious in direct-seeded tomatoes. Weeds can be removed from direct-seeded tomatoes during thinning, but by this time they have already delayed the growth of young seedlings. Furthermore, the cost of labor for thinning may be doubled or tripled by heavy annual-weed infestation. Lack of weed control delays tomato maturity and greatly reduces the efficiency of machine harvesting.

Some common annual weeds of tomatoes are barnyardgrass, crabgrass, foxtails, lambsquarters, purslane, mustard, smartweed, and nightshade. Fields containing perennial weeds such as quackgrass, johnsongrass, and field bindweed should be avoided, because there are no selective methods to control these weeds in tomatoes.

Stoddard solvent or paraquat may be used in direct-seeded tomatoes when annual weeds emerge before the tomatoes. Either is applied when 2–4% of the crop has emerged. Emerged tomato plants will be killed, but a considerable excess of tomato seeds is usually sown, and many seedlings are removed in the thinning operation. Uniform emergence of tomato plants is essential for this method to be effective.

Bensulide, CDEC, napropamide, and diphenamid can be applied preemergence to control annual weeds in direct-seeded tomatoes. In areas of limited rainfall these herbicides are usually incorporated into the soil before planting. Other preplant soil-incorporated herbicides for tomatoes are pebulate, diphenamid-plus-trifluralin (see Figure 25-5), diphenamid-plus-pebulate, and napropamide-plus-pebulate. EPTC can be used in certain California counties.

In general, transplants are more tolerant to herbicides than the direct-seeded crop. However, when the direct-seeded crop has four true leaves, it is generally tolerant to herbicides. Most of the following herbicidal treatments apply to transplants as well as to the established, direct-seeded crop. The size of the plant, method of application, and other critical factors vary for each of these herbicides; therefore, detailed information from the manufacturer's label, local extension staff, and company field representatives should be checked before use.

These herbicides can be used to control annual weeds in transplanted or established plants of tomatoes: bensulide, CDEC, chloramben, chlor-

Figure 25-5. Annual weed control in direct-seeded tomatoes with diphenamid plus trifluralin; band treatment on top of beds. *Foreground*: Not treated. (Elanco Products Company, a division of Eli Lilly and Company.)

propham, DCPA, diphenamid, napropamide, metribuzin, pebulate, and trifluralin. CDEC-plus-CDAA granules can be used in New York only. Certain combinations of the above herbicides are also used. Pebulate also suppresses the growth of yellow and purple nutsedge.

Except for trifluralin and napropamide, the preceding tomato herbicides have a relatively short period of soil activity; therefore, it is often necessary to apply both a preemergence or preplant treatment as well as a lay-by treatment to give seasonlong control.

OTHER VEGETABLE CROPS

Chickpeas, eggplant, horseradish, lentils, okra, southern peas, and sweet potatoes are covered in this section. As is the case with many other vegetable crops, annual broadleaf weeds and grasses are their main weed problem.

The following herbicides are registered for use in these specialty vegetable crops. In *chickpeas* (garbanzos) profluralin is applied as a preplant soil-incorporation treatment. In *eggplant,* DCPA is sprayed on the soil surface postemergence to the crop when the seedling is 4–6 in. high or four to six weeks after transplanting. DCPA does not control emerged weeds. In *horseradish,* DCPA is applied as a preemergence treatment.

Nitrofen is used as a preemergence treatment or postemergence when horseradish is in the two- to four-leaf stage.

In *lentils,* four herbicides are used to control wild oat: diallate, triallate, and propham as a preplant soil-incorporation treatment or barban postemergence to wild oat at the two-leaf stage. Propham also controls volunteer grain. In *okra,* diphenamid is applied as a preemergence treatment; profluralin or trifluralin is used as a preplant soil-incorporation treatment.

In *southern peas* (cowpeas) profluralin, trifluralin, and DCPA are applied as a preplant soil-incorporation treatment. DCPA can also be used preemergence. Chlorpropham is also applied as a preemergence treatment. Bentazon is applied postemergence.

In *sweet potatoes* the following herbicides may be used to control annual weeds: CDAA, chloramben, DCPA, diphenamid, and EPTC. EPTC also suppresses nutsedge growth. Most of these herbicides require application methods unique to this crop, and the rates may also vary; local authorities should be consulted for specific information.

SUGGESTED ADDITIONAL READING

Dallyn, S., and R. Sweet, 1970, *FAO International Conference on Weed Control,* WSSA, Champaign, IL. p. 210.

Danielson, L. L., 1970, *FAO Internat. Conf. on Weed Control,* WSSA, Champaign, IL. p. 245.

Menges, R. M., and T. D. Longbrake, 1970, *FAO International Conference on Weed Control,* WSSA, Champaign, IL. p. 229.

Orsenigo, J. R., 1970, *FAO International Conference on Weed Control,* WSSA, Champaign, IL. p. 198.

Romanowski, R. R., 1970, *FAO International Conference on Weed Control,* WSSA, Champaign, IL. p. 184.

Shadbolt, C. A., and L. H. Holm, 1956, *Weeds* **4**, 11.

USDA, SEA, 1980, *Suggested Guidelines for Weed Control,* U.S. Government Printing Office, Washington, D.C. (621-220/SEA 3619).

USDA, SEA, 1981, *Compilation of Registered Uses of Herbicides.*

Zimdahl, R. L., 1980, *Weed-Crop Competition,* International Plant Protection Center, Oregon State University, Corvallis.

For chemical use, see the manufacturer's label for method and time of application, rates to be used, weeds controlled, and special precautions. *Label recommendations must be followed–regardless of statements in this book.* Also see the Preface, Inclusion or omission of a product name does not constitute recommendation or nonrecommendation of a product.

26 Fruit and Nut Crops

Weeds can damage fruit and nut crops seriously and in many ways. In newly planted crops, weeds compete directly with young trees for soil moisture, soil nutrients, carbon dioxide, and perhaps light. In older plantings, weed competition may be less serious, but still may reduce yields noticeably.

In addition, weeds may harbor plant diseases, insects, and rodents such as field mice that may girdle the trees. Also, weeds such as poison ivy may interfere with harvest. When crops, such as some nut crops, are picked up from the ground, weeds may seriously interfere with harvesting. The wasteful use of water by weeds is always important, especially in arid regions.

Crops discussed in this chapter are perennial plants. They are propagated in nurseries and transplanted to fields. The discussion is limited to field-weed control. Although nurseries have serious weed problems, they are not covered here because tree tolerance to most herbicides varies according to age, soil type, and so on. The herbicide label should be consulted for more specific instructions.

Troublesome perennial weeds, such as quackgrass, johnsongrass, bermudagrass, nutsedge, field bindweed, or Canada thistle, should be brought under control before setting a new orchard or making a small-fruit planting. Perennial weeds can be controlled much easier and at less cost before planting than afterward.

In most perennial crops when perennial weeds are absent or controlled before planting the crop, the primary weed problem is annual weeds. However, perennial weeds may invade the area later. This is especially true if perennial weeds are tolerant to the herbicides used. Also, absence of competition provides an essentially noncompetitive environment for tolerant weeds, including perennials, and they flourish.

A total weed-control system in fruit and nut crops may combine several methods including *cultivation, mowing, mulching,* and *herbicides.*

Weed control in deciduous fruit and nut crops has been reviewed by Lange (1970). Jordan and Day (1970) reviewed weed control in citrus crops.

CULTIVATION

When considered as a single weeding operation in tree crops, cultivation is just as effective and economical as with other tilled crops. Growers know well the advantages of cultivation. As for the disadvantages, shallow feeder roots are torn up, soil structure is changed, soil erosion increases (especially on hillsides), weeds under trees are difficult to control by mechanized methods, and cultivation brings new weed seeds to the surface, where they may germinate and grow. For one or more of these reasons, growers are depending less and less on cultivation by itself.

MOWING

Mowing is popular in borders and areas between trees of many fruit and nut crops, especially where soil erosion is serious. These areas can be maintained as short turf, effectively controlling erosion while keeping plant or weed competition to a minimum. Mowed turf gives a clean and neat appearance. The area under trees and vines is kept weed-free by appropriate use of herbicides or mulches.

MULCHING

Mulching usually controls weeds by depriving them of light. Also, most mulches conserve soil moisture. Decomposable organic matter, such as straw, is used in many horticultural crops including strawberries and young fruit-tree plantings. Also, in recent years, both black plastic sheets and paper sheets impregnated with asphalt have been used. Plants are set through small holes made in these sheets.

HERBICIDES

Orchards

Orchards may be maintained with cultivation or noncultivation methods. Noncultivation methods are usually mowing or the use of herbicides. The specific method used depends somewhat on the crop, but perhaps more important is the slope of the land and soil type. Some type of ground cover is desirable on hillsides to prevent erosion; mowing and use of a herbicide is important. On relatively flat terrain, cultivation, mowing, and herbicides are used.

Often a combination of these methods is used. Frequently a herbicide band is used down the tree row and cultivation or mowing is used between rows. This permits mowing or cultivation and minimizes the amount of herbicide used (see Figure 26-1). Cultivation close to the trees may cause mechanical damage to the trunks and to the root systems.

Numerous herbicides are used in orchards to control annual weeds. Table 26-1 lists those registered for use in these crops. Most of these herbicides control only annual weeds or germinating seeds of perennial weeds. However 2,4-D, dalapon, dichlobenil, DSMA or MSMA, EPTC, glyphosate, and trifluralin are effective on certain perennial weeds. As for perennial weeds—(1) glyphosate is effective on most species, (2) 2,4-D acts only on broadleaf species, (3) dalapon is a grass herbicide, and (4) EPTC and DSMA or MSMA are particularly useful for nutsedge control. Dichlobenil and trifluralin may control several perennial weeds when appropriate application methods are used; labels should be consulted. For additional information, see page 422.

Understanding how these herbicides work should help each grower develop a comprehensive program giving best results from these practices (see Chapters 8-21).

Resistant Weed Species

The key to successful chemical weed control in orchards is closely related to the weed species present. Once a herbicide program has been

Figure 26-1. Strip application of a herbicide in an almond orchard. The untreated areas between the treated strip will later be mowed, cultivated, or both.

Table 26-1. Herbicides Used on Orchard Crops[1]

	Ametryn	AMS	Atrazine	Bromacil	Cacodylic acid	2,4-D amine	Dalapon	Dichlobenil	Dinoseb	Diphenamid	Diuron	DSMA and/or MSMA	EPTC	Glyphosate	Napropamide	Norflurazon	Oryzalin	Oxyfluorfen	Paraquat	Simazine	Terbacil	Trifluralin
Almond								×	×			×	×	×	×		×	×	×	×		×
Apple		×				×	×	×	×	×	×	×		×	×		×		×	×	×	
Apricot							×		×			×			×	×	×		×			×
Avocado								×						×			×		×	×		
Cherry								×	×	×		×		×	×	×	×		×	×		
Date									×													
Fig								×	×						×		×		×			
Filbert								×	×					×	×	×			×	×		
Grapefruit	×			×	×		×	×	×		×	×	×	×	×		×		×	×	×	×
Lemon				×	×			×	×		×	×	×	×	×		×		×	×	×	×
Lime					×		×	×	×	×		×		×					×			
Macadamia			×				×				×			×					×	×		
Nectarine								×	×						×	×	×	×	×			×
Olive									×		×						×		×	×		
Orange	×			×	×		×	×	×	×	×	×	×	×	×		×		×	×	×	×
Peach							×	×	×	×	×	×			×	×	×	×	×	×	×	×
Pear		×				×	×	×	×		×	×		×	×		×		×	×		
Pecan								×	×		×			×	×		×		×	×	×	×
Pistachio														×			×		×			
Plum and/or prune							×	×	×			×			×	×	×	×	×	×		×
Tangerine					×		×	×	×		×	×	×	×	×				×			×
Walnut, English								×	×		×	×		×	×	×	×		×	×		×

[1] Most of these control annual weeds. However, specific ones also control certain perennial weeds. Several also have restrictions as to soil type, geographic location, tree age, use on bearing or nonbearing crops, and minimum time between application and harvest. *Read labels carefully* and follow directions.

selected, it will usually have to be revised periodically to prevent an increase of resistant weed species; for example, if simazine or terbacil is being used, the annual grasses will soon be a problem. If diuron is being used, weeds such as groundsel, burclover, wild oat, and other resistant species will usually take over. Alternating herbicides on sequential years or using a combination of two or more herbicides usually stops resistant weeds from becoming a problem.

Grapes

Weeds can be serious competitors of the grapevine. This is especially true of deep-rooted perennial weeds in unirrigated vineyards of the West, where little, if any, rain falls during the growing season. Here the grapevine depends totally on moisture stored in the soil from winter rains, and this is usually not enough for both the grapevine and perennial weeds. Annual weeds also compete, especially in young plantings.

As with orchards, annual weeds in the crop row are often controlled with herbicides, and the interrow area is cultivated or mowed. Cultivation close to the grapevine often injures the plant. In some soils, repeated cultivations for weed control can hasten the development of a plow sole, which can impede water penetration. A mowed cover crop between rows with a herbicide-treated strip down the row is used in some areas (see Figure 26-2).

Annual weeds in vineyards can be controlled by contact sprays such as dinoseb or paraquat; and soil applications of diuron, napropamide, oryzalin, simazine, or trifluralin (see Figure 13-4). Some contact sprays should not touch the trunk and foliage.

Field bindweed, a serious perennial broadleaf weed, can be controlled by careful selective placement of 2,4-D amine in grapes, which are quite sensitive to 2,4-D. To place the herbicide correctly, a wax bar impregnated with 2,4-D amine is dragged over the weeds, or a hooded boom or nozzles that deliver large droplets are used (see Chapters 6 and 7). Only acid or amine forms of 2,4-D should be used; they are essentially nonvolatile. Contact with the grapevines should be avoided. The chemical should be applied when field bindweed is in the bloom stage and growing vigorously. This application is safest on grapes after shatter following bloom, but before shoots reach the ground.

Field bindweed and rhizome johnsongrass can be controlled by trifluralin using special application techniques; the label should be consulted. Dichlobenil granules applied to the soil surface, with rainfall or overhead irrigation, control several perennial weeds including artemisia, Canada thistle, fescue, orchardgrass, quackgrass, and Russian knapweed.

Perennial grasses in vineyards can be controlled with directed sprays of dalapon. Repeat applications to regrowth are required about once a

Figure 26-2. Strip application of a herbicide in a vineyard. Untreated areas between treated strips will be mowed or cultivated later.

month, or about five times during the growing season. The user should avoid spraying fruit or foliage. It should not be applied within 30 days before harvest.

Blueberries

Weed problems and general methods of control of weeds in blueberries are similar to those for grapes. The herbicides usually used are dinoseb, chlorpropham, dichlobenil, diuron, paraquat, pronamide, simazine, or terbacil.

Caneberries

Caneberries or brambles include mainly blackberries, boysenberries, loganberries, and raspberries, The herbicides registered for use in these crops are given in Table 26-2. Dinoseb and paraquat are contact sprays and are applied to foliage of annual weeds. The other herbicides listed are applied to the soil to control annual weeds as they germinate. They are not effective on emerged weeds. However, dichlobenil and pronamide also control certain perennials.

Table 26-2. Herbicides Used on Caneberries[1]

Herbicide	Blackberries	Boysenberries	Loganberries	Raspberries
Chlorpropham	X			X
Dichlobenil	X			X
Dinoseb	X	X	X	X
Diphenamid	X			X
Diuron	X	X	X	X
Oryzalin	X	X	X	X
Paraquat	X	X		X
Pronamide	X			X
Simazine	X	X	X	X
Terbacil	X	X	X	X

[1] Check labels for limitations such as geographic area, plant age, stage of growth and methods of application.

Strawberries

Weed control in strawberries may cost up to several hundred dollars per acre per year. By using proper herbicides and other management practices, this cost can be reduced drastically. The lack of available labor for hand weeding has stimulated the use of herbicides in this crop. Black polyethylene plastic or straw also controls weeds in strawberries, as well as conserving moisture and keeping berries clean (see Figure 26-3).

Strawberry beds are usually abandoned after one to three years as a result of either disease or serious weed infestation. Appropriate use of herbicides can considerably extend the productive life of many plantings, especially as more-effective disease controls become available.

In some areas, fields intended for strawberries are fumigated with a mixture of methyl bromide and chloropicrin before planting. This treatment not only kills many weed seeds but also soil-borne disease organisms, for example, *Verticillium* and nematodes. Methyl bromide is the primary weed killer of this mixture. Malva, burclover, and filaree seeds are often not killed by this treatment.

Methyl bromide and chloropicrin are gases under normal temperatures and pressures. They are sold in pressurized containers as a liquid. Both are injected into the soil about 6 in. deep through chisels spaced about 12 in. apart on a tool bar. Immediately after treatment, the treated area is sealed with a gas-tight tarpaulin (polyethylene) for at least 48 hr. Injection of the fumigant and covering with the tarpaulin is completed in one operation, with all equipment mounted on a single tractor.

Strawberries may be planted three days after removing the tarpaulin. If weeds resistant to this treatment emerge later, they can usually be

Figure 26-3. Black polyethylene plastic used as a mulch for weed control in strawberries.

controlled with a postemergence treatment of chloroxuron at least 60 days before harvest.

Other herbicides that control annual weeds in strawberries are chloroxuron, dinoseb, DCPA, and diphenamid. Varietal responses of strawberries to herbicides vary considerably and the degree of dormancy differs in various locations. The label and local extension agent or other authorities should be consulted before using any of these herbicides.

SUGGESTED ADDITIONAL READING

Jordan, L. S., and B. E. Day, 1970, *FAO International Conference on Weed Control,* WSSA, Champaign, IL., p. 128.

Lange, A. H., 1970, *FAO Internactional Conference on Weed Control,* WSSA, Champaign, IL., p. 143.

USDA, SEA, 1980, *Suggested Guidelines for Weed Control,* U. S. Government Printing Office. Washington, D. C. (621-220/SEA 3619).

USDA, SEA, 1981, *Compilation of Registered Uses of Herbicides.*

For chemical use, see the manufacturer's label for method and time of application, rates to be used, weeds controlled, and special precautions. *Label recommendations must be followed–regardless of statements in this book.* Also see the Preface. Inclusion or omission of a product name does not constitute recommendation or nonrecommendation of a product.

27 Brush and Undesirable Trees

Control of woody-plant growth is a problem affecting most types of property. This includes grounds surrounding homes and industrial plants; telephone, telegraph, highway, and railroad rights-of-way; and recreation and grazing areas.

There are about 1 billion acres of pasture, pasturelands, and grazing lands in the United States. On parts of nearly all of this area, woody plants present some problem. In range and pasture areas it is often desirable to eliminate all or most of the woody plants, leaving only grasses and legumes for livestock grazing.

On many Western dryland ranges, native grasses increase rapidly where brush is controlled. In Wyoming, range forage yields doubled the first year after sagebrush was treated and increased fourfold during a five-year control program. In Oklahoma, control of brush with chemicals plus proper management (principally keeping livestock off during the first summer) increased the growth of grass four to eight times during the first two years after treatment.

Figure 27-1 shows the relationship between mesquite control and increase of perennial grass forage on Southwestern rangeland. With 150 mesquite trees/acre, production of grass forage was reduced by about 85%, and total production including that produced by the mesquite was reduced by nearly 60%.

In the Southeastern United States, over 100 million acres of land well suited to loblolly and short-leaf pine are being invaded by less-desirable hardwoods and heavy brush undergrowth. The same is true in Northwestern United States and Canada, where dominant, but inferior, hardwoods may invade stands of Douglas fir, balsam fir, and spruce. See Figure 27-2.

Aside from ranges, pastures, and forests, woody plants must be controlled along electric power and telephone lines, firebreaks, highways,

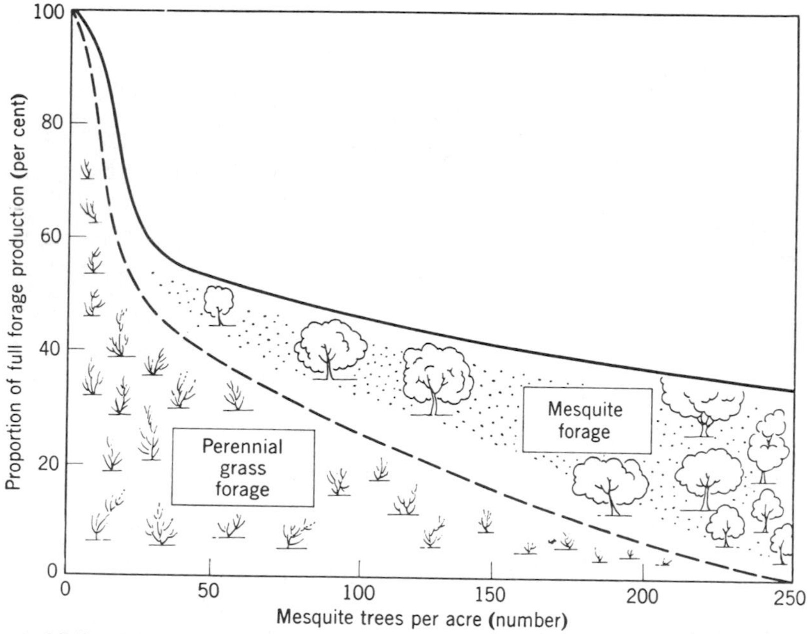

Figure 27-1. The effect of mesquite on forage production. (Adapted from USDA Leaflet No. 421, 1957.)

Figure 27-2. Red pine released from oak overstory by aerial spraying with 1 lb/acre of 2,4,5-T. Photograph was taken one year after treatment. (J. L. Arend, U.S. Forest Service.)

Figure 27-3. Woody plants controlled by 2,4-D and 2,4,5-T sprays. Forage grasses have increased (O. A. Leonard, University of California, Davis.)

railroads, farm fences, field borders, irrigation and drainage ditches, and in recreation areas. See Figure 27-3.

Some brush species act as alternate hosts for disease organisms affecting other plants. Common barberry harbors the organism causing stem rust in wheat and some other grasses. Other brush plants, such as poison ivy and poison sumac, are poisonous to man, and some, including wild cherry and locoweed, are poisonous to ruminant animals.

Control methods for woody plants vary with nature of the plant and size of the job. Some plants can be easily killed by one cutting. For example, many conifer trees have no adventitious buds on the lower parts from which new shoots or sprouts may develop. However, on large acreages even these species are controlled more efficiently by herbicidal sprays than by mechanical means.

In Oregon and Washington, studies compared manual brush control with herbicide control in forests. Brush cut close to the ground by hand reached an average height of over 4 ft in sixth months, and this brush overtopped more than 40% of the timber (crop) trees. Furthermore,

22% of small Douglas firs were damaged by this handwork. Costs for the manual control averaged \$230/acre compared with \$15–\$30/acre for aerial herbicide sprays. Workers did not want to cut any more brush by hand.

Plants hardest to control have underground buds from which new shoots or sprouts may develop. Some woody plants such as velvet and honey mesquite have these buds on the lower part of the trunk; other plants have their buds on underground, horizontal roots, or stems (see Figure 27-4). Most of the serious woody weeds are in one of these categories. As long as such plants have enough stored food and are not inhibited in some other way, they will send up new shoots. Some plants continue to send up new shoots for as long as three years or more, even when all top growth is repeatedly removed.

As discussed in Chapter 1, higher plants produce their own food. This food-building process starts with photosynthesis. If areas where photosynthesis takes place (mainly the leaves) are repeatedly destroyed, the plant will eventually die of starvation.

Control methods for woody species are similar to those for other weeds, except that they must be adapted to woody and heavy type of growth. Methods include repeated cutting or defoliation, digging or

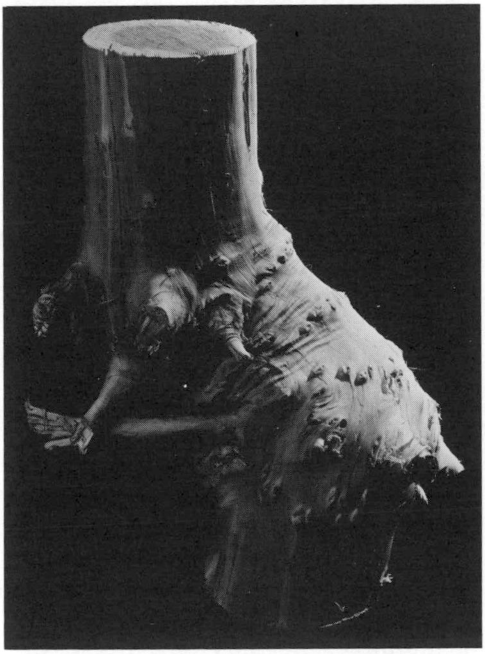

Figure 27-4. Section of mesquite stump. The wartlike structures are buds that produce sprouts. (Texas A & M University, Department of Range and Forestry.)

grubbing, chaining and crushing, burning, girdling, and herbicide treatments.

REPEATED CUTTING OR DEFOLIATION

The object of repeated cutting or repeated defoliation is to reduce food reserves within the plant until it dies. Plants commonly die during winter months if reserve foods have been reduced enough to make them more susceptible to winter injury (see Figure 2-3).

Most woody species bear their leaves well aboveground, making repeated cutting or defoliation effective. Most species must be defoliated several times each season if they are to be controlled. Plants may be cut by hand, by mowers, by special roller-type cutters, or by saws. They may be defoliated with special equipment or by chemicals. Species having palatable leaves may be sufficiently defoliated by livestock. Sheep and goats have eradicated undesirable shrubs and tree sprouts, and insects may also defoliate a plant, gaining control as a biological predator.

Plants have well-established, annual patterns of food usage and food storage. This knowledge of root reserves as they vary during the year is valuable in reducing the number of treatments needed. With most deciduous plants, root reserves are highest during the fall. These reserves gradually diminish through the winter months as the plant uses its energy to survive. In the spring the plant needs considerable energy to send up new stems and leaves. Thus food reserves in roots are usually lowest just before full-leaf development.

Many plants reach this stage at about spring flowering or slightly before; therefore, mowing or defoliation is best started at that time (see Figure 2-3). Repeat treatments are ordinarily needed for one or more seasons. Usually treatments are repeated when leaves reach full development. For more precise timing, root-reserve studies must be made on each plant species.

As leaves become full grown, they start synthesizing more food than they need. As the excess accumulates, at least part of it is moved back to replenish the depleted root reserves.

Apical Dominance

The terminal bud of a twig or a shoot may hold back development of buds lower on the stem or roots. When the apical bud is cut off, lateral buds are no longer inhibited and may develop shoots. This mechanism is controlled by hormones, or auxins, within the plant.

When apical dominance is destroyed, more stems may develop than were present originally. Following cutting, the stand may appear thickened rather than thinned. However, if cuttings are persistently continued so as to keep root reserves low, the stand will start to thin as its food supply is depleted. Under such conditions, the increase in number of shoots may actually speed reduction of root reserves (see "Mowing," page 21).

DIGGING OR GRUBBING

Plant species that sprout from an individual stem or root and remain as a "bunch" or "clump" can be effectively controlled by digging or grubbing. Whether this method is practical depends on the ease of removing the "clump," and density of stand. Mesquite is an example of this type of plant. Young plants can easily be removed by grubbing.

Giant bulldozers have been equipped with heavy steel blades or fingers (root plows) that run under the mesquite, lifting out the bud-forming crown along with the rest of the tree. The operation is often expensive and hard on desirable forage species. Other weeds may quickly invade the disturbed area. Reseeding the disturbed area with desirable grasses will help reduce invasion by weeds. Disadvantages of this method include the following: (1) Remaining roots may resprout, (2) the soil surface is severely disturbed, (3) the method is difficult in rocky soils, and (4) dormant weed seeds are brought to the surface to germinate and invade the area.

CHAINING AND CRUSHING

Chaining is a technique of pulling out trees without digging. A heavy chain is dragged between two tractors; this action tends to uproot most of the trees if the soil is moist. Then the chain may be pulled in the opposite direction to finish tearing trees loose from the soil. The method cannot be used on large, solidly rooted trees, and it is not effective on small, flexible brush, or on any species that sprouts from underground roots or rhizomes.

Crushing mashes and breaks solid stands of brush and small trees by driving over them with a bulldozer and/or with rolling choppers. Rolling choppers have been developed with drums 3–8 feet in diameter and weighing at least 2000 lb/ft of length. Chopper blades, about 8 in. wide, are attached across the entire length of the drum. This treatment crushes and breaks most of the branches and main stems. The chopper is usually pulled by a large, track-type tractor.

Crushing has disadvantages–it disturbs the soil, releasing buried, dormant weed seed to germinate and infest the area. It involves expensive equipment. In experiments in Texas, roller chopping before, during, or after herbicide application did not increase herbicide effectiveness in control of live oak. Roller chopping was followed by no increase in livestock grazing per acre (Meyer and Bovey, 1980). Similar results have been observed with other species including white brush, McCartney rose, and mesquite. The main advantage to chopping was a temporary increase in visibility.

BURNING

Prior to modern weed science (about 1950), many U. S. grasslands and woodlands were burned regularly. These fires were started by lightning, Indians, ranchers, and farmers. Some fires were accidental, others intentional.

Effective chemical methods are available to control many brush species on rangeland, yet burning is still used in many places. There are two reasons: cost is low and certain species are resistant to most herbicides but controlled by burning (e.g., red cedar). In Kansas experiments, late-spring burning controlled brush best and injured desirable forage species least.

Many people consider forest fires only as destructive. This is usually true of large, uncontrolled fires that occasionally ravage forests.

But fire in forest areas can produce benefits too. Since about 1950 "prescribed burning" or "controlled burning" has come to mean periodic burning with clear-cut purposes. When done properly, this method leaves desirable forest trees uninjured. Such burning removes some undesirable trees, brush, leaves, branches, and other debris, prepares a seedbed, and serves other useful purposes. By removing undesirable trash, (1) there is less competition from "weed" species, (2) the seedbed is improved for desirable tree seedlings and made less desirable for others, (3) less fuel remains for wild fires, and (4) some disease organisms may be held in check.

Thickness of bark is important in determining susceptibility to fire. If the phloem and cambium of the trunk are killed by heat, the tree may show an effect similar to that of girdling.

Most pine seedlings develop best with less than $1\frac{1}{2}$ in. of debris on the forest floor; oak seedlings need more debris. One reason for prescribed burning is to prepare seedbeds favorable to pine and unfavorable to oak.

All tree seedlings, when small, are susceptible to fire. Most pine seedlings become resistant after the first year and will withstand considerable fire after the trunk has reached a 2-in. diameter at breast height (DBH). Oak and most other hardwoods the same age are still susceptible. Therefore, where pine forests are desired, controlled fire can effectively and selectively kill the deciduous hardwoods. Frequency of prescribed burning will depend on local conditions; usually every two to three years is best.

Before setting a fire, local authorities should be consulted for regulations and precautions; firebreaks, fire-fighting equipment, wind direction and possibility of a shift, and other fire hazards must be considered. A heavy share of the cost of burning is for personnel and equipment on standby for emergency use. Obviously an escaped fire can be extremely costly.

During several years of protection from fire, considerable debris and undergrowth may accumulate. If so, the first burning may be hot enough to damage even large trees. Burning when woods are only partially dry will reduce the intensity of the fire.

GIRDLING

Girdling consists of completely removing a band of bark around the woody stem. The method is effective on most woody, dicotyledonous plants, but is too laborious to be efficient on thick stands of smaller-stemmed species. In most species, sprouting near the base of the tree is eliminated or at least considerably reduced on large, mature trees. Young trees may sprout profusely following girdling.

Plant foods move down to the roots through the phloem, and water and nutrients move upward through the xylem (see Chapter 4). When bark is removed, phloem tissue is destroyed, and plant foods can no longer move to the roots. Since roots depend on tops for food, the roots eventually die of starvation.

This process usually takes from one to three years. The length of time depends upon the rate of food usage and amount of food stored in the roots at the time of girdling. If girdling is properly done, the top of the tree shows little injury until the root is near death. As soon as the root dies, the top also succumbs.

The girdle must be wide enough to prevent healing and deep enough to prevent the cambium from developing a new phloem. The girdled area should dry in a short time. Chemicals similar to those used for frill treatment or heat may help kill an active cambium. The xylem should not be injured. If the top is killed too rapidly through injury to the xylem (sapwood), root reserves may not be depleted enough to prevent shoot or sprout growth.

If sprouts or lateral branches develop below the girdle, these will soon provide enough food for the roots to keep the tree alive. Thus any sprouts that develop may be cut, girdled, or treated with suitable chemical sprays to prevent food from returning to the roots.

HERBICIDES

Use of herbicides to control brush and undesirable trees has expanded rapidly since 1950. Proper use of chemicals has generally proven to be more effective and less costly than most other methods.

Herbicides can be applied in many different ways to accomplish different purposes. These can be classified as foliage sprays, bark treatment, trunk injections, stump treatment, and soil treatment. See Tables 27-1 and 28-2.

Table 27-1. Herbicides Used to Control Some Common Woody Plants[1,4]

	2,4-D	2,4,5-T	2,4-D plus 2,4,5-T	Dicamba[2]	Picloram[2,3]	Fosamine	Amitrole[3]	Glyphosate	Tebuthiuron[3]
Alder	F, B	F, B, C	F, B, C	C, S		F			S
Ash		F, B, C	F, B, C	C, S	S	F	F	F	S
Blackberry		F, B	F, B	F, S		F	F	F	S
Buckbrush	F, B	F, B	F, B						S
Elderberry	F, B	F, B	F, B						
Elm	C	F, B, C	F, B, C		F	F		F	S
Locust	F, B	F, B, C	F, B, C		S	F	F		S
Manzanita	F, B	F, B	F, B		F				
Maple		F, B, C	F, B, C	C, S	S	F	F	F	
Mesquite		F, B, C		F, S	F				S
Oak	F, C	F, B, C	F, B, C	C, S	S	F		F	S
Poison ivy or Oak	F, B	F, B	F, B	F, S	F		F		S
Rabbitbrush	F, B		F, B						S
Rose		F, B	F, B		S	F		F	S
Sagebrush	F, B	F, B	F, B						S
Saltcedar	F, B, C		B, C	S	S				S
Sassafrass		F, B	F, B			F			
Sumac	F, B	F, B	F, B	F, S		F	F		S
Willow	F, B	F, B, C	F, B, C	F, S		F		F	S

[1] F=foliar, B=stems, C=cutsurface (frill injection or stump treatment), S=soil.

[2] Dicamba, picloram, and dichlorprop are often combined with 2,4-D, 2,4,5-T, or both.

[3] Noncropland only.

[4] **All formulations of these herbicides are not suitable for all the uses indicated. *Check manufacturer's label* for specific species controlled with various formulations and for use precautions. *Follow label instructions.***

The effectiveness of herbicides should not be evaluated too quickly; one should wait at least until the next growing season. In some cases, trees may die two to three years after herbicide treatment, after repeatedly leafing out followed by defoliation.

Foliage Sprays

Foliage sprays (see Figure 27-5) are applied with the intent of getting uniform coverage. The need for thorough wetting will depend on the chemical used and the species treated. In most cases, uniform coverage is more important than thorough wetting; the latter is more often simply a means of obtaining uniform coverage.

Effectiveness of foliage sprays varies considerably among species and with size of woody plants. Several chemicals kill aboveground parts, but sprays have been less effective in killing roots. A resurgence of sprouts is common with some species. Repeat treatments are often needed as new sprouts develop.

Absorption by leaves is the first major problem of herbicide effectiveness. Many woody plants have coarse, thick leaves with a heavy cuticle. On such plants a nonpolar substance is absorbed more effectively than a polar substance. Thus esters of 2,4-D, 2,4,5-T, silvex, dichlorprop, and other oil-soluble herbicides are absorbed better than salt formulations. For the same reasons, an effective wetting agent added to phenoxy salts increases absorption. Absorption is also discussed in Chapters 4, 6, 7.

Figure 27-5. Applying a foliage spray. (F. A. Peevy, Agricultural Research Service, U.S. Department of Agriculture.)

Once absorbed, water-soluble or polar forms (especially of 2,4-D) are translocated more readily through the phloem than the esters. Esters appear to be hydrolyzed to their respective acids upon entering the plant.

A second important problem is acute toxicity versus chronic toxicity. Sometimes for promotional reasons, a "quick show" of herbicide effects is needed. Therefore, chemicals and rates may be used to "burn" the plant's leaves quickly. This may be contrary to physiological principles of effective translocation. For example, the excessively high application rate may inhibit translocation and the chemical will not be translocated to the roots. Thus roots will not be killed.

For foliar application this suggests three practical steps: (1) low-acute-toxicity herbicides and wetting agents should be used for maximum absorption and translocation of growth substances through phloem tissues, (2) herbicides should be applied at rates that cause chronic toxicity, but not acute toxicity, and (3) low rates of treatment and repeated treatment should be used for some species as they often give superior final results.

Foliage sprays of growth substances (such as 2,4-D and 2,4,5-T) have usually given best root kills when large quantities of food are being translocated to the roots. With species that go somewhat dormant in summer, this peak translocation period usually occurs about the time of full-leaf development in the spring; a second peak period may occur in the fall. Other plants continue rapid growth and translocation throughout the summer; these can be treated anytime during the summer.

Chemicals are applied by hand sprayers; mist nozzles; power-mist applicators; truck-, trailer-, or tractor-mounted power-spray equipment; and airplanes. With ground equipment, water is normally used as the carrier.

Late-season plants develop a thickened, waxy cuticle. On such plants oil–water mixes are more effective. Airplane application may use water, oil, or oil–water emulsions depending on the type of foliage to be treated and the relative humidity of the atmosphere.

Herbicides such as 2,4-D, 2,4,5-T and dichlorprop are applied at rates of 1–4 lb/acre. Amitrole, 2–4 lb/100 gal of water, is effective on poison ivy, poison oak, kudzu, ash, locust, blackberry, dewberry, and sumac.

Picloram and dicamba are effective as a foliar or soil application for certain woody species. They are used on species resistant to phenoxy-type herbicides. They are often combined with 2,4-D, 2,4,5-T, or both, to control a broad range of woody species.

Ammonium sulfamate is an effective foliage treatment. It is usually suggested where phenoxy compounds are too hazardous to use because of spray drift. One should use 60 lb/100 gal and apply it as a wetting spray.

Fosamine is foliar-applied in late summer or early fall. Response is not normally observed until the next spring. Susceptible treated plants fail to refoliate and later die. It is applied at the rate of 6–12 lb/acre.

Glyphosate is a nonselective, foliar-applied herbicide. It will translocate from foliage to roots, resulting in kill of many species. It is applied when plants are actively growing and when most plants are at or beyond the full-bloom stage of growth. It is effective against blackberries, honeysuckle, kudzu, maple, multiflora rose, trumpetcreeper and willow.

Bark Treatment

In bark treatments the chemical is applied to the stem near the ground and is absorbed through the bark. Two methods are used: broadcast bark treatment and basal bark treatment.

Broadcast Bark Treatment

In broadcast bark treatment, the herbicide is applied to the entire stem area of the plant. This treatment is especially adapted to thick stands of small-stemmed woody plants. Spray is usually applied between fall leaf drop and midwinter. Phenoxy herbicides are effective on several species; from 4 to 12 lb/100 gal are applied, mixed in oil. The spray is applied so that the greatest amount of it will cover the stems evenly. For example, treating upright stems with a horizontally directed spray exposes the greatest stem surface to the spray.

Basal Bark Treatment

Here the herbicide is applied so as to wet the stem base, 10–15 in. aboveground, until rundown drenches the stem at groundline. The chemical can be applied from a sprayer or from a container that simply lets the chemical "trickle" on the stem base. A small number of trees can easily be treated with a tin can to pour the chemical on the base of the tree trunk. Brush or large trees are left standing. This method is especially effective on brush and trees less than 6 in. in diameter (see Figure 27-6). Also, it controls some plants that sprout from horizontal, underground roots and stems, especially if applied during early summer.

This method has the advantage that the operator can selectively treat only those trees he wishes to kill, without injury to other trees.

Herbicides used most often are phenoxy compounds such as 2,4-D or 2,4,5-T. From 8 to 16 lb active ingredient of herbicide are mixed in enough diesel oil to make 100 gal of mixture. The lower rate is suggested when trees are growing actively and the higher rate for dormant work. The herbicide mixture is usually applied at the rate of 1 gal/100

Figure 27-6. Basal bark application is effective on trees up to 6 in. in diameter. (Amchem Products, Inc., Ambler, PA.)

in. of tree diameter; for example, 50 trees 2-in. in diameter or 33 trees 3-in. in diameter (see Figure 27-7).

Scientists do not fully understand the way basal bark treatment kills plants. It is possible that treatment either kills the phloem or immobilizes it to the point that the tree is chemically girdled. In addition, the herbicide may inhibit bud formation and sprout development. If the tree is chemically girdled, the tree root dies from starvation, as it does with girdling.

Trunk Injections

Trunk injections often help the herbicide penetrate through the bark. In some cases, injections may serve as a girdle, with the herbicide acting as a chemical girdle. Techniques used are frilling or notching with an ax or tree injector.

Frill Treatment

Frill treatment consists of a single line of overlapping, downward ax cuts around the base of the tree. The herbicide is then sprayed or squirted into the cut entirely around the tree (see Figure 27-8). The method is effective on trees too large in diameter for basal bark treatment.

Phenoxy compounds, such as 2,4-D and 2,4,5-T, are the principal herbicides; they are used at rates of from 8 to 16 lb diluted in enough diesel oil to make 100 gal of spray mixture.

Figure 27-7. Basal bark application of 4% 2,4,5-T solution was dribbled to saturate the lower 15 in. of this red oak tree trunk in November. Eleven months later, the tree is dead. (D. B. Cook, Cooxrox Forest, Stephenson Center, NY.)

Ammonium sulfamate is applied by saturating the frill with water containing 3–5 lb of chemical per gallon.

Notching

The tree trunk may be notched with an axlike frill treatment, except that one notch is cut for every 6 in. of trunk circumference. Notches filled with one teaspoonful of ammonium sulfamate crystals will kill many kinds of broadleaved trees (see Figures 27-9 and 27-10). This method is less effective than frill treatment, which uses a continuous notch around the tree.

The tree-injector tool speeds notch treatment, and when properly used does a satisfactory job. The oil-soluble amine or ester form of the phenoxy compounds plus diesel oil (2:9 ratio) should be used; one cut for every 2 in. in trunk diameter. This tool is especially effective against elm, postoak, white oak, live oak, and willow. More-resistant trees are treated by spacing injections closer together. Trees that are more resistant include ash, cedar, hackberry, hickory, blackjack oak, red oak, persimmon, and sycamore.

Figure 27-8. Frill application is effective on all trees. It is used primarily on trees that are too large for basal bark applications. (Dow Chemical Company, Midland, MI.)

Figure 27-9. Application of ammonium sulfamate in notches. (F. A. Peevy, Agricultural Research Service, U.S. Department of Agriculture.)

Dicamba is also injected. The undiluted commercial liquid concentrate is added to water at a ratio of 1:1 for resistant trees and 1:4 for susceptible trees, with 0.5 to 1.0 ml applied per injection cut.

Stump Treatment

Stumps of many species may quickly sprout after trees or brush are cut. Most trees can be prevented from sprouting by proper stump treatment (see Figure 27-11). However, weedy trees that can develop sprouts from underground roots or stems are difficult to control by stump treatment alone. With most such species, stump treatment is effective if followed by a midsummer basal bark treatment of the sprouts when they develop.

Phenoxy herbicides (such as 2,4-D or 2,4,5-T) are usually used in stump treatment at the rate of 8–16 lb of acid in enough diesel oil to make 100 gal of spray solution. Enough spray should be applied to "wet" the tops and sides of the stump so that rundown drenches the stem at groundline.

Ammonium sulfamate crystals are also highly effective on many species. Crystals are usually applied to the cut top of the stump at $\frac{1}{4}$ oz (about $1\frac{1}{2}$ teaspoonfuls)/in. of diameter.

Figure 27-10. The tree injector speeds notch application of herbicides, and permits selection of the tree to be killed. (W. C. Elder, Oklahoma State University.)

Figure 27-11. The stump will not sprout if the top and sides are saturated with the proper herbicide. (Dow Chemical Company, Midland, MI.)

Stump treatments are most effective if applied immediately after the tree is cut. Sprouting from older cuts can usually be controlled by increasing the amount of chemical.

Soil Treatments

Some herbicides control woody plants when applied to the soil. Most of these herbicides are used as dry pellets. They require rainfall to leach them into the soil as deep as the feeder roots. Therefore, they are usually applied just before or early in the rainy season. Herbicides used this way usually persist in the soil for more than one year for maximum efficacy. Effects may develop slowly and not be apparent for one to two years after treatment. See Figure 27-12.

Research has shown that dicamba, hexazinone, karbutilate, picloram, and tebuthiuron can be used effectively by this method. The species to be controlled determines which of these herbicides is used. Sometimes combinations of these have been more effective than either one alone.

Since the early 1970s, soil-applied herbicides have gained increasing

Figure 27-12. *Right*: Tebuthiuron was applied two years prior to taking this picture. The brush and trees are dead. Reseeding the area with grass has converted the area to a highly productive grazing area. *Left*: Not treated. (Elanco Products Company, a division of Eli Lilly and Company.)

attention as a method to control undesirable brush and trees, thus improving grazing capacity on rangelands. Some of these herbicides are formulated as pellets or granules. They can be applied by hand, ground equipment, or airplane.

Many woody plants have extensive finely branched root systems, whereas many desirable grasses and forbs have root systems restricted to a small area. Some herbicides are phytotoxic to desirable grasses and forbs. Therefore, some questions arise. Can a wide spacing of herbicide pellets control the woody plants with spaces between pellets large enough to *not* injure most of the desirable grasses and forbs? What might be the most efficient spacing of pellets?

In Oklahoma, a grid application of tebuthiuron spaced 6 × 6 ft was equally or more effective than a broadcast application for control of oak (Stritzke, 1976). Doses were placed in a small area 6 ft apart in each direction. In Texas, tebuthiuron placed in rows from $4\frac{1}{2}$ to 18 ft apart was equally effective as broadcast treatment for control of several woody species. Researchers concluded that this spacing principle makes application of herbicides from airplanes in more or less grid patterns commercially possible (Meyer, Bovey, and Baur, 1978).

With highly concentrated herbicide application in a 6 × 6 ft grid, a bare area 8–24 in. in diameter developed around the dosage sites about

90 days after application. Bare spots covered from 1.5 to 4.1% of the total area. After one year, in most cases, the bare areas had disappeared. Desirable grasses and forbs in the interspaces were not injured (Scifres, Mutz, and Meadors, 1978). Obviously for successful use of this method for each herbicide, the percent concentration of active ingredient, size of pellet, and area covered by each pellet must be determined through research.

Hexazinone is formulated as a Gridball® to be distributed in a random grid pattern for control of woody plants. To apply 10 lb/acre, pellets are spaced about 6 ft apart in each direction on the soil surface; for 20 lb/acre, $4\frac{1}{2}$ ft apart; and for 40 lb/acre, 3 ft apart. The herbicide controls elm, hawthorn, maple, red oak, post oak, white oak, sweetgum, and wild plum.

SUGGESTED ADDITIONAL READING

Meyer, R. E., and R. W. Bovey, 1980, *Weed Science* **28**(1), 51.

Meyer, R. E., R. W. Bovey, and J. R. Baur, 1978, *Weed Science* **26**(5), 444.

Scifres, C. J., J. L. Mutz, and C. H. Meadors, 1978, *Weed Science* **26**(2), 139.

Stritzke, J. F., 1976, *Abstract, Proc. South Weed Sci. Soc.* **29**, 255.

Weed Science published by WSSA, 309 West Clark St., Champaign, IL 61820. See index at end of each volume.

Weeds Today. Same availability as above.

FAO International Conference on Weed Control, 1970. Same availability as above.

Regional Weed Science research reports and proceedings. Available in the libraries of most land grant universities. Published annually.

Obtain weed control bulletins, leaflets, and so forth, from your state university or college of agriculture on woody plant control.

For chemical use, see the manufacturer's label for method and time of application, rates to be used, weeds controlled, and special precautions. *Label recommendations must be followed—regardless of statements in this book.* Also, see the Preface. Inclusion or omission of a product name does not constitute recommendation or nonrecommendation of a product.

28 Pastures and Ranges

Hundreds of kinds of weeds infest pastures and ranges. These include trees, brush, broadleaf herbaceous weeds, poisonous plants, and undesirable grasses. Control of trees and brush in pastures and range was discussed in chapter 27. We will limit this chapter primarily to control of broadleaf weeds and grasses.

Almost half of the total land area of the United States is used for pasture and grazing (D. Klingman, 1970). Nearly all of this forage land is infested with weeds, some of it seriously. Weeds interfere with grazing, lower the yield and quality of forage, increase costs of managing and producing livestock, slow livestock gains, and reduce quality of meat, milk, wool, and hides. Some weeds are poisonous to livestock (see Figure 1-4). The total cost of these losses is hard to estimate. Losses from undesirable woody plants are discussed in Chapter 27. Controlling heavy infestations of some woody weeds has increased forage yields from two to eight times.

Grass yields increased 400% after the removal of sagebrush in Wyoming. Forage consumed by cattle increased 318% on native Nebraska pasture after perennial broadleaved weeds were controlled by improved agronomic practices and use of 2,4-D. This 318% increase was a result of better pasture species, deferred and rotational grazing, and effective weed control (Klingman, D., and McCarty, 1958). Spraying with 2,4-D on high-level-fertility weedy pastures increased consumption of forage by cattle 1000 lb/acre over the no-weed-control, high-fertility treatment (Peters and Stritzke, 1971).

These results emphasize the usual need for improved management practices along with a better weed-control program. Both must be used together.

Weed-control programs for pastures and ranges often include a combination of management, mechanical, fire, and chemical methods. Biological control has been effective on certain species (see Figure 2-4).

CHEMICAL COMPOSITION OF GRASSLAND WEEDS

Nutrient or chemical composition of grassland weeds is important to livestock farmers for two reasons. First, weeds contribute to the livestock ration. See the next paragraph. Second, weeds compete for nutrients and water needed by more-palatable and more-desirable species and thus cut down yields of desirable forage.

Scientists collected forage and weed samples before mowing in an intensive agricultural area of the Connecticut River Valley in Massachusetts. Table 28-1 shows nutrient content of various plants as determined by chemical analysis. In addition to chemical composition, one must evaluate freedom from toxins, palatability, yield, and persistence in determining the worth of a species for forage. Also see Table 2-1.

MANAGEMENT

Choice of the most desirable forage species for the area is perhaps the first step in pasture improvement.

Proper management will control some weeds by itself. Proper management favors heavy plant growth, and many annual weeds are crowded out. For example, broomsedge disappears from pastures of the southeastern United States when the pastures are properly fertilized and seeded to high-yielding species such as Ladino clover, orchardgrass, or fescue.

Some weeds are favored by the recommended agronomic program. For example, dock responds to high soil fertility and favorable moisture. It is favored as much or more than the desired species.

In some cases weedy pastures will best be plowed, fertilized if needed, and reseeded. These are usual steps in areas favored with adequate moisture. In many other areas, desirable species will flourish if grazing pressure is temporarily removed or time of use adjusted, and weedy species brought under control. This latter program is often most practical where rainfall is limited. Reseeding may also hasten establishment of desirable species in dryland areas. A weed-control program is usually necessary to allow new seedlings to become established.

MOWING

Mowing in the past was often recommended to control pasture weeds. Now, mowing is of little importance for rangeland weed control, and it is steadily becoming less important on more intensively grazed areas.

Table 28-1. Chemical Composition of Grassland Weeds Compared to Timothy and Red Clover (Sampling Date June 5–10) (Vengris et al., 1953)

			Mean Percentage Composition (Air-Dry Basis)				
Plant	Growth Stage	Number of Samples	N	P	K	Ca	Mg
Timothy	Early heading	19	1.55	0.26	2.17	0.34	0.10
Red Clover	In buds, before bloom	19	2.84	0.25	1.09	1.88	0.42
Tufted vetch	Early bloom	3	3.58	0.30	1.52	1.52	0.30
Yarrow	In buds, before bloom	8	1.56	0.31	2.35	0.82	0.18
Oxeye daisy	50% heads in bloom	7	1.63	0.34	2.48	0.94	0.21
Daisy fleabane	In buds, before bloom	11	1.47	0.38	2.12	1.12	0.20
Common dandelion	Mostly leaves	14	2.25	0.44	3.39	1.21	0.43
Yellow rocket	After bloom	7	1.44	0.24	1.55	1.23	0.17
Plantain	Mostly leaves	11	1.48	0.30	2.10	2.55	0.46
Narrowleaf plantain	Mostly leaves	5	1.85	0.37	1.90	1.90	0.33
Yellow dock	50% heads in bloom	13	1.84	0.30	2.29	1.11	0.42
Tall buttercup	In bloom	7	1.45	0.31	1.98	0.94	0.25
Wild carrot	Vegetative growth leaves	5	2.52	0.54	2.37	1.92	0.44
Mouseear chickweed	In bloom	11	1.73	0.41	3.14	0.70	0.26
Cinquefoil	Early bud stage	2	1.49	0.28	1.31	2.08	0.33
Common milkweed	Vegetative growth	2	3.02	0.47	3.08	0.80	0.45
Sensitive fern	Vegetative growth	6	2.27	0.48	2.50	0.65	0.39
Quackgrass	Before heading	11	1.82	0.28	2.14	0.36	0.10

Mowing has an effect on some kinds of weeds, but not on others; it is more effective on upright-growing annuals, but is ineffective on those with leaves and seed heads close to the ground. Mowing during early flowering usually slows down or stops seed production of the upright types. Repeated mowing will kill most tall-growing annual and some tall-growing perennial weeds. With most perennials mowing is best done during early flowering and repeated as needed (see Figure 2-3).

Mowing is often disappointing. It tends to improve the appearance of the area, but kills few perennial weeds. In North Carolina, five years of monthly mowing was required to control horsenettle. Weekly mowing for three years (about 18 times during each growing season) reduced wild garlic plants in bermudagrass turf by only 52%.

In Nebraska, mowing a native grass pasture in either June or early July for three years left 65% of the perennial broadleaved weeds still living at the end of the experiment. After 20 years of mowing, 24–38% of the ironweed plants still persisted.

Mowing may be used to good advantage in new grass-legume seedings to lessen weed competition. Clipping the tops off of broadleaved weeds may sufficiently reduce weed competition to permit survival of seedling grasses and legumes. However, mowing also clips the tops off the forage species, setting them back to some extent.

HERBICIDE CONTROL

Pasture and grazing areas are well suited to chemical weed control. It is often possible to control a weed with little or no injury to desirable forage species. These forage species then respond with increased ground cover and larger yields (see Figures 27-12, 28-1, 28-2, and 28-3).

Figure 28-1. *Left*: Before treatment with herbicides. *Right*: After treatment, showing weed-free pasture. (Fisons Pest Control, Ltd, Chestford Park, England.)

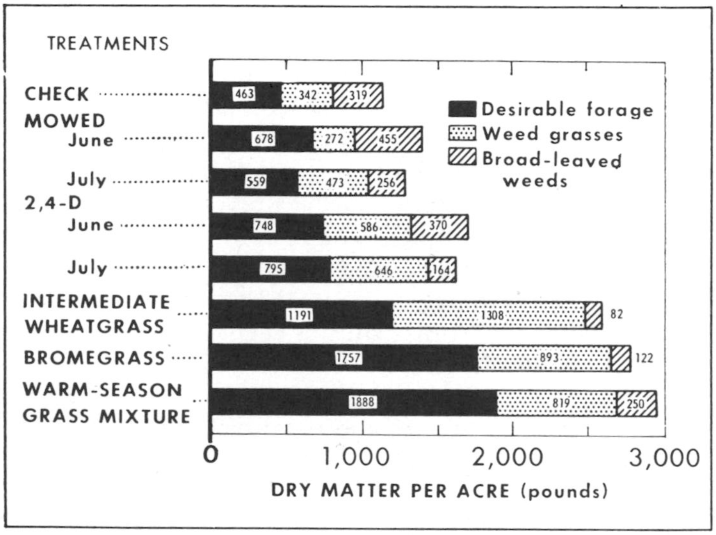

Figure 28-2. The effect of weed-control treatments on average pounds of dry matter per acre eaten by cattle. The seedbed for wheatgrass, bromegrass, and the warm-season grass mixture was plowed, killing many weeds. In addition, these plots were sprayed annually with 2,4-D for weed control. (D. L. Klingman and M. K. McCarty, USDA and University of Nebraska.)

Most broadleaf weeds can be controlled with phenoxy-type sprays. Control of undesirable grasses, especially annual grasses such as downy brome, has become an increasingly serious problem. Use of a preemergence herbicide, such as atrazine or diuron, before weed-seed germination may provide control.

Control of brush and woody weeds is covered in Chapter 27, and weed control in small-seeded legumes is discussed in Chapter 23. Control of these weed problems in pastures is essentially the same.

Table 28-2 lists the herbicides and methods of treatment for weed control, including woody plants, in permanent pastures and rangeland.

Figure 28-3. *Left*: Tebuthiuron removed weeds, leaving desirable grass forage. *Right*: No herbicide treatment, with heavy weed infestation. (Dr. J. F. Stritzke, Oklahoma State University.)

Table 28-2. Weed Control, Including Woody Plants, on Permanent Pastures and Rangeland by Various Herbicide Applications[1]

Herbicide	Foliage Spray	Basal Bark Spray	Stump Spray	Frill, Notching	Dormant Stem Spray	Tree Injection	Soil Treatment
Atrazine	X						X
Amitrole	X						
AMS	X		X	X			
2,4-D	X	X	X	X	X	X	
Dichlorprop	X						
Dicamba	X					X	X
Diuron	X						X
Fosamine	X						
Glyphosate	X						
Hexazinone	X						X
Karbutilate	X						X
MCPA	X						
Picloram	X						X
Silvex	X	X	X	X	X		
2,4,5-T	X	X	X	X	X	X	
Tebuthiuron							X
Triclopyr	X						

[1]Weeds controlled by a number of herbicides are given on page 364 and 418. Also, commercial herbicide labels list different weeds controlled, give instructions for use, and precautions. *Follow the label instructions.*

LIVESTOCK POISONING AND ABORTION

Most herbicides in common use in grazing lands are relatively nonpoisonous. If recommendations on the label are followed, no poisoning should occur.

Some herbicides, such as 2,4-D and similar compounds, are thought to temporarily increase the palatability of some poisonous plants. Livestock may eat these plants after treatment, whereas before spraying animals would have avoided them. Under such conditions livestock may be poisoned after plants have been sprayed, even though the herbicide is nonpoisonous.

Hydrocyanic acid (HCN) or prussic acid may be produced from glucosides that are found in various species of the sorghum family and in wild cherry. If cattle, sheep, and other animals with ruminant stomachs eat these plants, HCN is produced, and animals may be poisoned. Studies of wild cherry indicated no increase in HCN content following treatment with 2,4-D or 2,4,5-T.

"Tift" Sudan grass was treated when 8 in. tall with 1 lb/acre of 2,4-D, 2,4,5-T, and MCPA. Samples were taken 32 days after treatment. Chemical analysis showed that HCN increased after all three chemical treatments (Swanson and Shaw, 1954). The use of phenoxy herbicides evidently does not change the usual precautions necessary to prevent HCN poisoning. Ruminant animals should not be permitted access to plants containing HCN, whether treated or not.

Livestock may also be poisoned by eating plants having a high nitrate content. Nitrate is reduced to nitrite by microorganisms in the animal's intestinal tract. Nitrite in the bloodstream interferes with effective transport and use of oxygen, and the animal dies of asphyxia (suffocation). The lethal dose of potassium nitrate is 25 g/100 lb of animal weight.

In Mississippi a number of pasture weeds were analyzed for chemical content. The scientists reported that redroot pigweed and horse nettle contain enough nitrate nitrogen to be toxic to livestock. Even though many weeds were found to have sufficient crude protein to meet animal needs, the authors concluded that most weeds should be considered detriments and should be removed. Factors cited as detrimental were bitterness, spines, toxic mineral levels, other toxic components of weeds, and the fact that weeds caused lower yields of forage (Carlisle, Watson, and Cole, 1980).

The effect of herbicides on nitrate content of 14 weed species was studied in Michigan. 2,4-D, 2,4,5-T, MCPA, dinoseb, chlorpropham, and MH were tested. Before treatment, 10 of the 14 weeds contained enough nitrate to cause poisoning if consumed in considerable quantities. Following herbicide treatment, four species showed no change in nitrate content, five species showed increases, and one a definite decrease.

Variation in the other species prevented drawing definite conclusions. Scientists concluded that many weeds contain enough nitrate to cause poisoning if eaten by livestock, whether sprayed by herbicides or not (Frank and Grigsby, 1957).

A high nitrate content in plants has been associated with abortion in cattle in Wisconsin. In Portage County, 400 abortions in cattle were reported in 1954. Reproductive diseases and pathogens accounted for only a very small number of abortions. "Poisonous weeds" were considered a possible explanation. Further study revealed that the muck soils were high in nitrogen but lacking in phosphorus and potassium; this condition is conducive to nitrate storage in plants. Weed species were analyzed for nitrate nitrogen and classified according to nitrate content (Table 28-3).

Next, pastures were treated with 2,4-D to eliminate weeds thought to contribute to the high abortion rate. Pasture areas were divided for experimental purposes. On one pasture, 2,4-D was applied in both 1956 and 1957. The area was weed-free in 1957. Ten heifers that grazed on this area calved normally in 1957, but all 11 heifers that grazed on non-treated and weedy pastures aborted in the same year.

A feeding trial was conducted to test the effectiveness of dosing pregnant cattle with nitrate to induce abortion. Three 700-lb heifers that were given 3.56 oz of potassium nitrate each day aborted after three to five days. The aborted fetuses and placentas were similar to those aborted on the weedy pastures (Simon et al., 1958).

In summary, herbicides are generally nonpoisonous to livestock if used as directed on the label. Poisoning may occur if palatability is increased following spraying so that livestock consume larger-than-usual quantities of poisonous weeds. Killing poisonous weeds with herbicides may reduce the poisoning hazard.

Table 28-3. Nitrate Nitrogen Content of Plants (Sund and Weight, 1959)

High NO_3 Content (above 1000 ppm)	Medium NO_3 Content (300–1000 ppm)	Little, or No NO_3 Content (below 300 ppm)
Elderberry	Goldenrod	Linaria
Canada thistle	Cinquefoil	Meadow rue
Stinging nettle	Boneset	Yarrow
Lambsquarters	Mints	Vervain
Redroot pigweed	Foxtail	Dandelion
White cockle	Aster	Milkweed
Burdock	Groundcherry	Willow
Smartweed	Toadflax	Dogwood
		Spirea

PREVENTING LIVESTOCK POISONING BY WEEDS

Poisonous plants should be immediately isolated from livestock to prevent livestock poisoning. You may need to fence the infested area or remove livestock from the area. A small number of poisonous plants may be cut and removed from the pasture or killed by herbicide treatment.

Hundreds of plants cause livestock poisoning. Usually eradication, or at least very effective control, is needed. Cutting and removal of plants, or treatment with an effective herbicide may be most desirable. If the area is small, a soil sterilant may be the best answer, as discussed in Chapter 30. Small, isolated areas can probably be treated best with hand equipment or with granular materials. Large areas may be treated best by broadcast-type equipment, either ground operated or aerial.

SUGGESTED ADDITIONAL READING

Bovey, R. W., 1977, *Agriculture Handbook No. 493,* USDA, ARS.

Carlisle, R. J., V. H. Watson, and A. W. Cole, 1980, *Weed Science* **28**(2), 139.

Frank, P. A., and B. H. Grigsby, 1957, *Weeds* **5**(3), 206.

Klingman, D. L., 1970, *Int. Conf. on Weed Cont.,* p. 401, WSSA, Urbana, IL.

Klingman, D. L., and M. K. McCarty, 1958, USDA Bulletin 1180.

Peters, E. J., and J. F. Stritzke, 1971, USDA Technical Bulletin, 1430.

Simon, J., J. M. Sund, M. J. Wright, and A. Winter, 1958, *J. A. Vet. Med. Assoc.* **132**, 164.

Sund, J. M., and M. J. Weight, 1959, *Down to Earth* (Dow Chem) **15**(1), 10.

Swanson, C. R., and W. C. Shaw, 1954, *Agron. Jour.* **46**(9), 418.

Vengris, J., M. Drake, W. G. Colby, and J. Bart, 1953, *Agron. Jour.* **45**(5), 213.

Weed Science published by WSSA, 309 West Clark St., Champaign, IL 61820. See index at end of each volume.

Weeds Today. Same availability as above.

FAO International Conference on Weed Control, 1970. Same availability as above.

Regional Weed Science research reports and proceedings. Available in the libraries of most land grant universities. Published annually.

Obtain weed control bulletins, leaflets, and so forth, from your state university or college of agriculture on weed control on pastures and ranges.

For chemical use, see the manufacturer's label for method and time of application, rates to be used, weeds controlled, and special precautions. *Label recommendations must be followed—regardless of statements in this book.* Also, see the Preface. Inclusion or omission of a product name does not constitute recommendation or nonrecommendation of a product.

29 Aquatic-Weed Control

Aquatic plants, as the term is used here, include those plants that normally start in water and complete at least part of their life cycle in water.

Excessive growth of aquatic weeds causes many serious problems for people who use ponds, lakes, streams, and irrigation and drainage systems. Excess weeds (1) obstruct water flow and increase water losses; (2) interfere with navigation, fishing, and other recreational activities; (3) destroy wildlife habitats; (4) cause undesirable odors and flavors; (5) lower real-estate values; (6) create health hazards; and (7) speed up the rate of silting by increasing the accumulation of silt and debris.

On the plus side, aquatic weeds may reduce erosion along shorelines, and some plant species provide food and protection for aquatic invertebrates, fish, fowl, and game. Algae are the original source of food for nearly all fish and marine animals; and swamp smartweed, wildrice, wild millet, and bulrush provide food and protection for waterfowl, especially ducks.

Controlling aquatic weeds sometimes causes problems other than those of the chemical itself. For example, the rapid killing of dense, weedy growth may kill fish, which happens even though the chemical is nontoxic to the fish. During photosynthesis, living plants release oxygen, and fish depend on this oxygen for respiration. When plants are killed, they produce no more oxygen. Worse yet, dead plants are decomposed by microorganisms that require oxygen for respiration. These two actions may reduce oxygen content in the water, causing the fish to suffocate. The answer is to treat only a part of very heavily infested areas at one time; fish will move to the untreated part.

Suppose you owned a recreation area suitable for both swimming and fishing. The right fertilization favors microscopic plants, used as food by fish. This heavy growth of microscopic plants makes the water appear cloudy or dirty and may give it an undesirable odor; hence it is less desirable for swimming. Here you can choose swimming or fishing, but not both—at least not at their best.

METHODS OF CONTROLLING AQUATIC WEEDS

Methods of controlling aquatic weeds include the following:

1. Proper construction of pond.
2. Competition for light.
3. Pasturing.
4. Drying.
5. Mowing.
6. Hand cleaning.
7. Chaining.
8. Dredging.
9. Burning.
10. Biological control.
11. Chemical control.

Proper Construction of Pond

Proper pond construction is highly important in controlling pond weeds. Many rooted aquatic plants are not easily established in deep water. The pond should be built so that as much water as possible is at least 3 ft deep. You can have water 3 ft deep only 9 ft from shoreline if all the edges of the pond have a slope of 3 to 1. Such a slope greatly reduces the area where cattails, rushes, and sedges first start growing. However, steep banks are hazardous for swimming. Gentle slopes should be provided for swimming areas.

Competition for Light and Fertilization

Ponds adequately fertilized develop millions of tiny plants and animals that give the water a cloudy appearance (bloom). If the water has a bloom and is at least 3 ft deep, submerged aquatic weeds have almost no chance to grow because of inadequate light. The benefits of fertilization include the following:

1. Increased growth of beneficial microscopic life, including phytoplankton and zooplankton.
2. Increased food supply for fish from the food chain that develops from the above.
3. Effective weed control by shading. Plants that do reach the surface should be cut off; otherwise they will be stimulated by the fertilizer.

A 16-20-4 or similar fertilizer is suggested at about 50 lb/acre. The first application is put on in early spring and is repeated as needed to maintain the cloudy appearance. A light-colored object should not be visible 1.5 ft below the surface. Fertilization of ponds is practical only where there is little loss of water from the pond, because fertility is lost with the overflow.

Pasturing

Pasturing is economical and effective in controlling marginal aquatic grasses, weeds, and some woody species. A good legume-grass pasture mixture, if properly managed and grazed, will give the banks and dam a lawnlike appearance. A good sod also protects the banks against erosion and helps to control undesirable species.

Excessive trampling may destroy the banks and muddy the water. Also, leeches in the water may attack animals, and some diseases are spread in the water to livestock, especially to dairy animals.

Drying and/or Freezing

Drying is a simple way to control many submerged aquatics. If the water can be drawn from the pond or ditch, leaf and stem growth of submerged weeds may be killed after 7–10 days' exposure to sun and air. Drying usually must be repeated to control regrowth from roots or propagules in the bottom mud or sand. In ditches this operation may be repeated several times per season.

Especially in cold climates, if the water is drawn down in late fall and the lake not allowed to refill until early spring, many aquatic weeds will be killed. As the lake refills, reinfestation may occur from weed pieces from the deeper part of the lake.

Mowing

Mowing effectively controls some ditch-bank weeds. Power equipment can be most easily used where the banks are relatively smooth and not too steep. Also, underwater power-driven weed saws and weed cutters are available. The effects usually are short-lived. Mowing is usually required at rather frequent intervals, and disposal of mowed weeds is often difficult.

Hand Cleaning

In lightly infested areas, hand cleaning may be the most practical method of control. A few hours spent in pulling out an early infestation may prevent the weed from spreading. The method is particularly effective on new infestations of emergent weeds such as cattail, arrowhead, and willow.

Chaining

Chaining aquatic weeds resembles chaining woody plants (see Chapter 27). A heavy chain, attached between two tractors, is dragged in the ditch. The chain tears loose the rooted weeds from the bottom. The method is effective against both submerged and emergent aquatics.

Chaining should be started whenever new shoots of emersed weeds rise about 1 ft above the water or when submersed weeds reach to the water surface. It should be repeated at regular intervals. Dragging the chain both ways may be effective in tearing loose most of the weeds.

The method is limited primarily to ditches of uniform width, accessible from both sides with tractors, and free of trees and other obstructions. After chaining it is usually necessary to remove plant debris from the ditch to keep it from collecting and stopping the flow of water.

Dredging

Dredging is a common method of cleaning ditches that are accessible from at least one side. The dredge may be equipped with the usual bucket, or a special weed fork may be used. Dredging may solve two problems: removal of weeds and removal of silt and debris.

Dredging has been tried in ponds from specially built pontoons, but in general the pontoon dredge has not proven practical. Dredging is an expensive operation, because of high equipment costs and the large amount of labor involved.

Burning

Burning may control ditch-bank weeds such as cottonwoods, willows, perennial grasses, and many annual weeds. Green plants are usually given a preliminary searing. After 10–14 days, vegetation may be dry enough to burn from its own heat.

Large trucks have been used in irrigation districts in the West. Each truck is equipped with a 30–35-ft maneuverable boom with oil-spraying nozzles at the end; lighter-weight equipment is also available.

Burning can also be combined with chemical- or mechanical-control programs. Burning the previous year's debris allows better spray coverage of regrowth. It may be desirable to burn the dead debris after chemical treatment. Mowing followed by burning the dried weeds may increase the effectiveness of the mowing.

Biological Control

Aquatic weeds have been controlled by fish, snails, insects, microorganisms, and higher plants. Ducks often effectively control duckweed in small ponds. Biological control has appeal because of the continuing control potential and the nonuse of chemicals in the water. However, as with other biological-control methods, care must be taken not to introduce a biological-control organism that will have undesirable side effects; for example, a fish that reduces the population of game fish.

Some freshwater fish will eat aquatic vegetation. The white amur (Chinese grass carp), tilapia, and silver dollar fish are used to control aquatic weeds in certain areas of the world. The white amur has been used in the People's Republic of China, Czechoslovakia, Poland, and the Soviet Union. It is now being used in Arkansas. See Figure 2-5.

For alligatorweed control a fleabeetle (*Agasicles hygrophila*), and a moth (*Vogtia malloi*) are providing control in the southeastern United States.

Although biological control holds great potential, its actual use has been limited. The federal government now supports considerable research on this method. Hydrilla and eurasian watermilfoil are two aquatic weeds receiving considerable research effort from the federal research program.

Chemical Control

Chemicals effectively control many aquatic and ditch-bank weeds. To control these you need to know (1) the name or names of the weed species, (2) the appropriate chemical, recommended rate, and time of treatment, and (3) the amount of water or size of area to be treated.

Water surface areas are usually measured in acres, like field areas. One acre is 43,560 ft^2, or an area 208.7 ft square (on each side). One acre-ft of water means 1 acre of water, 1 ft deep—that is, 43,560 ft^3 of water; 325,828 gal, or 2,719,450 lb. Thus, to produce a chemical concentration of 1 ppm, one would add 2.7 lb of the chemical (active ingredient) per acre-ft of water; 5 ppm would require 13.5 lb, and 5 ppm to an average depth of 3 ft would require 40.5 lbs. A closely related technique involves treating the "bottom acre-foot". Two methods of application are used. Using formulations that are heavier than water, the chemical can be applied to the water surface as either a spray or granule. The chemical sinks to the bottom. The second method involves drop hoses or weighted hoses dragged behind a boat, releasing the chemical at the lake bottom. Rates of application are usually based on a bottom acre (43,560 ft^2), much the same as with field-crop application. For conversion factors to other measurements see page 429.

Running water is measured by several methods. Rates are usually given as *cubic feet per second;* 1 ft^3/sec is equal to 450 gal/min. Usually the rate of water flow is determined by the use of a weir and gauge.

CHEMICALS USED IN AQUATIC-WEED CONTROL

The more important chemicals used in aquatic-weed control are discussed below. The use of specific chemicals for control of certain aquatic weeds is given in Table 29-1. For more complete discussions of chemicals see Chapters 8 through 21. Trade names are on page 422.

Restrictions on the use of water that has been treated with an aquatic herbicide are extremely important. In potable domestic water supplies, copper has a residue tolerance of 1.0 ppm; 2,4-D, 0.1 ppm; and endothall, 0.1 ppm. Detailed instructions and restrictions are printed on the label. *Those instructions must be followed.*

Acrolein

Acrolein is useful for treating weed-infested drainage ditches and irrigation canals. It controls most submersed water weeds and many snails (see Table 29-1). In small irrigation canals in western states at rates of 1–2.5 gal/ft^3/sec flow of water, this chemical controls weeds from 6 to 20 miles downstream. However, the principal use of acrolein is in large canals where the usual treatment is 0.1–0.6 ppm on a volume basis (ppmv) for 8–48 hr. Weed control with these treatments may extend 20–50 miles downstream.

Treated water did not harm crops when used for irrigation at low concentrations. Higher concentrations may cause injury to susceptible crops, such as cotton. Acrolein is not effective in the control of waterplantain.

Acrolein is a potent irritant and lacrimator. It must be applied by licensed applicators. It may be toxic to fish.

Amitrole

Amitrole is safe to fish at normal rates. It is especially effective on cattails and bulrushes. A special form of amitrole and sodium thiocyanate known as amitrole-T is very effective on waterhyacinth and certain

Table 29-1. Herbicides[1] Used to Control Several Common Aquatic Weeds

	Class[2]	Acrolein	Amitrole	Aromatic Solvents	Copper Ion	2,4-D	Dichlobenil	Diquat	Endothall	Fenac	Fluridone	Simazine
Algae												
Unicellular	S				X				X			X
Filamentous	S				X			X	X			X
Chara	S				X		X					X
Alligatorweed	E					X						
Arrowhead	E					X					X	
Bladderwort	S	X		X		X	X	X	X	X	X	
Bulrush	E		X			X						
Cattail	E		X			X						
Coontail	S	X		X		X	X	X	X	X	X	X
Duckweed	F					X		X			X	X
Elodea	S	X		X		X	X	X		X	X	
Hydrilla	S				X			X	X		X	
Naiad	S	X		X		X	X	X	X	X	X	X
Pickerelweed	E					X					X	
Pondweed	S	X		X		X	X	X	X	X	X	X
Rush	E					X						
Spikerush	S, E					X				X		
Waterhyacinth	F		X			X		X				
Waterlily	E					X	X				X	
Watermilfoil	S, E	X		X		X	X	X	X	X	X	X
Waterprimrose	S, F					X	X		X	X	X	

[1] See the manufacturer's label for rates, method of usage, species susceptibility, and precautions. Follow the label—regardless of statements in this book.

[2] F = Floating (unattached, tops above water). E = Emersed (rooted underwater or growing on wet soil, tops above water). S = Submersed (usually rooted in soil, tops mostly underwater).

other emergent and floating species. Amitrole is applied at 7–9 lb/acre active ingredient as a foliage spray. It should be applied on cattails between the time of flowering and seed maturity.

Aromatic Solvents

Aromatic oils have been particularly effective in controlling weeds in ditches with running water. These chemicals are deadly to fish, but the movement of water in irrigation canals usually has higher priority than saving fish in the canals. In the West, hand-weeding and dredging costs have been cut 80% with the aromatic solvents.

Aromatic solvents are sold as Varsol, Stoddard's solvent, paint thinners, general solvents, and for dry-cleaning clothes. Commercial forms specially developed for aquatic-weed control are available. The most effective ones had a flash point of not less than 80°F, distillation ranging between 278 and 420°F, and an aromatic content of not less than 85% (Bruns et al., 1955).

Velocity, temperature, exposure time, and rate all affect results in treating ditches. But one should not wait until the ditch is clogged seriously. Water velocities of $\frac{3}{4}$ to $1\frac{1}{4}$ ft/sec produced the best results. Water temperatures should be 70°F or above; in fact, the warmer the water, the lower the concentration required. Aromatic solvents are applied at the rate of 6–10 gal/ft^3 of water per second during 30–60 min. This provides a solvent concentration of 450–750 ppm.

Aromatic solvents strong enough to kill most submersed aquatic weeds caused no important injury when used to irrigate alfalfa, beans, lima beans, carrots, cotton, sorghum, white clover, lettuce, oats, sweet corn, potatoes, sugar beets, or potatoes (Bruns, et al., 1955).

Aromatics are very toxic to fish, crayfish, snails, plankton, and mosquito larvae. These chemicals often kill fish at concentrations as low as 5–10 ppm. If given the opportunity, fish will usually move from treated areas. Aromatics cause strong and persistent off-flavor in fish.

Livestock tend to avoid treated water, so there is no apparent danger to them. Guinea pigs forced to drink treated water showed no ill effects.

Copper Sulfate

Copper sulfate (bluestone, blue copperas, blue vitriol) and chelated copper (such as ethylenediamine, triethanolamine, and alkanolamine) are very effective against most kinds of algae, including chara. Soon after treatment the algae's color changes to a grayish white. Several days after treatment, the water should be nearly free of algal growth (see Figure 29-1).

2,4-D

Many aquatic plants are susceptible to 2,4-D dissolved in water at rates of 1–5 ppm by weight (see Table 29-1). To provide a concentration of 1.0 ppm in 1 acre-ft of water, 2.7 lb of 2,4-D (acid equivalent) are required. Water-soluble liquid forms of 2,4-D or granular forms may be used.

2,4-D should be exposed in water to susceptible weeds at full concentration for at least 10 hr. In *nonmoving* water, 2,4-D in water-soluble form will tend to *distribute itself equally* over several days' time for dis-

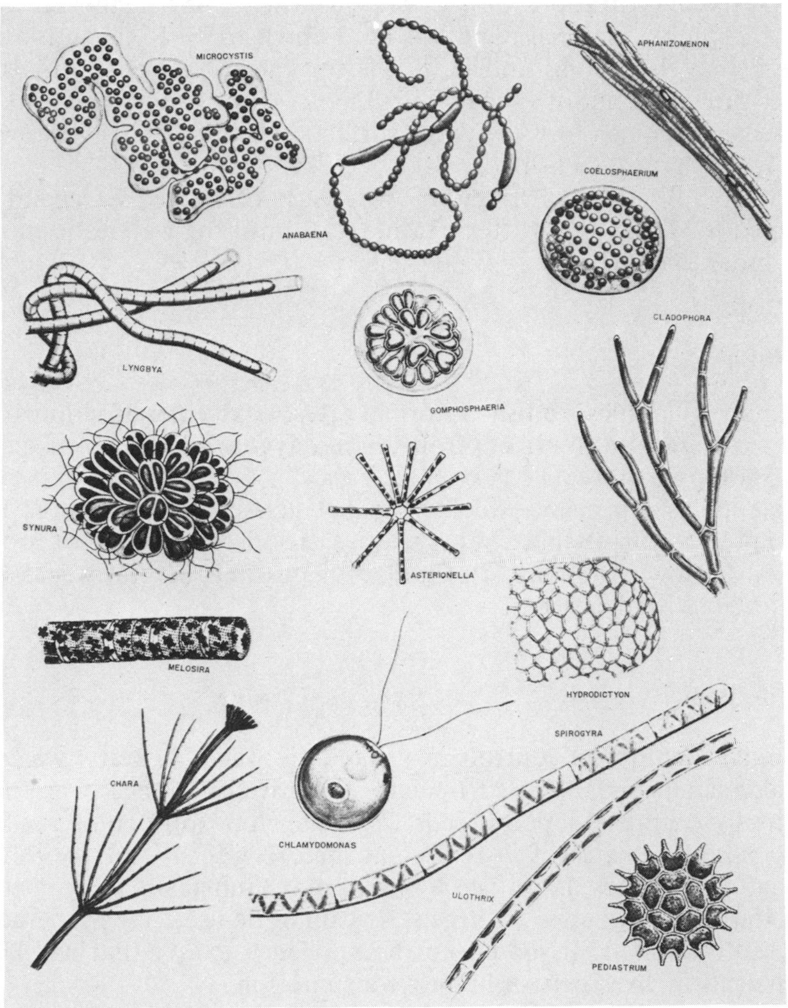

Figure 29-1. Algae often responsible for creating nuisance blooms. (C. M. Palmer and R. A. Taft, U.S. Public Health Service.)

tances up to 40 ft. *In contrast*, the granular form falls to the pond bottom, where a relatively high concentration may develop at the soil–water interface.

Fish were kept in 2,4-D sodium salt solution for one week. The LD_{50} concentration of 2,4-D for minnows 1.5–2.5 in. long was about 2000 ppm; for sunfish 2.5–4 in. long, 1000 ppm; and for catfish 5–6 in. long, 2000 ppm (Harrison and Rees, 1946). Thus 2,4-D sodium salt is considered low in fish toxicity. The same is true of most amine salts of 2,4-D.

Other work has shown ester forms of 2,4-D to be toxic to fish. Fingerling bluegills suffered losses of 40–100% from the butyl ester of

2,4-D at concentrations of 1–5 ppm (Snow, 1948). Esters are oil-like and oil-soluble materials, and oils are known to be toxic to most fish. Esters may be acting similar to oils, or the oil solvents, emulsifying agents, or other additives may be killing the fish. When pure 2,4-D ester is used in granular form, without surface-active agents, there is considerably less hazard to fish.

For potable water or irrigation water, use of treated water should be delayed for three weeks after treatment or until the water does not contain more than 0.1 ppm of 2,4-D acid.

Dalapon

Dalapon is harmless to fish at normal rates. Lake emerald shiners apparently suffered no ill effects from three days in 3000 ppm of dalapon, but 5000 ppm was fatal.

Dalapon is especially effective against grasses and cattails. It is best applied as a foliage spray. Most species are susceptible if not rooted in standing water. Draining is advisable, if possible, several weeks before treatment with dalapon.

Dichlobenil

Granular dichlobenil controls many aquatic weeds in static water such as lakes and ponds. It is particularly effective on elodea, watermilfoil, coontail, chara, and pondweeds (*Potamogeton* spp.). It is applied in early spring at rates of 7–10 lb/acre before weeds start to grow. Although it does not harm fish at these rates, humans may not eat these fish until 90 days after treatment. It cannot be used in commercial fish or shellfish waters. It should not be applied to water that will be used for irrigation, livestock, or human consumption.

Diquat

Diquat controls many submersed aquatic weeds and algae in static water. Bladderwort, coontail, elodea, naiad, pondweeds, watermilfoil, spirogyra, and pithophora are among the species controlled. It is applied at rates of 1–4 lbs/surface acre.

Diquat can be applied by pouring directly from the container into the water while moving slowly in a boat. Early in the season strips 40 ft apart are suggested, later in the season 20 ft. For best results, it should be applied before weeds reach the surface of the water. Because diquat is a contact spray, repeated applications may be necessary to give seasonlong control.

It should not be used in muddy water because diquat is rendered ineffective by its tight adsorption on soil particles. Treated water should not be used for irrigation, agricultural sprays, swimming, or domestic purposes within 10 days after treatment. For human consumption, treated water should not be used for at least 14 days. It is not harmful to most fish at the recommended rate. In heavy weed infestations only $\frac{1}{3}$–$\frac{1}{2}$ of the area should be treated to avoid fish kill through oxygen depletion.

Endothall

Endothall is available as a number of derivatives and in several liquid and granular formulations. It controls most algae and submersed aquatic weeds in static and in some flowing waters. However, elodea is not controlled. It is applied at a broad range of rates—from 0.3 to 14 lb/acre-ft of water. Lower rates are for algae control and higher rates for coontail, watermilfoil, pondweeds, naiad, bass weed, and burreed control. Treated water should not be used for irrigation, agricultural sprays, animal consumption, or domestic purposes within seven days. Fish may be used as food three days after treatment.

Fenac

Fenac controls most submersed aquatic weeds where water can be removed from the area. Unlike other aquatic herbicides, fenac is applied to the soil surface, not to the water. Water is drawn down to expose the area where weeds root. Only exposed soil is treated. Water is kept down for three weeks after treatment. It is particularly useful in drainage ditches, ponds, and margins of lakes. However, it is not permitted where these waters are used for irrigation, livestock consumption, or domestic needs. Applications can be made anytime, but a treatment in the fall before freezing weather gives the best results. The rate is 15–20 lb/acre.

Fluridone

Fluridone is a new herbicide for aquatic-weed control. It controls at least 14 submersed and emersed plants. After application, chlorosis and a pinkish color show up in 7–10 days, but full herbicidal effects may take up to 60 days. About four weeks after treatment, susceptible weeds begin to sink to the bottom. This slow herbicidal response reduces the potential of fish kill from oxygen depletion.

Fluridone is adsorbed by the hydrosoil (lake-bottom soil), reaching a maximum one to four weeks after treatment. These hydrosoil residues declined to a nondetectable level after 16–52 weeks.

Simazine (Aquazine®)

Simazine may be used in ponds for control of several submerged and floating aquatic weeds and algae. Control of weeds is usually seasonlong. Algae may be controlled for one to three months or may require retreatment later in the season. Simazine may be used in ponds containing fish. It should be used only in ponds that will have little or no outflow after treatment. Simazine should not be used as a spot treatment.

Ponds are treated after seasonal flow has ceased early in the weed and algae growth period. They should be treated when 5–10% of the pond surface is covered with scum (algae mats) or floating weeds, and/or while submerged aquatic weeds are actively growing and before they reach the surface of the water. As a general rule, ponds in northern areas may be treated between May 1 and June 15. In southern areas, where water warms up and weed growth is earlier, ponds may be treated between April 1 and May 15.

High water temperatures cause more-rapid natural decay of dead weeds and algae that can cause fish distress; therefore, simazine should be applied before water temperatures exceed 75°F.

Ponds having an extremely heavy infestation of weeds and algae such as occurs in mid and late summer should not be treated, since rapid decomposition of heavy growth greatly reduces the oxygen content of the water, and this can cause fish distress and/or death.

Algae control requires 1.40–3.40 lb/acre-ft; submerged weeds, 2.7–6.8 lb; and floating weeds 2.7–5.4 lb.

Use of Fish and Water Following Use of Simazine

1. Fish taken from treated ponds may be used for human consumption.
2. Treated ponds may be used for swimming.
3. Water from treated ponds may not be used for irrigation or spraying of agricultural crops, lawns, or ornamental plantings, or for watering cattle, goats, hogs, horses, poultry, or sheep, or for human consumption until 12 months following treatment.
4. *Precaution:* Ponds that have bordering trees with roots visibly extended into the water must not be treated, since injury to these trees may occur. Usually, trees 50 ft or more from the pond's edge will not be injured.

AQUATIC WEEDS OF SPECIAL IMPORTANCE

Aquatic weeds are commonly classified by their growth habits: (1) *floating,* (2) *emersed,* and (3) *submersed.* Floating aquatic weeds are not attached to soil, and their tops are above water. Emersed aquatic weeds are rooted in soil with their tops above water. Submersed aquatic weeds are usually rooted in soil with their tops entirely or mostly under water. Table 29-1 gives the classification of 20 common aquatic weeds and herbicides that effectively control them. See Figures 29-2 and 29-3. Following are descriptions of some of the more important aquatic weeds.

Algae

Algae are so tremendously important as aquatic weeds that you might read more about them in an encyclopedia or botany book. Algae may annoy bathers, causing a type of dermatitis and symptoms of hay fever. Blue-green algae have been known to cause poisoning of horses, cattle, sheep, hogs, dogs, and poultry. Odors and fishy tastes often result from decaying algae in water reservoirs. Extremely heavy algae growth may suffocate fish by depleting the supply of oxygen in the water at night.

Algae are thallus-like plants with chlorophyll (green color). They may consist of a single cell, a single filament, branched filaments, or they may grow 200 ft long like the giant algae of the Pacific (known as kelp) (see Figure 29-1). They are often called pond scums or seaweeds.

Plankton algae are microscopic, free-floating (unattached) organisms. They frequently cause "water bloom" of mid to late summer. They may become so numerous that they give the water a thickened, green, pea-soup appearance.

Algae are extremely important synthesizers (producers) of food. They are the "original" sources of food and energy for most fish and aquatic

Figure 29-2. Several submersed aquatic weeds. *Left to right*: Watermilfoil, coontail, eelweed, pondweed, and largeleaf pondweed. (King and Penfound, 1946.)

Figure 29-3. Emersed and floating aquatic weeds. *Left to right*: Burreed, duckweed, white waterlily, floating pondweed. (King and Penfound, 1946.)

animals. Algae live under moist conditions on land, in the soil, or in water—both salt and fresh.

Chara (*Chara* spp.)

Chara is a large, green algae characterized by branched, erect stems that have cylindrical whorls of branches along the stem. It is anchored to the bottom. In general appearance it resembles a higher plant. There are many different species (see Figure 29-1). These plants reproduce both vegetatively and by sexual fruiting bodies. After chara becomes encrusted with calcareous deposits, the plant takes on a coarse, gritty feeling and a gray-green shade. Chara, a submersed aquatic weed, thrives best in clear, hard water, and is found in water up to 20 ft deep.

Alligatorweed

Alligatorweed (*Alternanthera philoxeroides*) grows mostly along the coast from Texas to Virginia, reaching 25–150 miles inland, and it has been reported in California. It is widely distributed in the tropical areas of Central and South America.

It is a coarse, fleshy plant, forming dense mats in shallow water and on mud flats. It has narrow, opposite leaves and short spikes of whitish flowers (see Figure 29-4).

Alligatorweed depends on vegetative reproduction (stems) for its spread. Broken-off branches root easily and spread rapidly.

Figure 29-4. Alligatorweed.

Arrowhead (*Sagittaria* spp.)

Leaves of this group of plants give it its general name of arrowhead (see Figure 29-5). These plants are usually perennial, reproducing by seed and tuber-bearing rootstocks. Indians used the tubers as food. The stems usually have a milky juice. The plant is best adapted to shallow water or muddy shores. It can usually be found near the edges of the pond as a marginal weed.

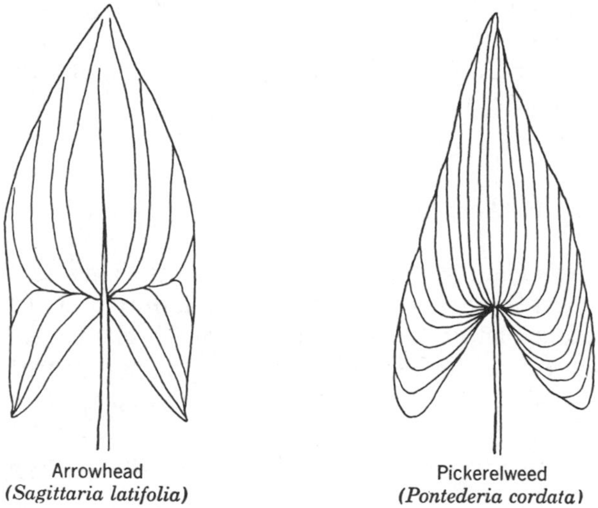

Figure 29-5. *Left*: Common arrowhead. *Right*: Pickerelweed.

Bulrush (*Scirpus* spp.)

The Bulrush genus has about 150 species, widely distributed. The genus *Scirpus* is a member of the sedge family and is not a true rush. Most grow in marshes, ponds, and lakes. The stems are either triangular or rounded and sheathed at the base.

Cattails (*Typha* spp.)

Cattails dominate extensive marshy areas of the world. They are tall, grasslike plants with fleshy leaves having no midvein. They spread rapidly by rhizomes and by small, airborne seeds. Seeds remain viable for five years or more.

Small areas of cattails can be eliminated by pulling. Repeated underwater mowing is also effective. They should be cut when the first spikes reach two-thirds full size; this should be repeated when the cattail is again 2 ft tall. Most kinds of cattail are killed if submerged in 4 ft of water; *T. domingensis* requires deeper flooding.

Common Coontail (*Ceratophyllum demersum*)

Coontail is a submersed aquatic herb with whorls of finely dissected leaves (see Figure 29-2). In the spring the lower stem may anchor in the mud with the plant growing upright. Later the plant may become a tangled mass with filamentous algae. Leaves are crowded toward the tip of the stem, giving it the coontail appearance. The plant spreads by both seed and vegetative growth.

Coontail is common in shallow lakes, ponds, and sluggish streams. It is favored by water rich in organic material.

Duckweed Family (*Lemnaceae*)

Duckweeds are small, stemless, floating plants. They lack true leaves, but have thickened or rounded fronds. The family includes the smallest of the flowering plants. Duckweeds propagate mainly from vegetative growth that develops at the edge or base of the parent frond. Also, during the fall, small bulblets may develop at the edge of the leaf. These sink to the pond bottom, live through the winter, and germinate the next spring. Seed is usually produced sparingly, or not at all (see Figure 29-3).

Elodea (*Elodea canadensis*)

Elodea, also known as waterweed, is a common plant of home aquariums. The leaf is veinless; leaf margins have microscopic teeth. Its leaves are attached to the stem in a whorled arrangement. Stems are floating and usually rooted to the bottom.

The plant is dioecious. Male flowers break loose or grow to the surface to shed their pollen. Female flowers grow to the surface after fertilization and produce up to five spindle-shaped seeds. See Figure 29-6.

Hydrilla (*Hydrilla verticillata*)

Hydrilla was first found in the United States near Miami, Florida, in 1960. It is now widely spread and has become a troublesome aquatic weed in southern United States and in California.

Figure 29-6. Egeria (*left*), hydrilla (*center*), and American elodea (*right*) have similar growth habits. Although individual plants of all three species vary in length and number of leaflets in whorls and in the distance between whorls, the scratchy feel of hydrilla readily distinguishes it. (Jack K. Clark, University of California, Davis.)

Figure 29-7. Small thornlike projections, occasionally black on the tips, that protrude from the midrib on the lower side of the leaf are characteristic of hydrilla. (Jack K. Clark, University of California, Davis.)

Hydrilla is a submersed vascular aquatic weed, rooted to the bottom with long branching stems. The lower leaves are opposite and small, whereas the medium and upper leaves are in whorls of fours and eights. Leaves are verticillate and narrow lanceolate. Hydrilla is easily confused with elodea and egeria. Hydrilla has sharply toothed leaf margins, small thornlike projections from the lower side of the leaf midrib, and a "harsh" feel. The flowers have a threadlike pedicel. See Figures 29-6 and 29-7. Vegetative reproduction from broken shoots, subterranean shoots, and turions are more important in its spread than seeds. Hydrilla grows in water up to 50 ft deep.

Pickerelweed (*Pontederia cordata*)

Pickerelweed has fleshy, heart-shaped to lance-shaped leaf blades that are easily confused with arrowhead (see Figure 29-5). It has thick, creeping rootstocks, strongly rooted to the bottom. The flowers are blue spikes.

Pickerelweed grows primarily as a marginal weed in shallow water, and it reproduces by creeping rootstocks and from seed.

Pondweed (*Potamogeton* spp.)

Potamogeton, with 60 species, is the largest genus of true aquatic seed plants of temperate regions (Muenscher, 1944).

Pondweed plants grow in both fresh and salt water. They grow best in water 1–2 ft deep, but some species grow in water 6 ft deep or more. Some species have two different kinds of leaves. Those remaining submerged are usually thick, transparent, long, and slender. Those emerging are firm and leathery and are usually attached to the stem by a petiole, with leaves alternately arranged. The stems are jointed, with roots forming at the lower nodes (see Figure 29-2).

Pondweeds reproduce by seed, by turions and tubers that develop in the leaf axils of some species, and from rootstock, which may produce tubers.

Deep water with appropriate fertilization will help to control pondweed.

Waterhyacinth (*Eichornia crassipes*)

Waterhyacinth includes five species, but only one is of major importance in the United States. It is a tropical plant, confined primarily to the Southern Coastal States and to California. It is found principally in ponds and quiet streams.

Waterhyacinth has a blue flower that is somewhat funnel-shaped and two-lipped. The plant is normally free-floating, buoyed by bladderlike, inflated leaf petioles. It forms a matlike growth. The leaf blade may be kidney-shaped to somewhat rounded.

The plant reproduces mainly by vegetative offsets. Several thousand new plants may develop during one season from one original plant. Some reproduction results from seed.

Watermilfoil (*Myriophyllum* spp.)

Watermilfoil includes 20 different species with parrotfeather (*Myriophyllum brasiliense*) and eurasian watermilfoil (*M. spicatum*) being troublesome species. The lower leaves are finely divided and threadlike, each leaf resembling a small, green feather, whereas the upper leaves may resemble bracts (see Figure 29-2). The plant usually grows submersed in water, but stems often emerge above water during the last half of the summer. Flowers and seeds are produced sessile in the leaf axils.

Reproduction is from seed, creeping rhizomes, and broken stem parts that take root.

Pond fertilization will help to control watermilfoil species. Light infestations and small areas can be removed by hand.

SUGGESTED ADDITIONAL READING

Bruns, V. F., J. M. Hodgson, H. F. Arle, and F. L. Timmons, 1955, USDA Circular No. 971.

Burkhalter, A. P., L. M. Curtis, R. L. Lazor, M. L. Beach, and J. C. Hudson, 1974, *Aquatic Weed Identification and Control Manual,* Florida Department of Natural Resources, Tallahassee.

Harrison, J. W. W., and E. W. Rees, 1946, *Am. J. Phar.* **118**(12),4.

King, J. E., and W. T. Penfound, 1946, *Ecology,* **27**(4), 372.

Muenscher, W. C., 1944, *Aquatic Plants of the United States,* Comstock, Ithaca, NY.

Snow, J. R., 1948, Thesis, Alabama Polytechnic Institute.

Timmons, F. L., 1970, *FAO International Conference on Weed Control,* WSSA, Urbana, IL. p. 357. Available from WSSA. See address below.

USDA, 1980, Agricultural Handbook 565.

Yeo, R. R., and T. W. Fisher, 1970, *FAO International Conference on Weed Control,* WSSA, Urbana, IL. p. 450. Available from WSSA. See address below.

Weed Science published by WSSA, 309 West Clark St., Champaign, IL 61820. See index at end of each volume.

Weeds Today. Some availability as above.

FAO International Conference on Weed Control, 1970. Same availability as above.

Regional Weed Science research reports and proceedings. Available in the libraries of most land grant universities. Published annually.

Obtain weed control bulletins, leaflets, and so forth, from your State University or college of agriculture on weed control in aquatic areas.

For chemical use, see the manufacturer's label for method and time of application, rates to be used, weeds controlled, and special precautions. *Label recommendations must be followed–regardless of statements in this book.* Also see the Preface. Inclusion or omission of a product name does not constitute recommendation or nonrecommendation of a product.

30 Total Vegetation Control

Total control of vegetation is the removal of all higher green plants and maintenance of these areas vegetation-free. Complete absence of vegetation is desirable on many sites such as railway roadbeds, industrial areas, highway brims, fencerows, and irrigation and drainage-ditch banks. Industrial sites include storage and work areas, lumberyards, utility transmission stations and railroad right-of-ways (see Figures 30-1, 30-2, and 30-3).

Vegetation can be controlled totally by mechanical or chemical methods. Sometimes both methods are used. Brush and trees may be removed by mechanical methods at first clearance, although this method adds considerable cost. These areas are then maintained weed-free with herbicides. Herbaceous species can be eliminated by either mechanical or chemical methods. Discing or other mechanical means may need to be repeated several times a season to keep the area weed-free. But persistent herbicides at high rates are needed only annually or less often. Herbicide Trade names are on page 422.

FOLIAR HERBICIDES

To obtain vegetation-free areas, certain nonselective foliar-applied herbicides are used together with persistent, nonselective, soil-applied herbicides. These foliar herbicides include contact herbicides such as paraquat, cacodylic acid, dinoseb, weed oil, or the translocated herbicide amitrole. Also, to broaden the spectrum of weed species controlled, other herbicides are added. For example, 2,4-D, 2,4,5-T, silvex, picloram, dicamba, and fenac are used in this way. Glyphosate readily translocates from the foliage to other parts of the plant, including roots. Seven to 10 days may be required to show total effect. Gly-

Figure 30-1. Pipeline and storage-tank areas kept weed-free by an annual application of monuron. (E. I. du Pont de Nemours and Company.)

phosate is nonselective and controls many difficult-to-control weeds, including perennial weeds. Foliar-applied herbicides are often applied to areas treated with the persistent nonselective herbicides discussed below.

SOIL HERBICIDES

Persistent nonselective herbicides used for total vegetation control are given in Table 30-1. Chemical and physical properties of these compounds, as well as other uses, have been discussed in previous chapters. Although some of these herbicides are also used as selective herbicides, the rate of application is usually lower for selective uses than for nonselective uses.

Soil-applied herbicides may remain toxic to higher green plants for more than one year. Factors affecting the length of time that a herbicide remains toxic in the soil were discussed in Chapter 5. In general, dry weather with little or no leaching, cool or cold temperatures, and heavy soils tend to lengthen the time that a herbicide will remain toxic.

Under any given condition the length of time that a herbicide will remain toxic can be predicted with reasonable accuracy (see Table 5-1).

Annual applications of most of these herbicides at rates that persist somewhat longer than one year are usually more economical than massive rates that will persist for two years or more.

Figure 30-2. Bermudagrass control with three annual treatments of karbutilate on noncropland. *Foreground*: Untreated. *Background*: Treated. (W. B. McHenry, University of California, Davis.)

These persistent, nonselective soil-applied herbicides must be leached into the rooting or seed-germination zone of the weeds to be effective. Therefore, they are usually applied just before or during the rainy season.

Persistent soil-applied herbicides gradually lose phytotoxicity. With loss of phytotoxicity weeds reinfest the area. Usually new plants are stunted and grow slowly at first. At that time, use of a broad-spectrum foliar herbicide will usually extend the period of total vegetation control.

Figure 30-3. Tebuthiuron, at increased rates, provides effective total vegetation control for industrial areas and railroads. (Elanco Products Company, a division of Eli Lilly and Co.)

Table 30-1. Herbicides Commonly Used for Total Vegetation Control of Weeds on Noncropland[1]

Herbicide[2] Soil-Applied	Rate (lb/acre)	Herbicide[2] Foliar-Applied	Rate (lb/acre)
Atrazine	4.8-40	Ametryn	1-2
Bromacil	2.4-24	Amitrole	1.8-10
Dicamba	2-8	AMS	95
Diuron	3.2-48	Asulam	3.3-6.6
Fenac	4.5-18	Bromoxynil	1.0
Picloram	2-8.5	Cacodylic acid	1.25-5.0
Prometon	10-15	Dalapon	15-30
Simazine	10-40	Dicamba	1-3
Sodium chlorate	190-760	2,4-D	1-3
Tebuthiuron	1.2-16	Dichlorprop	0.5-12
		Dinoseb	1.25-3.75
		Diquat	0.5
		Fosamine	6-12
		Glyphosate	0.75-4.0
		Hexazinone	1-12
		Linuron	1-3
		MSMA; DSMA	2.5
		Paraquat	0.5-1
		Triclopyr	1-9

[1] Various combinations of these herbicides are often used to give greater persistence and/or broader spectrum of weeds controlled. Trade names are given on page 422.

[2] Some herbicides have considerable foliar activity as well as soil activity. See the manufacturer's label for rates, method of usage, species susceptibility, and precautions. Follow the label—regardless of statements in this book.

PREVENTING INJURY TO TREES, SHRUBS, ORNAMENTALS, AND LAWNS

Nearly all nonselective herbicides will kill all types of plant growth. As one would expect, some plants are more susceptible to some herbicides than to others. At high rates they should all be considered very effective against trees, shrubs, ornamental flowers, and lawns. Soil sterilants should never be applied to the rooting zone of such plants or so that the chemical is washed into their rooting zone.

SUGGESTED ADDITIONAL READING

Weed Science published by WSSA, 309 West Clark St., Champaign, IL 61820. See index at end of each volume.

Weeds Today. Same availability as above.

FAO International Conference on Weed Control, 1970. Same availability as above.

Regional Weed Science research reports and proceedings. Available in the libraries of most land grant universities. Published annually.

Obtain weed control bulletins, leaflets, and so forth, from your state university or college of agriculture on total vegetation control or soil sterilants.

For chemical use, see the manufacturer's label for method and time of application, rates to be used, weeds controlled, and special precautions. *Label recommendations must be followed–regardless of statements in this book.* Also see the Preface. Inclusion or omission of a product name does not constitute recommendation or nonrecommendation of a product.

31 Lawn, Turf, and Ornamentals

Lawn- and turf-weed problems have been largely underestimated. Nearly every home and apartment house has a weed problem. Other turf areas that have weed problems are golf courses, public and private parks, other recreation areas, grounds surrounding many commercial and governmental buildings, and roadsides.

No other type of weed control directly affects so many people. In the United States over 5 million acres are in home lawns, and an additional 10 million acres are in other types of turf.

Because there are millions of consumers, numerous turf species, and a multitude of ornamental plants, the job of educating users is more complex than in other areas of weed control.

The importance of management practices that produce a strong and vigorous turf cannot be overemphasized. In general these practices are well understood for each area of the United States. They are specific for each geographic area. They include choice of an adapted lawn grass, proper grading and seedbed preparation, fertilization, mowing, and watering, as well as control of insects, diseases, and weeds.

LAWN

Weed control alone cannot guarantee a beautiful lawn. Other recommended practices must also be followed. For example, ridding a lawn of crabgrass may leave a bare yard unless plans are made to encourage desirable turf grasses to become established. However, desirable turf grasses are usually present in areas where crabgrass predominates. With elimination of crabgrass and other weeds, proper fertilization, mowing, and watering, lawn grasses such as Kentucky bluegrass or bermudagrass soon cover the area.

Topsoil Added to New Lawns

After new homes or buildings are finished, only subsoil may remain for starting a lawn. Often topsoil is hauled in to cover the area 2–4 in. deep. It usually contains weed seeds and weedy plants that soon infest the area.

The owner can (1) use the topsoil and hope to control the weeds later (also, he should attempt to get topsoil from fields known to be free of serious perennial lawn weeds, (2) use a soil fumigant such as methyl bromide to rid the soil of weeds, or (3) not use the topsoil. *Proper fertilization,* including thorough mixing into the upper 3–4 in. of soil, makes it possible to grow many turf plants on subsoil. Turf plants are favored by adding peat moss at the rate of 1 bale (7 ft^3) per 200 ft^2 of surface plus liberal fertilization, with both worked into the surface 3–4 in. of subsoil. *Well-rotted* sawdust is as good as peat; however, the fertilization program depends upon the degree of sawdust decomposition. Well-decomposed sawdust will require less fertilizer than fresh sawdust. Also, *well-rotted* manure (fresh manure will have weed seeds) can be used in place of the peat moss. A surface mulch of peat moss at seeding time may discourage weeds, reduce soil erosion, and keep the surface from drying rapidly.

Before Planting

Serious weeds that cannot be controlled after turf is established should be eliminated before planting. Such methods may delay planting, but will reduce the work required after the turf is planted.

Weed control before seeding may include shallow cultivation after emergence of most annual weeds, or the use of herbicides. To be effective the herbicide needs to control all weedy growth and to have no residual toxicity to the turf to be seeded or sodded later (see Table 5-1).

Two herbicides are especially useful for this purpose. Glyphosate is a nonselective, broad-spectrum, foliar-applied herbicide. Glyphosate kills annual and most perennial plants. It is absorbed principally through foliage, requiring 7 to 10 days for translocation to the roots in some species. Application to foliage of desirable plants should be avoided. Glyphosate has no residual soil activity. It does not control dormant seeds in the soil. Glyphosate reacts with galvanized steel to form hydrogen gas, which is explosive. Stainless steel, aluminum, fiberglass, or plastic are suitable sprayer containers.

Cacodylic acid is a contact, nonselective herbicide. It kills the tops of plants but seldom kills the roots of perennial plants. At usual rates of application it has no residual herbicidal activity through the soil.

Following the above herbicide treatment and after seeding or sodding, appropriate herbicide treatment should be made to continue the weed control program (see Table 31-1).

Table 31-1. Herbicides for Established Turf, Lawns, [1,3]

Weeds[2] To Be Controlled	Bentgrass	Bermudagrass	Bluegrass	Fescue	St. Augustine	Zoysia
Barnyardgrass	Bensulide Siduron	Benefin Bensulide DSMA, MSMA	Benefin Bensulide DSMA, MSMA Siduron	Benefin Bensulide DSMA, MSMA Siduron	Asulam Benefin Bensulide	Benefin Bensulide DSMA, MSMA Siduron
Bindweed, field	2,4-D	2,4-D	2,4-D	2,4-D	–	2,4-D
Bluegrass, annual	Bensulide	Benefin Bensulide DCPA Oxadiazon Pronamide	Benefin Bensulide DCPA Oxadiazon	Benefin Bensulide DCPA	Benefin Bensulide DCPA Oxadiazon	Benefin Bensulide DCPA
Carpetweed	2,4-D Dicamba	Benefin 2,4-D Dicamba Oxadiazon	Benefin 2,4-D Dicamba Oxadiazon	Benefin 2,4-D Dicamba	Benefin Dicamba Oxadiazon	Benefin 2,4-D Dicamba
Chickweed, common	Dicamba Mecoprop	Benefin Dicamba DCPA DSMA, MSMA Mecoprop	Benefin Dicamba DCPA DSMA, MSMA Mecoprop	Benefin Dicamba DCPA DSMA, MSMA Mecoprop	Benefin Dicamba DCPA	Benefin Dicamba DCPA DSMA, MSMA Mecoprop
Chickweed, mouseear	Dicamba Mecoprop	Dicamba Mecoprop	Dicamba Mecoprop	Dicamba Mecoprop	Dicamba	Dicamba Mecoprop

Clover, white	2,4-D Dicamba Mecoprop	2,4-D Dicamba Mecoprop	2,4-D Dicamba Mecoprop	2,4-D Dicamba Mecoprop	Dicamba	2,4-D Dicamba Mecoprop
Crabgrass	Bensulide DSMA, MSMA Siduron	Benefin Bensulide DCPA DSMA, MSMA Oxadiazon	Benefin Bensulide DCPA DSMA, MSMA Oxadiazon Siduron	Benefin Bensulide DCPA DSMA, MSMA Siduron	Asulam Bensulide DCPA Oxadiazon	Benefin Bensulide DCPA DSMA, MSMA Siduron
Dallisgrass	–	DSMA, MSMA	DSMA, MSMA	–	–	DSMA, MSMA
Dandelion	2,4-D	2,4-D	2,4-D	2,4-D	–	2,4-D
Fennel, dog	2,4-D Dicamba	2,4-D Dicamba	2,4-D Dicamba	2,4-D Dicamba	Dicamba	2,4-D Dicamba
Dock, curly	2,4-D Dicamba	2,4-D Dicamba	2,4-D Dicamba	2,4-D Dicamba	Dicamba	2,4-D Dicamba
Foxtail	Bensulide Siduron	Benefin Bensulide DCPA	Benefin Bensulide DCPA Siduron	Benefin Bensulide DCPA Siduron	Asulam Benefin Bensulide DCPA	Benefin Bensulide DCPA Siduron
Garlic, wild	2,4-D	2,4-D	2,4-D	2,4-D	–	2,4-D
Goosegrass	Bensulide	Benefin Bensulide DCPA Oxadiazon	Benefin Bensulide DCPA Oxadiazon	Benefin Bensulide DCPA	Asulam Benefin Bensulide DCPA Oxadiazon	Benefin Bensulide DCPA

Table 31-1 (Continued)

Weeds[2] To Be Controlled	Bentgrass	Bermudagrass	Bluegrass	Fescue	St. Augustine	Zoysia
Henbit	2,4-D Dicamba	2,4-D Dicamba	2,4-D Dicamba	2,4-D Dicamba	Dicamba	2,4-D Dicamba
Ivy, poison	2,4-D	2,4-D	2,4-D	2,4-D	–	
Knotweed, prostrate	2,4-D Dicamba Mecoprop	Benefin 2,4-D Dicamba Mecoprop	Benefin 2,4-D Dicamba Mecoprop	Benefin 2,4-D Dicamba Mecoprop	Benefin Dicamba	Benefin 2,4-D Dicamba Mecoprop
Lambsquarters	Bensulide 2,4-D Mecoprop	Benefin Bensulide 2,4-D Mecoprop	Benefin Bensulide 2,4-D Mecoprop	Benefin Bensulide 2,4-D Mecoprop	Benefin Bensulide	Benefin Bensulide 2,4-D Mecoprop
Nutsedge, purple	2,4-D	2,4-D DSMA, MSMA	2,4-D DSMA, MSMA	2,4-D DSMA, MSMA	– –	2,4-D DSMA, MSMA
Nutsedge, yellow	Bentazon 2,4-D	Bentazon 2,4-D	Bentazon 2,4-D	Bentazon 2,4-D	Bentazon	Bentazon 2,4-D

Onion, wild	2,4-D	2,4-D	2,4-D	2,4-D	–	2,4-D
Pennywort, lawn	2,4-D	2,4-D	2,4-D	2,4-D	–	2,4-D
Pepperweed	2,4-D Dicamba	2,4-D Dicamba	2,4-D Dicamba	2,4-D Dicamba	Dicamba	2,4-D Dicamba
Pigweed, redroot	Bensulide 2,4-D Mecoprop	Benefin Bensulide 2,4-D DSMA, MSMA Mecoprop Oxadiazon	Benefin Bensulide 2,4-D DSMA, MSMA Mecoprop Oxadiazon	Benefin Bensulide 2,4-D DSMA, MSMA Mecoprop	Benefin Bensulide Oxadiazon	Benefin Bensulide 2,4-D DSMA, MSMA Mecoprop

[1] Herbicide activity. Commercial names of herbicides given on page 422. Preemergence or soil-applied: Benefin, bensulide, 2,4-D, dicamba, DCPA, oxadiazon, siduron. Postemergence or foliar-applied: asulam, bentazon, 2,4,-D, dicamba, mecoprop, DSMA, MSMA. Repeat treatments are often needed.

[2] Weeds names established by WSSA; *Weed Science,* 1971, **19**(4), 435–476.

[3] Use 2,4-D amine to avoid volatile drift. Also avoid spray drift. See Chapters 6 and 7.

For chemical use, see the manufacturer's label for method and time of application, rates to be used, weeds controlled, and special precautions. *Label recommendations must be followed–regardless of statements in this book.* Also, see the Preface. Inclusion or omission of a product name does not constitute recommendation or nonrecommendation of a product.

At Planting

Siduron can be applied at seeding of cool-season turf (bluegrass, ryegrass, fescue, and some bent grasses) to control annual grasses. For about one month siduron selectively controls annual grasses such as crabgrass, foxtail, and barnyardgrass. Diphenamid can be used at planting of dichondra for control of annual grasses and some broadleaf weeds.

After Planting

Postplanting control is needed after turfgrass species have emerged, but before they become well-established. This can be done by hand weeding, mowing, use of herbicides, or a combination. Mowing of newly planted turf will control some erect broadleaf weeds, but is not effective on grasses or prostrate broadleaf weeds. While the lawn is young, mowing height should be kept high so that a minimum of foliage of the turf species is removed.

Herbicides must be used with care on young turfgrass species to avoid injury. Uniform distribution and proper rate of application are essential. Many emerged broadleaf weeds can be controlled by low rates of bromoxynil, 2,4-D, or dicamba once the grass seedlings have reached the three- to four-leaf stage. At this stage, dosage rates should be cut to about one-fourth to one-half of those recommended for established turf. As the grass becomes older, these rates can be gradually increased.

ESTABLISHED TURF

With good management practices, weeds in established turf often can be controlled by hand pulling or cutting the occasional weed out of a home lawn.

Since an established turf tolerates herbicides much better than a new planting, many herbicides may be used. A turf is established when the grasses have developed an extensive root system and are well tillered or when the rhizome (runner) system is well grown.

Herbicides used on established turf are either soil-active or foliar-active. Generally, soil-active compounds are applied about two to three weeks before germination of weed seeds to be controlled. They are preemergence-type herbicides. Some require immediate sprinkle irrigation to minimize foliar injury to desirable turf species. All must be leached into the soil within a few days after application by sprinkle irrigation or rainfall to be effective.

Foliar-active herbicides are applied to emerged weeds. Because it usually takes several hours or sometimes a few days for maximum absorption, they should not be applied if rain is expected or if sprinkle irrigation is to be applied during that time.

Table 31-1 shows herbicides available to control many common turf weeds in six grass species. Because formulations of these herbicides available on the market vary considerably, and formulations have considerable influence on their activity, application rates are not given. The manufacturer's label gives the recommended rate. Many formulations contain two or more herbicides to increase the number of weed species controlled.

ORNAMENTALS

Weeds in ornamentals present serious problems for nurserymen as well as professional and home gardeners. Weed control methods include *hand weeding, mulches, cultivation,* and *herbicides.* Selecting the most suitable method or methods of controlling weeds in ornamentals is a difficult task.

Hand weeding, although tedious, is often very useful for the home gardener; however, for the nurseryman or professional gardener it is usually too expensive. *Mulches* are commonly used quite effectively. To control most annual weeds, a mulch 2 or 3 in. thick is sufficient. Good mulching material includes wood or bark chips, sawdust, peatmoss, small grain straw free of weed seeds, pine needles, and gravel or stones. An ideal mulch allows free passage of moisture and air, but smothers growth of young weeds and prevents germination of seeds that require light. *Cultivation* is used widely by the nurseryman, somewhat by the professional gardener, but little by the home gardener.

Woody ornamentals around the home, in nurseries, forest plantings, Christmas trees, and cemeteries can be safely treated with herbicides. It is possible to spray the soil and small weeds around the stem and below the foliage of woody ornamentals with postemergence herbicides such as amitrole, glyphosate, DSMA, and MSMA.

Since woody ornamentals are perennials, it is possible to apply a soil-residual type of herbicide preemergence to the germination of weed seeds. Some may be tank-mixed and applied at the same time as the postemergence spray. Preemergence herbicides should be applied just before periods of weed-seed germination. One should avoid spraying the leaves of woody ornamentals. Such herbicides include most of the preemergence types listed in Table 31-1.

Also, if turf maintenance is not involved, other herbicides useful for woody-ornamental-weed control include alachlor, atrazine, chloramben, chlorpropham, dichlobenil, diphenamid, diuron, EPTC, naptalam, oryzalin, oxyfluorfen, pronamide, simazine, and trifluralin.

Nonselective sprays such as glyphosate, cacodylic acid, DSMA, and MSMA are especially effective for killing *all vegetation* in brick walks along borders or under woody ornamentals. For best results, weeds should be treated when 1–2 in. tall. One should be careful not to spray the foliage of valued ornamentals.

To help select proper herbicides for weed control in ornamentals, local county extension agents or other public agencies and local landscape architects should be consulted. One should always read the herbicide manufacturer's literature and labels.

SUGGESTED ADDITIONAL READING

Weed Science published by WSSA, 309 West Clark St., Champaign, IL 61820. See index at end of each volume.

Weeds Today. Same availability as above.

Regional Weed Science research reports and proceedings. Available in libraries of most land grant universities.

Obtain weed control bulletins, leaflets, and so forth, on weed control, from your state university or college of agriculture.

For chemical use, see the manufacturer's label for method and time of application, rates to be used, weeds controlled, and special precautions. *Label recommendations must be followed–regardless of statements in this book.* Also, see the Preface. Inclusion or omission of a product name does not constitute recommendation or nonrecommendation of a product.

Appendix

Table A-0. Selective Crop Herbicides[1], Including Selective Application, and the Weeds[2] Controlled as Listed on the Manufacturer's Label[3]

Weed	Acifluorfen	Alachlor	Ametryn	Asulam	Atrazine	Benefin	Bensulide	Bentazon	Bifenox	Bromoxynil	Butylate	CDAA	CDEC	Chloramben	Chlorpropham	Cyanazine	Cycloate	2,4-D	2,4-DB	DCPA	Desmedipham	Diallate, triallate	Dicamba	Dichlorprop	Difenzoquat	Dinoseb	Diphenamid
1. Barley, wild	X				X										X												
2. Barnyardgrass	X	X	X	X	X	X	X		X		X	X	X	X			X			X						X	X
3. Beggarweed, Florida		X												X													
4. Bermudagrass											X																
5. Bindweed, field																		X					X				
6. Bluegrass, annual	X				X	X			X			X	X		X		X			X							X
7. Brome, downy	X				X										X												
8. Buckwheat, wild					X			X		X					X						X						
9. Carpetweed	X	X				X			X			X		X	X			X		X							X
10. Cheat	X				X							X			X												X
11. Chess, hairy					X										X												
12. Chickweed, common	X		X			X			X				X	X	X					X	X		X	X		X	X
13. Chickweed, mouseear																							X	X			X
14. Cocklebur			X		X			X		X						X		X	X				X			X	
15. Crabgrass(s)	X	X	X	X	X	X	X				X	X	X	X			X			X						X	X
16. Dallisgrass			X																								
17. Dock				X											X			X	X				X				
18. Dodder															X					X							
19. Dogfennel					X					X								X									
20. Foxtail(s)	X	X	X	X	X	X	X				X	X	X	X			X			X	X					X	X
21. Garlic, wild																		X									
22. Goosegrass(s)	X	X	X	X	X	X	X				X		X	X						X							X
23. Gromwell										X																	
24. Groundcherry																				X	X						
25. Guineagrass																											
26. Hemp																		X									
27. Henbit	X		X							X			X				X										
28. Jimsonweed					X			X	X	X								X	X								
29. Johnsongrass				X		X					X			X						X							X
30. Junglerice			X		X	X	X																				
31. Knawel										X																	X
32. Knotweed	X					X									X			X			X			X			X
33. Kochia			X		X			X	X	X				X				X	X		X						
34. Lambsquarters	X	X	X		X	X	X	X	X	X				X			X	X	X	X	X		X	X		X	X
35. Lettuce, prickly					X				X									X	X								
36. Morningglory	X		X		X				X	X						X		X	X				X	X		X	
37. Mustard, wild	X		X		X			X		X				X	X			X	X		X					X	
38. Nightshade	X	X	X							X				X	X		X		X		X					X	
39. Nutsedge, purple			X								X						X										
40. Nutsedge, yellow		X	X		X			X			X						X										
41. Oat, wild				X											X		X					X			X		X
42. Panicum, fall	X	X	X	X	X	X	X				X				X					X							X
43. Pennycress, field									X	X								X	X								

Table A-0 (Continued)

Weed	Diuron	DSMA, MSMA, MAA, MAMA	EPTC	Fenac	Fluchloralin	Glyphosate	Isopropalin	Linuron	MCPA	Mefluidide	Methazole	Metolachlor	Metribuzin	Molinate	Napromide	Naptalam	Nitrofen	Norea	Norflurazon	Oryzalin	Oxyfluorfen	Pebulate	Pendimethalin	Prometryn	Pronamide	Propachlor	Propanil	Propazine	Propham	Pyrazon	Simazine	Terbutryn	Trifluralin	Vernolate
1.															X					X														
2.	X	X	X	X	X		X	X			X	X	X	X	X			X	X	X	X	X	X	X	X	X	X				X	X	X	X
3.								X					X																					
4.			X			X																X												
5.				X		X										X																	X	
6.	X		X	X		X									X		X			X					X				X		X		X	
7.				X		X																			X				X		X	X	X	
8.	X												X													X	X							
9.			X		X		X	X				X	X			X	X		X	X	X		X		X	X		X			X	X	X	X
10				X									X												X				X		X	X	X	
11.																																		
12.	X	X	X	X				X					X		X	X	X	X		X	X				X				X		X	X	X	
13.		X																																
14.	X	X				X		X	X		X		X			X		X	X					X										
15.	X	X	X	X	X	X	X	X				X	X		X		X	X	X	X	X	X	X	X	X	X	X				X	X	X	X
16.		X				X																												
17.				X		X																					X							
18.																									X									
19.	X							X																								X		
20.	X	X	X	X	X	X	X	X				X	X		X					X	X	X	X	X		X	X	X	X		X	X	X	X
21.																																		
22.		X	X	X	X		X	X			X	X					X	X	X	X	X	X	X	X	X	X	X				X		X	X
23.	X																															X		
24.	X															X								X								X		
25.						X									X																		X	
26.																																		
27.		X	X	X									X							X		X			X					X	X	X	X	
28.													X					X																
29.		X	X	X	X	X	X			X		X	X		X				X	X			X										X	X
30.			X	X																				X							X	X	X	
31.	X							X																							X			
32.															X					X					X								X	
33.				X		X			X				X										X				X				X	X	X	
34.	X	X	X	X	X	X	X	X	X		X		X		X	X	X			X	X	X	X	X	X	X	X	X		X	X	X	X	X
35.	X					X									X																X	X		
36.	X	X	X	X	X			X			X						X	X	X	X	X			X	X			X			X	X		X
37.	X			X				X	X				X			X								X	X		X			X	X	X		
38.			X	X		X											X			X		X			X					X	X			
39.		X	X																X			X												X
40.		X	X									X	X						X			X												X
41.		X	X	X									X		X					X		X		X	X				X		X			
42.		X	X		X	X	X	X				X	X		X				X	X		X		X	X	X					X		X	
43.	X								X																					X		X		

Table A-0 (Continued)

Weed	Acifluorfen	Alachlor	Ametryn	Asulam	Atrazine	Benefin	Bensulide	Bentazon	Bifenox	Bromoxynil	Butylate	CDAA	CDEC	Chloramben	Chlorpropham	Cyanazine	Cycloate	2,4-D	2,4-DB	DCPA	Desmedipham	Diallate, triallate	Dicamba	Dichlorprop	Difenzoquat	Dinoseb	Diphenamid
44. Pigweed(s)		X	X		X	X	X	X	X			X	X	X		X	X	X	X	X	X		X	X		X	X
45. Puncturevine					X				X									X									
46. Purslane, common	X	X	X			X	X		X			X	X	X	X		X			X	X			X		X	X
47. Pusley, Florida		X	X			X														X						X	X
48. Quackgrass					X																						
49. Radish, wild										X								X	X								
50. Ragweed		X	X		X			X				X		X				X	X		X			X		X	
51. Ryegrass, annual	X					X									X		X										X
52. Sandbur(s)			X			X					X									X							X
53. Sesbania								X	X					X													
54. Shattercane			X								X																
55. Shepherdspurse	X								X						X		X	X	X		X						X
56. Signalgrass		X	X	X																							
57. Sida, prickly		X						X	X					X		X			X							X	
58. Smartweed	X	X	X		X			X	X	X				X	X			X	X		X		X			X	X
59. Sowthistle, annual			X															X			X		X				
60. Speedwell																								X			
61. Spurge, annual														X		X				X			X				
62. Spurry, corn									X																		X
63. Sunflower			X		X			X		X								X					X				
64. Thistle, Canada					X			X										X					X				
65. Thistle, Russian					X									X				X	X								
66. Velvetleaf			X		X			X	X	X				X				X	X				X				

[1]Herbicide common names as listed on back cover of *Weed Science*. Herbicide commercial names given on page 422 and in text.

[2]Common and scientific weed names given in *Weed Science,* 1971 19(4), 457-468.

[3]For chemical, see the manufacturer's label for method and time of application, rates to be used, weeds controlled, and special precautions. Label recommendations must be followed—regardless of statements in this book. Also, see the preface. Inclusion or omission of product name does not constitute recommendation or nonrecommendation of a product.

Table A-0 (Continued)

Weed	Diuron	DSMA, MSMA, MAA, MAMA	EPTC	Fenac	Fluchloralin	Glyphosate	Isopropalin	Linuron	MCPA	Mefluidide	Methazole	Metolachlor	Metribuzin	Molinate	Napromide	Naptalam	Nitrofen	Norea	Norflurazon	Oryzalin	Oxyfluorfen	Pebulate	Pendimethalin	Prometryn	Pronamide	Propachlor	Propanil	Propazine	Propham	Pyrazon	Simazine	Terbutryn	Trifluralin	Vernolate
44.	X	X	X	X	X	X	X	X	X		X	X	X		X	X	X		X	X	X	X	X	X		X	X	X		X	X	X	X	X
45.		X		X					X											X													X	
46.	X	X	X	X	X		X	X	X		X	X	X		X	X	X	X	X	X	X	X	X	X	X	X				X	X		X	X
47.			X		X		X	X				X	X					X	X	X		X	X	X		X					X		X	X
48.			X	X		X																			X									
49.								X	X																X									
50.	X			X		X		X	X				X			X		X	X	X				X		X		X		X	X	X		
51.			X	X			X	X																	X	X			X		X	X		
52.		X	X			X	X					X			X					X				X									X	
53.	X							X			X		X																					X
54.			X		X	X																											X	X
55.	X		X						X				X			X	X			X		X			X					X	X	X		
56.		X	X		X							X	X					X	X	X		X	X	X		X	X				X		X	
57.	X		X					X			X		X						X		X			X										
58.				X		X		X					X				X	X	X		X	X	X	X	X	X		X		X	X	X		
59.		X		X					X						X																			
60.																															X	X		
61.											X		X						X	X			X											
62.	X		X																													X		
63.						X							X																					
64.				X		X																												
65.				X		X																									X	X	X	
66.						X		X					X			X		X				X	X					X		X		X	X	

Table A-1. Alphabetical Listing of Herbicides by Common Name[1] and Corresponding Trade Name, and Manufacturer

Common Name	Trade Name	Manufacturer
Acifluorfen	Blazer	Rohm and Haas
Acrolein	Aqualin	Shell
Alachlor	Lasso	Monsanto
AMA	(Many)	(Many)
Ametryn	Evik, Gesapax	Ciba-Geigy
Amitrole	Amizol, Amino Triazole Weedazol	American Cyanamid, Union Carbide
AMS	Ammate	duPont
Asulam	Asulox	Rhone-Poulenc
Atrazine	AAtrex, Atrazine	Ciba-Geigy, Shell
Barban	Carbyne	Gulf
Benazolin	Benazolin	Boots
Benefin	Balan	Lilly/Elanco
Bensulide	Betasan, Prefar	Stauffer
Bentazon	Basagran	BASF
Benzadox	Topcide	Gulf
Bifenox	Modown	Mobil
Borate (meta)	(Many)	Occidental
Borate (octa)	Polybor	U. S. Borax
Borax	(Many)	Occidental
Bromacil	Hyvar	duPont
Bromoxynil	Brominal, Buctril, Bronate	Rhone-Poulenc Union Carbide
Butachlor	Machete	Monsanto
Buthidazole	Ravage	Velsicol
Butralin	Amex	Union Carbide
Butylate	Sutan	Stauffer
Cacodylic acid	Rad-E-Cate	Vineland
Calcium cyanamide	Cyanamide	American Cyanamid
CDAA	Randox	Monsanto
CDEC	Vegadex	Monsanto
Chloramben	Amiben, Vegiben	Union Carbide
Chlorbromuron	Maloran	Ciba-Geigy
Chlorflurenol	Maintain	U. S. Borax
Chloropicrin	Chloropicrin	Dow, Monsanto
Chloroxuron	Tenoran	Ciba-Geigy
Chlorpropham	Chloro IPC, Furloe	PPG
Copper chelate	Cutrine	Applied Biochemists
Copper sulfate	Copper Sulfate, Bluestone	(Many)
Copper ethylenediamine	Komeen	Sandoz

Table A-1 (Continued)

Common Name	Trade Name	Manufacturer
Copper triethanolamine	K-Lox	Sandoz
Cyanazine	Bladex	Shell
Cycloate	Ro-Neet	Stauffer
2,4-D	(Many)	(Many)
2,4-DB	Butoxone, Butyrac, Embutox	Rhone-Poulenc, Union Carbide, many others
Dalapon	(Many)	(Many)
Dazomet	Mylone, DMTT	Union Carbide
DCPA	Dacthal	Diamond Shamrock
Desmedipham	Betanex	Nor-Am
Diallate	Avadex	Monsanto
Dicamba	Banvel	Velsicol
Dichlobenil	Casoron	Thompson-Hayward
Dichloropicolinate	Lontrel	Dow
Dichlorprop	(Many)	BASF, Bayer, Union Carbide
Diclofop	Hoelon	American Hoechst
Diethatyl	Antor	Hercules
Difenzoquat	Avenge, Finaven	American Cyanamid
Dinitramine	Cobex	U. S. Borax
Dinoseb	(Many)	Dow, Hoechst, others
Diphenamid	Enide	Upjohn
Dipropetryn	Sancap	Ciba-Geigy
Diquat	Diquat, Reglone	ICI, Chevron
Diuron	Karmex	duPont
DSMA	(Many)	(Many)
Endothall	(Many)	Pennwalt
EPTC	Eptam	Stauffer
Ethalfluralin	Sonalan	Lilly/Elanco
Ethofumesate	Norton	Fisons
Fenac	Fenac	Union Carbide
Fenuron TCA	Dozer, Urab	Hopkins
Fluchloralin	Basalin	BASF
Fluometuron	Cotoran, Lanex	Ciba-Geigy, Nor-Am
Fluorodifen	Preforan	Ciba-Geigy
Fluridone	Brake, Sonar	Lilly/Elanco
Fosamine	Krenite	duPont
Glyphosate	Round-Up	Monsanto
Hexaflurate	Nopalmate	Pennwalt
Hexazinone	Velpar	duPont
Ioxynil	Actril, Oxytril, Totril	May & Baker
Isopropalin	Paarlan	Lilly/Elanco

Table A-1 (Continued)

Common Name	Trade Name	Manufacturer
Karbutilate	Tandex	Ciba-Geigy
KOCN	Aerocyanate	American Cyanamid
Linuron	Lorox	duPont
MAA, MAMA	(Many)	(Many)
MCPA	(Many)	Dow, Rhone-Poulenc, Union Carbide
MCPB	(Many)	Rhone-Poulenc, Union Carbide
Mecoprop	(Many)	Cleary, Diamond Shamrock, Rhone-Poulenc
Mefluidide	Embark, Vistar	3M
Metham	Vapam	Stauffer
Methazole	Probe	Velsicol
Metolachlor	Dual	Ciba-Geigy
Methyl bromide	Methyl bromide fumigant	Dow, Velsicol
Metribuzin	Sencor, Lexon	Mobay, duPont
MH	(Many)	Uniroyal, Drexel
Molinate	Ordram	Stauffer
MSMA	(Many)	(Many)
Napropamide	Devrinol	Stauffer
Naptalam	Alanap	Uniroyal
Neburon	Kloben	duPont
Nitralin	Planavin	Shell
Nitrofen	TOK	Rohm and Haas
Norea	Herban	Hercules
Norflurazon	Evitol, Solicam, Zorial	Sandoz
Oryzalin	Ryzelan, Surflan	Lilly/Elanco
Oxadiazon	Ronstar	Rhone-Poulenc
Oxyfluorfen	Goal	Rohm and Haas
Paraquat	Gramoxone, Paraquat	ICI, Chevron
Pebulate	Tillam	Stauffer
Pendimethalin	Prowl	American Cyanamid
Perfluidone	Destun	3M
Phenmedipham	Betanal	Nor-Am
Picloram	Tordon	Dow
PMA	PMA	Linck, Cleary, Scott
Potassium azide	Kazoe	PPG
Prodiamine	Rydex	U. S. Borax
Profluralin	Tolban	Ciba-Geigy
Prometon	Pramitol	Ciba-Geigy
Prometryn	Caparol	Ciba-Geigy

Table A-1 (Continued)

Common Name	Trade Name	Manufacturer
Pronamide	Kerb	Rohm and Haas
Propachlor	Ramrod, Bexton	Monsanto, Dow
Propanil	Stam, Stampede	Rohm and Haas
Propazine	Milogard	Ciba-Geigy
Propham	Chem-hoe	PPG
Prosulfalin	Sward	Lilly/Elanco
Pyrazon	Pyramin	BASF
Secbumeton	Sumitol	Ciba-Geigy
Sethoxydim	Poast	BASF
Siduron	Tupersan	duPont
Silvex	Kuron, Weedone	Dow, Union Carbide
Simazine	Princep, Aquazine	Ciba-Geigy
Sodium azide	Smite	PPG
Sodium chlorate	Sodium chlorate	Pennwalt
2,4,5-T	(Many)	(Many)
2,3,6-TBA	Benzac, Trysben	duPont, Union Carbide
TCA	Sodium TCA	Dow
Tebuthiuron	Graslan, Spike	Lilly/Elanco
Terbacil	Sinbar	duPont
Terbutryn	Igran	Ciba-Geigy
Thiobencarb	Bolero, Saturn	Chevron, Kumiai
Triallate	Far-go	Monsanto
Triclopyr	Garlon	Dow
Tricamba	Banvel-T	Velsicol
Trifluralin	Treflan	Lilly/Elanco
Vernolate	Vernam	Stauffer

[1] Common names and chemistry given on the back cover page of any current *Weed Science* publication.

Table A-2. Alphabetical Listing of Herbicides by Trade Name and Corresponding Common Name

Trade Name	Common Name
Actril	Ioxynil
Aerocyanate	KOCN
Alanap	Naptalam
Amex	Butralin
Amiben	Chloramben
Amino Triazole	Amitrole
Amixol	Amitrole
Ammate	AMS

Table A-2 (Continued)

Trade Name	Common Name
Antor	Diethatyl
Aqualin	Acrolein
Aquazine	Simazine
Asulox	Asulam
AAtrex	Atrazine
Atrazine	Atrazine
Avadex	Diallate
Avenge	Difenzoquat
Balan	Benefin
Banvel	Dicamba
Banvel-T	Tricamba
Basagran	Bentazon
Basalin	Fluchloralin
Benazolin	Benazolin
Benzac	2,3,6-TBA
Betanal	Phenmedipham
Betanex	Desmedipham
Betasan	Bensulide
Bexton	Propachlor
Bladex	Cyanazine
Blazer	Acifluorfen
Bluestone	Copper sulfate
Bolero	Thiobencarb
Brake	Fluridone
Brominal	Bromoxynil
Buctril	Bromoxynil
Bronate	Bromoxynil
Butoxone	2,4-DB
Butyrac	2,4-DB
Caparol	Prometryn
Carbyne	Barban
Casoron	Dichlobenil
Chloro IPC	Chlorpropham
Chloropicrin	Chloropicrin
Chem-hoe	Propham
Cobex	Dinitramine
Copper Sulfate	Copper sulfate
Cotoran	Fluometuron
Cutrine	Copper Chelate
Cyanamide	Calcium Cyanamide
Dacthal	DCPA
Destun	Perfluidone

Table A-2 (Continued)

Trade Name	Common Name
Devrinol	Napropamide
Diquat	Diquat
DMTT	Dazomet
Dozer	Fenuron TCA
Dual	Metolachlor
Embark	Mefluidide
Embutox	2,4-DB
Enide	Diphenamid
Eptam	EPTC
Evik	Ametryn
Evitol	Norflurazon
Far-go	Triallate
Fenac	Fenac
Finaven	Difenzoquat
Furloe	Chlorpropham
Garlon	Triclopyr
Gesapax	Ametryn
Goal	Oxyfluorfen
Gramoxone	Paraquat
Graslan	Tebuthiuron
Herban	Norea
Hoelon	Diclofop
Hyvar	Bromacil
Igran	Terbutryn
Karmex	Diuron
Kazoe	Potassium azide
Kerb	Pronamide
Kloben	Neburon
K-Lox	Copper triethanolamine
Komeen	Copper ethylenediamine
Krenite	Fosamine
Kuron	Silvex
Lanex	Fluometuron
Lasso	Alachlor
Lexon	Metribuzin
Lontrel	Dichloropicolinate
Lorox	Linuron
Machete	Butachlor
Maintain	Chlorflurenol
Maloran	Chlorbromuron
Methyl bromide fumigant	Methyl bromide
Milogard	Propazine

Table A-2 (Continued)

Trade Name	Common Name
Modown	Bifenox
Mylone	Dazomet
Nopalmate	Hexaflurate
Norton	Ethofumesate
Ordram	Molinate
Oxytril	Ioxynil
Paarlan	Isopropalin
Paraquat	Paraquat
Planavin	Nitralin
PMA	PMA
Poast	Sethoxydim
Polybor	Borate (octa)
Pramitol	Prometon
Prefar	Bensulide
Preforan	Fluorodifen
Princep	Simazine
Probe	Methazole
Prowl	Pendimethalin
Pyramin	Pyrazon
Rad-E-Cate	Cacodylic acid
Ramrod	Propachlor
Randox	CDAA
Ravage	Buthidazole
Reglone	Diquat
Ro-Neet	Cycloate
Ronstar	Oxadiazon
Round-Up	Glyphosate
Rydex	Prodiamine
Ryzelan	Oryzalin
Rydex	Prodiamine
Sancap	Dipropetryn
Saturn	Thiobencarb
Sencor	Metribuzin
Sinbar	Terbacil
Smite	Sodium azide
Sodium chlorate	Sodium chlorate
Sodium TCA	TCA
Solicam	Norflurazon
Sonalan	Ethalfluralin
Sonar	Fluridone
Spike	Tebuthiuron

Table A-2 (Continued)

Trade Name	Common Name
Stam	Propanil
Stampede	Propanil
Sumitol	Secbumeton
Surflan	Oryzalin
Sutan	Butylate
Sward	Prosulfalin
Tandex	Karbutilate
Tenoran	Chloroxuron
Tillam	Pebulate
TOK	Nitrofen
Tolban	Profluralin
Topcide	Benzadox
Tordon	Picloram
Totril	Ioxynil
Treflan	Trifluralin
Trysben	2,3,6-TBA
Tupersan	Siduron
Urab	Fenuron TCA
Vapam	Metham
Vegadex	CDEC
Vegiben	Chloramben
Velpar	Hexazinone
Vernam	Vernolate
Vistar	Mefluidide
Weedazol	Amitrole
Weedone	Silvex
Zorial	Norflurazon

Table A-3. Conversion Factors

Liquid Measure

1 gallon (U.S.) = 3785.4 milliliters (ml); 256 tablespoons; 231 cubic inches; 128 fluid ounces; 16 cups; 8 pints; 4 quarts; 0.8333 imperial gallon; 0.1337 cubic foot; 8.337 pounds of water

1 liter = 1000 milliliters; 1.0567 liquid quarts (U.S.)

1 gill = 118.29 milliliters

1 fluid ounce = 29.57 milliliters; 2 tablespoons

3 teaspoons = 1 tablespoon; 14.79 milliliters; 0.5 fluid ounce

1 cubic foot of water = 62.43 pounds; 7.48 gallons

Table A-3 (Continued)

Weight
1 gamma = 0.001 milligram (mg)
1 grain (gr) = 64.799 milligrams
1 gram (g) = 1000 milligrams; 15.432 grains; 0.0353 ounce
1 pound = 16 ounces; 7000 grains; 453.59 grams; 0.45359 kilogram
1 short ton = 2000 pounds; 0.097 metric ton
1 long ton = 2240 pounds; 1.12 short ton
1 kilogram = 2.2046 pounds

Linear Measure
12 inches = 1 foot; 30.48 centimeters
36 inches = 3 feet; 1 yard; 0.914 meter
1 rod = 16.5 feet; 5.029 meters
1 mile = 5280 feet; 1760 yards; 160 rods; 80 chains; 1.6094 kilometers (km)
1 chain = 66 feet; 22 yards; 4 rods; 100 links
1 inch = 2.54 centimeters (cm)
1 meter = 39.37 inches; 10 decimeters (dm); 3.28 feet
1 micron (μm) = $\frac{1}{1000}$ millimeter (mm)
1 kilometer = 0.621 statute miles; 0.5396 nautical miles

Area
1 township = 36 sections; 23,040 acres
1 square mile = 1 section; 640 acres
1 acre = 43,560 square feet; 160 square rods; 4840 square yards; 208.7 feet square; an area $16\frac{1}{2}$ feet wide and $\frac{1}{2}$ mile long; 0.4047 hectare
1 hectare = 2.471 acres; 100 are

Capacity (Dry Measure)
1 bushel (U.S.) = 4 pecks; 32 quarts; 35.24 liters; 1.244 cubic feet; 2150.42 cubic inches

Pressure
1 foot lift of water = 0.433 pound pressure per square inch (psi)
1 pound pressure per square inch will lift water 2.31 feet
1 atmosphere = 760 millimeters of mercury; 14.7 pounds; 33.9 feet of water

Geometric Factors (π = 3.1416; r = radius; d = diameter; h = height)
Circumference of a circle = $2\pi r$ or πd
Diameter of a circle = $2r$
Area of a circle = πr^2 or $\frac{1}{4}\pi d^2$ or $0.7854d^2$
Volume of a cylinder = $\pi r^2 h$
Volume of a sphere = $\frac{1}{6}\pi d^3$

Table A-3 (Continued)

Other Conversions

Multiply	by	To Obtain
Gallons per minute	2.228×10^{-3}	Cubic feet per second
Gallons per acre	9.354	Liters per hectare
Kilograms per hectare	0.892	Pounds per acre
Liters	1.05	U.S. quarts
Liters	0.2642	U.S. gallons
Liters per hectare	0.107	Gallons per acre
Miles per hour	88.0	Feet per minute
Miles per hour	1.61	Kilometers per hour
Pounds per gallon	0.12	Kilograms per liter
Pounds per square inch	70.31	Grams per square centimeter (Atm)
Pounds per 1000 square feet	0.489	Kilograms per acre
Pounds per acre	1.12	Kilograms per hectare
Square inch	6.452	Square centimeter
Parts per million	2.719	Pounds acid equivalent per acre foot of water

Temperature, degrees

$F^\circ = C^\circ + 17.78 \times 1.8$

$C^\circ = F^\circ - 32.00 \times \frac{5}{9}$

°C	°F	°C	°F
100	212	30	86
90	194	20	68
80	176	10	50
70	158	0	32
60	140	-10	14
50	122	-20	-4
40	104	-30	-22

Table A-4. Weight of Dry Soil

Type	Pounds per Cubic Foot	Pounds per Acre, 7″ Deep
Sand	100	2,500,000
Loam	80-95	2,000,000
Clay or silt	65-80	1,500,000
Muck	40	1,000,000
Peat	20	500,000

Table A-5. Length of Row Required for One Acre

Row Spacing	Length or Distance
24 in.	7260 yards = 21,780 ft
30 in.	5808 yards = 17,424 ft
36 in.	4840 yards = 14,520 ft
42 in.	4149 yards = 12,445 ft
48 in.	3630 yards = 10,890 ft

Table A-6. Available Commercial Materials in Pounds Active Ingredients per Gallon Necessary to Make Various Percentage Concentration Solutions[1]

Pounds of Active Ingredient in 1 gal of Commercial Product	Pounds of Active Ingredient/Pint[1]	Liquid Ounces of Commercial Product to make one gallon of:				
		$\frac{1}{2}$%	1%	2%	5%	10%
2.00	0.25	2.68	5.36	10.72	26.80	53.60
2.64	0.33	2.02	4.05	8.10	20.25	40.50
3.00	0.375	1.78	3.56	7.12	17.80	35.60
3.34	0.42	1.59	3.18	6.36	15.90	31.80
4.00	0.50	1.34	2.68	5.36	13.40	26.80
6.00	0.75	0.89	1.78	3.56	8.90	17.80

[1] Based on 8.4 lb/gal (weight of water) and 128 liquid oz = 1 gal, 16 liquid oz = 1 pint.

Table A-7. Equivalent Quantities of Liquid Materials When Mixed by Parts

Water	1-400	1-800[1]	1-1600
100 gal	1 qt	1 pt	1 cup
50 gal	1 pt	1 cup	$\frac{1}{2}$ cup
5 gal	3 tbs	5 tsp[1]	$2\frac{1}{2}$ tsp
1 gal	2 tsp	1 tsp	$\frac{1}{2}$ tsp

[1] Example: If a recommendation calls for 1 part of the chemical to 800 parts of water, it would take 5 tsp in 5 gal of water to give 5 gal of a mixture of 1-800.

Table A-8. Equivalent Quantities of Dry Materials (Wettable Powders) for Various Quantities of Water

Water	Quantity of Material					
100 gal[1]	1 lb	2 lb	3 lb	4 lb[1]	5 lb	6 lb
50 gal	8 oz	1 lb	24 oz	2 lb	$2\frac{1}{2}$ lb	3 lb
5 gal[1]	3 tbs	$1\frac{1}{2}$ oz	$2\frac{1}{2}$ oz	$3\frac{1}{4}$ oz[1]	4 oz	5 oz
1 gal	2 tsp	3 tsp	$1\frac{1}{2}$ tbs	2 tbs	3 tbs	3 tbs

[1] Example: If a recommendation calls for a mixture of 4 lb of a wettable powder to 100 gal of water, it would take $3\frac{1}{4}$ oz (approximately 12 tsp) to 5 gal of water to give 5 gal of spray mixture of the same strength.

[2] Wettable materials vary considerably in density. Therefore, the teaspoonful (tsp) and tablespoonful (tbs) measurements in this table are not exact dosages by weight but are within the bounds of safety and efficiency for mixing small amounts of spray.

Table A-9. Pressure Drop in Hose Due to Friction; These Figures Are for 25 ft of Clean Hose With No Couplings

$\frac{1}{4}''$ ID Hose		$\frac{3}{8}''$ ID Hose		$\frac{1}{2}''$ ID Hose		$\frac{5}{8}''$ ID Hose		$\frac{3}{4}''$ ID Hose		1″ ID Hose		$1\frac{1}{4}''$ ID Hose	
Flow (gpm)	Pr. Drop (psi/ 25 ft)	Flow (gpm)	Pr. Drop (psi/ 25 ft)	Flow (gpm)	Pr. Drop (psi/ 25 ft)	Flow (gpm)	Pr. Drop (psi/ 25 ft)	Flow (gpm)	Pr. Drop (psi/ 25 ft)	Flow (gpm)	Pr. Drop. (psi/ 25 ft)	Flow (gpm)	Pr. Drop (psi/ 25 ft)
0.2	0.8												
0.3	1.5												
0.4	2.5												
0.5	4.0	0.5	0.5										
0.6	5.0	0.6	0.8										
0.8	9.0	0.8	1.3										
		1.0	1.8	1.0	0.5								
		2.0	6.0	2.0	1.5								
		3.0	13.0	3.0	3.1	3.0	1.0						
				4.0	6.0	4.0	1.8						
				5.0	8.5	5.0	2.5	5.0	1.0				
				6.0	12.0	6.0	3.7	6.0	1.5				
						8.0	6.5	8.0	2.5	8.0	0.6		
						10.0	9.5	10.0	3.7	10.0	1.0		
								15.0	8.0	15.0	2.0	15.0	0.7
								20.0	14.0	20.0	3.4	20.0	1.2
										25.0	5.0	25.0	1.8
										30.0	6.5	30.0	2.5
										40.0	12.0	40.0	4.4
												50.0	6.0
												60.0	9.0
												70.0	13.0

Index